Information Fusion and Data Science

Series Editor

Henry Leung, University of Calgary, Calgary, AB, Canada

This book series provides a forum to systematically summarize recent developments, discoveries and progress on multi-sensor, multi-source/multi-level data and information fusion along with its connection to data-enabled science. Emphasis is also placed on fundamental theories, algorithms and real-world applications of massive data as well as information processing, analysis, fusion and knowledge generation.

The aim of this book series is to provide the most up-to-date research results and tutorial materials on current topics in this growing field as well as to stimulate further research interest by transmitting the knowledge to the next generation of scientists and engineers in the corresponding fields. The target audiences are graduate students, academic scientists as well as researchers in industry and government, related to computational sciences and engineering, complex systems and artificial intelligence. Formats suitable for the series are contributed volumes, monographs and lecture notes.

More information about this series at http://link.springer.com/series/15462

Michel Barès • Éloi Bossé

Relational Calculus for Actionable Knowledge

 Springer

Michel Barès
Data Science Department
Expertises Parafuse Inc.
Jouy en Josas, France

Éloi Bossé
Image and Information
Processing Department
IMT Atlantique
Quebec City, QC, Canada

ISSN 2510-1528   ISSN 2510-1536   (electronic)
Information Fusion and Data Science
ISBN 978-3-030-92432-4   ISBN 978-3-030-92430-0   (eBook)
https://doi.org/10.1007/978-3-030-92430-0

This Springer imprint is published by the registered company Springer Nature Switzerland AG
The registered company address is: Gewerbestrasse 11, 6330 Cham, Switzerland

Preface

One of the major challenges of a newly created scientific domain called "data science" is to turn data into actionable knowledge to exploit the increasing data volumes and deal with their inherent complexity (Big data and IoT). The advances in networking capabilities have created the conditions of complexity by enabling richer, real-time interactions between and among individuals, objects, systems, and organizations. Networking involves relations of all kinds and presents challenges of complexity especially when the objective is to provide technological supports to human decision-making.

Actionable knowledge has been qualitatively and intensively studied in management, business, and social sciences but for computer sciences and engineering, recently, there has been a connection with data mining and its evolution "Knowledge Discovery and Data Mining (KDD)." The ambition of our book is to present advanced knowledge concepts and its formalization to support the analytics and information fusion (AIF) processes that aim at delivering actionable knowledge. The book offers four major contributions: (1) the concept of "relation" and its exploitation (relational calculus) for the AIF processes, (2) the formalization of certain dimensions of knowledge to achieve a semantic growth along the AIF processes, (3) the modeling of the interrelations within the couple (knowledge, action) to gain sense, and finally (4) the exploitation of relational calculus to support the AIF core technological processes that allow to transform data into actionable knowledge.

This book addresses two main poles: computations with relations (relational calculus) and creation of actionable knowledge. In the first three chapters, we explore basic properties of knowledge, knowledge representations, and knowledge processes from scientific and practical perspectives emphasizing existing directions and areas in knowledge studies. We also examine the fundamental role of information and define the relationship that exists between data, information, and knowledge. We discuss the need for formalization. Any automatic process geared to support human decision-making must be indeed endowed with reasoning ability, depending on the circumstances and the context of its employment. A suitable formalism to represent knowledge and information remains the required essential for any subsequent artificial reasoning that is achieved throughout a knowledge processing chain, the AIF processes.

The subsequent three chapters (4–6) address the understanding of the couple (knowledge, action) and how to support the processing chain in the creation of actionable knowledge using relational calculus. Chapter 4 presents preliminaries of crisp and fuzzy relational calculus to support the discussion in the subsequent chapters. The question of how to deal with knowledge imperfections is addressed. Chapter 5 examines the couple (knowledge, action). Knowledge is a prerequisite to taking any reasoned action or course of action according to rational rules. The questions are what facilitates the relevant decision-making and what are the modalities that can make the action (effect) more efficient. There is a strong dependency between the notion of knowing about a given world and the decisions that can be made and consecutively the potential actions that can be undertaken. The notion of *mastering knowledge* for efficient actions is treated. Analysis and synthesis of information, a prerequisite to any decision-making and action, is supported by AIF technologies. Chapter 6 addresses the usage of relational calculus when applied to the AIF core processes that perform the multiple transformations required along the processing chain from data to actionable knowledge.

Jouy en Josas, France   Michel Barès
Quebec City, QC, Canada   Éloi Bossé

Acknowledgement

The authors wish to express their gratitude to Prof. Basel Solaiman, IMT-Atlantique, for the numerous discussions and comments on earlier versions of several chapters of this book. We would also like to thank the Springer staff for their support in the production of this book.

Contents

About the Authors

Michel Barès Ph.D. is an independent researcher in Artificial Intelligence (AI). He is a former researcher at DRET (Direction des Recherches Et Technologies) under the French Ministry of Defence. There, he conducted different innovative research projects relevant to military motivations concerning symbolic reasoning, decision support, and distributed AI. He has been a pioneer to introduce distributed AI in military command decision-making aids. In parallel to his military research work, he acted as professor for several universities and specialized French schools. Since 2017, he is a scientific advisor for Expertises Parafuse inc., a small research firm located in Québec City, Canada. He holds a Ph.D. degree from the university Paris 6, France.

Michel Barès, Ph.D., received the degree Expert en Traitement de l'Information (ETI) from Institut de Programmation (computer sciences) (71), a DEA in maths-physics (75), and a Ph.D. in mathematics (78) from university Paris 6. In addition, he received the informatics engineering degree from Conservatoire National des Arts et Métiers Paris (76) and an '*Habilitation à Diriger des Recherches*' (HdR) from the University of Nancy (97). In 1970, he joined '*l'Institut en Informatique et en Automatique (INRIA)*' where he worked on operational research problems related to large organizations. In 1975, he joined '*La Documentation Française*' to work on database management. In 1984, he joined DRET (Direction des Recherches et Technologies under the French Ministry of Defence) to hold a head position for the division of computer sciences and numerical analysis. There, he conducted different innovative research projects related to symbolic reasoning, intelligent interfaces, and distributed AI. He was one of the pioneers to introduce distributed AI in military command decision-making aids. As professor, Dr. Barès has taught at several universities and specialized French schools: Ecole Nationale Supérieure des Techniques Avancées (ENSTA), Ecole Supérieure de l'Aéronautique (Sup Aéro), Université de Versailles, Université de Bretagne, Université de Rouen, Ecole Militaire et Centre Interarmées de Défense (CID), and George Mason University (visiting professor, Washington, DC). Over several years, Dr. Barès has been responsible for specialized sessions and acted as chair of the international Avignon conference on AI and expert systems. He represented France in several NATO Research Technology Organization (RTO) panels and research groups on information sciences: NATO RSG 10 workshop "speech processing," TG 006 "modelling of organization and decision architecture," ET 014 C4ISRS interoperability," the NATO Data Fusion Demonstrator, and others. He has published 8 academic books on computer sciences and AI and more than 150 scientific publications in conference proceedings, technical reports, and journal papers. The application domains of Dr. Barès' research activities range from civilian to military ones: data sciences, complex systems, analytics and information fusion (AIF), and finally, knowledge systems.

Éloi Bossé Ph.D. is a researcher on decision support and analytics and information fusion (AIF). He possesses a vast research experience in applying them to defense- and security-related problems and more recently to civilian domains. He is currently president of Expertise Parafuse inc., a consultant firm on AIF and decision support. He holds an academic position as an associate researcher at IMT-Atlantique, Brest, France, and a Ph.D. degree from Université Laval, Québec City, Canada.

Éloi Bossé, Ph.D., received B.A.Sc. (79), M.Sc. (81), and Ph.D. (90) degrees in Electrical Engineering from Université Laval, QC. In 1981, he joined the Communications Research Centre, Ottawa, Canada, where he worked on signal processing and high-resolution spectral analysis. In 1988, he was transferred to the Defence Research Establishment Ottawa to work on radar target tracking in multipath. In 1992, he moved to Defence Research and Development Canada Valcartier (DRDC Valcartier) to lead a group of four to five defense scientists on information fusion and resource management. He has published over 200 papers in journals, book chapters, conference proceedings, and technical reports. Dr. Bossé has held adjunct professor positions at several universities from 1993 to 2013 (Université Laval, University of Calgary, and McMaster University). He headed the C2 Decision Support Systems Section at DRDC Valcartier from 1998 till 2011. Dr. Bossé was the Executive Chair of the 10th International Conference on Information Fusion (FUSION`07), held in July 2007 in Québec City. He represented Canada (as DRDC member) in numerous international research fora under various cooperation research programs (NATO, TTCP, and bi- and tri-laterals) in his area of expertise. He is coauthor and coeditor of five to six books on analytics and information fusion. He left DRDC in September 2011. Since then, he has conducted some research activities under NATO Peace and Security Program, as researcher in Mathematics and Industrial Engineering Department at Polytechnic of Montreal, as associate researcher at IMT-Atlantique, and as researcher at McMaster University, Canada. Since 2015, concurrently with the activities just mentioned, he acts as president of Expertise Parafuse inc., a consultant research firm on analytics and information fusion (AIF) technologies, a great component of data sciences.

List of Abbreviations

AD	Archetypal Dynamics
AI	Artificial Intelligence
AIF	Analytics and Information Fusion
Big Data 5Vs	(Velocity, Volume, Veracity, Value, Variety)
C2	Command Control
C4ISR	Command Control, Communications, Computer, Intelligence, Surveillance and Reconnaissance
CI	Contextual Information
CoA	Course of Action
CPS	Cyber-physical Systems
CPSS	Cyber-physical and Social Systems
CSE	Cognitive System Engineering
DF	Data Fusion
DIK	Data-Information-Knowledge
DM	Decision-Making
DQ	Data Quality
DSS	Decision Support Systems
DS	Dempster–Shafer
DST	Dempster–Shafer's Theory
EBDI	Entity-Based Data Integration
ER	Entity Resolution
ES	Epistemic Structure
ETURWG	Evaluation of Technologies for Uncertainty Representation Working Group
GIT	Generalized Information Theory
GTI	General Theory of Information
HLIF	High-Level Information Fusion
H2M	Human-to-Machine
H2S	Human-to-System
IBM	International Business Machine

ICN	Information Centric Networking
ICT	Information and Communication Technology
ID	IDentification
IF	Information Fusion
IFS	Intuitionistic Fuzzy Set
IG	Interoperable Groups
Intel	Military Intelligence cycle
IoT	Internet of Things
IoE	Internet of Everything
IS	Information Systems
ISIF	International Society of Information Fusion
JDL	Joint Directors of Laboratories
JDL DIFG	Joint Directors of Laboratories' Data and Information Fusion Group
KDD	Knowledge Discovery in Databases
KID	Knowledge, Information and Data
KIME	Knowledge-Information-Matter-Energy
KS	Knowledge System
MAPE	Monitor-Analyze-Plan-Execute
MAS	Multi-agent Systems
MCDA	Multi-criteria Decision Analysis
MCDM	Multi-criteria Decision-Making
MS	Management Science
M2M	Machine-to-Machine
NATO	North Atlantic Treaty Organization
NATO SAS RG	NATO Systems Analysis and Studies Research Group
ORBAT	ORder of BATtle
OODA	Observe-Orient-Decide-Act
QoI	Quality of Information
SA	Situation Analysis
SAW	Situation Awareness
SM	Sense-Making
STO	Socio-technical Organizations
TER	Total Entity Resolution
TQM	Total Quality Management
TU	Total Uncertainty
UMM	Uncertainty Management Methods
UN	United Nations
URREF	Uncertainty Representation and Reasoning Evaluation Framework
WoT	Web of Things

List of Symbols

$\mathbb{N} = \{1, 2, 3, \ldots\}$	The set of natural numbers
$\mathbb{R}$	The set of real numbers
$|A|$	The cardinality of a set A
A^C	The complement of A
$\in$	Membership sign; belongs to
$\subseteq$	Subset; inclusion sign
$\subset$	Proper subset; strict inclusion
$\varnothing$	Empty set
$\cup$	Union
$\cap$	Intersection
$\times$	Cartesian product
$<$	Less than
$\leq$	Less than or equal to
$>$	Greater than
$\geq$	Greater than or equal to
sup	Supremum
inf	Infimum
max	Maximum
min	Minimum
$::$ or $\stackrel{\text{def}}{=}$	Defined as; given by
$\therefore$	Therefore
$\sim$ or $\neg$	Negation
$\Rightarrow$	Implication
$\rightarrow$	Correspond to
$\forall$	Universal quantifier; for all
$\exists$	Existencial quantifier; there exists
$asc_\uparrow$	Ascendant of
$desc_\downarrow$	Descendent of
$\vdash$	Conclusion; turnstile symbol; assertion sign
$\Leftrightarrow$	Equivalence
$\mathrm{Dom}(R)$	Domain of relation R
$\mathrm{Rng}(R)$ or $\mathrm{Im}(R)$	Range or image of relation R
coR	The *complement* relation coR of R
R^{-1}	The reverse or inverse of relation R
$\neq$	Not equal
P-relation	A relation with property P
aRb	a is related to b
poset P	*a partial ordering or a partial order of P*
lub	or *sup* or $\sqcup$
	Least upper bound
glb or *inf* or $\sqcap$	*Greatest lower bound*
$P \oplus Q$	The ordinal sum of two posets

$\mathbb{L} = (P, \bigsqcup, \sqcap, 0, 1)$	A lattice as a *poset P*
$R \circ S$	Composition of relations R and S
$R \lhd S$	Subcomposition of relations R and S
$R \rhd S$	Supercomposition of relations R and S
$R \diamond S$	Ultracomposition of relations R and S
$x \notin A$	x is not element of A
μ	membership function
ν	non-membership function
$\mu_{A(x)}$	degree of membership of element x in A
$\upsilon_{A(x)}$	degree of non-membership of element x in A
${}^{\alpha}A$	The α-cut of A
${}^{\alpha+}A$	The strong α-cut of A
${}^{0+}A$	The *support* of A
${}^{1}A$	The *core* of A
hgt(A)	The *height* of A
plth(A)	The *plinth* of A

Chapter 1
Introduction to Actionable Knowledge

Abstract This chapter presents an introduction to actionable knowledge, its related notions, and to what general context actions are going to take effect? What is actionable knowledge? From what angle, this book is approaching it? Where and how do we position relational calculus with respect to actionable knowledge? The context of Cyber-Physical and Social Systems is briefly described. Important related notions of knowledge, dynamic decision-making, situations and situation awareness, and analytics and information fusion are being introduced. These notions are necessary to position relational calculus in the processes of creating actionable knowledge.

1.1   Actionable Knowledge

In a very recent book on knowledge and action [1], the authors start by quoting a widely accepted idea [2] that *"parts of knowledge can be defined as ability, aptitude, or 'capacity for social action' and that the production and dissemination of knowledge are always embedded in specific environments (spatial context, spatial relations, and power[1] structures)."*

Knowledge, learning, and information-processing determine how objectives of actions are set:

- How are situations, opportunities, and risks assessed?
- How are patterns interpreted?
- How are problems solved?

They are links between action and environment as pictured in Fig. 1.1. For instance, acting under conditions of uncertainty, people must rely on knowledge acquired from various situations and environments, and they must gather new

---

[1] The close relationship between knowledge and power is evident by the very fact that they have the same etymological roots. The word *power* derives from the Latin *potere* (to be able). The Latin noun *potentia* denotes an ability, capacity, or aptitude to affect outcomes, to make something possible. It can therefore be translated as both knowledge and power. (*from the same reference*)

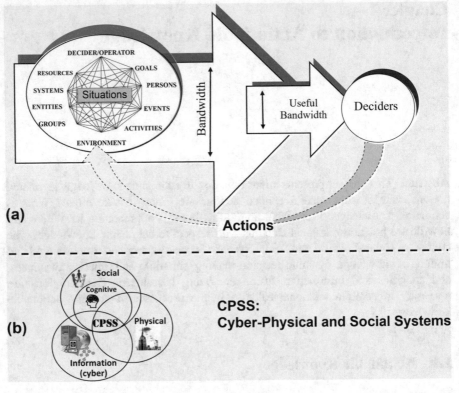

Fig. 1.1 Actionable knowledge in Cyber-Physical and Social Systems (CPSS): (**a**) a useful *information bandwidth*—a metaphor of the signal processing community and (**b**) CPSS as the intersection of various worlds

information, gain new knowledge, and develop new skills to cope with unexpected situations and unfamiliar challenges. The following points are generally accepted for people to achieve their goals:

- Goal setting is impacted by knowledge, skills, experience, and the search for new information.
- Experience rests upon former actions in specific settings.
- There are multiple relationships between knowledge and action.
- Learning processes are shaped by the social and material environment.
- The spatial dimension plays a key role in the acquisition of knowledge and implementation of actions.

Figure 1.1a illustrates the flow of information required to assess situations that are occurring in real world. Using an analogy borrowed from the signal processing community, one can imagine a useful "information" bandwidth where the overall goal would be to provide useful information to deciders, i.e., actionable knowledge. That conceptual multidimensional "useful bandwidth" could be defined by

Fig. 1.2 Analytics and information fusion (AIF) as a computer-support system

assembling appropriate smart filtering and metadata-based technologies to perform analysis and synthesis of information that we refer as analytics and information fusion (AIF), pictured in Fig. 1.2.

Relationships between knowledge, action, and environment are quite complex, some of them are still not fully understood. To understand the interrelations of knowledge and action, it yields to pose the following questions:

- To what extent is knowledge a precondition for action? How much knowledge is necessary for action?
- To what extent do various types of knowledge influence aspirations, attention, evaluation of situations, search for alternatives, implementation of intentions, decision-making, and problem-solving?
- How do different representations of knowledge shape action?
- How does the digital revolution change the formation of knowledge?
- How can one measure an environment's impact on action and knowledge production?
- ...?

These and other questions indicate that relations between knowledge, action, and environment are not simple and affect the modeling of decision and risk. The questions above suggest a need to explore the interdependencies of knowledge, action, and environment from a multidisciplinary perspective. With the evolution of societies, the so-called "*environment*" has become quite complex. Information overload and complexity are core problems that both military and civilian organizations are facing today. With the increase in networking, these large military and civilian organizations are referred to as Cyber-Physical and Social Systems (CPSS) (see Fig. 1.1b and next section).

Executives or commanders want better ways to communicate complex insights so they can quickly absorb the meaningful information (actionable knowledge) to decide and act. *Big data* [3] is contextual to CPSS complex dynamic environments. *Big data*, being at the same time a problem or an opportunity, has emerged with its 5Vs (*volume, veracity, variety, velocity, and value*) dimensions and it is the main object of a newly created scientific domain named *"data science."*[2] It requires that new technology be developed to provide the analysis (e.g., analytics) and synthesis (e.g., information fusion) support for the decision-makers to make sense out of data-information to create actionable knowledge for efficient actions.

A nice factual description of what CPSS is dealing with is given in [3] as follows:

Every day, around 20 quintillion (10^{18}) bytes of data are produced. This data includes textual content (unstructured, semi-structured, and structured) to multimedia content (images, video, and audio) on a variety of platforms (enterprise, social media, and sensors). The growth of physical world data collection and communication is supported by low-cost sensor devices, such as wireless sensor nodes that can be deployed in different environments, smartphones, and other network-enabled appliances. This trend will only accelerate, as it's estimated that more than 50 billion devices are currently connected to the Internet Extending the current Internet and providing connections and communication between physical objects and devices, or 'things,' is described under the general term of Internet of Things (IoT). Another often used term is Internet of Everything (IoE), which recognizes the key role of people or citizen sensing, such as through social media, to complement physical sensing implied by IoT. Integrating the real-world data into the Web and providing Web-based interactions with the IoT resources is also often discussed under the umbrella term of Web of Things (WoT)

Data science [4] is facing the following major challenges:

1. Developing scalable cross-disciplinary capabilities.
2. Dealing with the increasing data volumes and their inherent complexity.
3. Building tools that help to build trust.
4. Creating mechanisms to efficiently operate in the domain of scientific assertions.
5. Turning data into actionable knowledge units.
6. Promoting data interoperability.

Actionable knowledge is not a new term. It has been qualitatively and intensively studied in management and social sciences. It illustrates the relationship between theory and practice. Actionable knowledge is linked with its user: the practitioner. It has been positioned as a response to the relevance of management research to management practice. Is the generated knowledge actionable by the users whom it is intended to engage (business practitioners, policymakers, researchers)? Actionable knowledge should advance our understanding of the nature of action as a phenomenon and the relationship between action and knowledge (modes of knowing) in organizations.

Actionable knowledge is explicit symbolic knowledge that allows the decision-maker to perform an action, such as select customers for a direct marketing campaign or select individuals for population screening concerning high disease risk. Its main connection is with data mining, an emerging discipline that has been booming for the

---

[2]**Data science** is an interdisciplinary field that uses scientific methods, processes, algorithms, and systems to extract knowledge and insights from structured and unstructured data and apply knowledge and actionable insights from data across a broad range of application domains. Data science is related to data mining, machine learning, and big data. (https://en.wikipedia.org/wiki/Data_science)

last two decades. Data mining seeks to extract interesting patterns from data. However, it is a reality that the so-called interesting patterns discovered from data have not always supported meaningful decision-making actions. This shows the significant gap between data mining research and practice, and between knowledge, power, and action. This has motivated the evolution of data mining next-generation research and development from data mining to actionable knowledge discovery and delivery [4–6].

Although big data is highly linked to business and social sciences and the genesis of actionable knowledge is from there, a CPSS environment present cases where the actions are much more complex than the implementation of conventional one-time decisions. A CPSS environment presents situations where interdependent decisions take place in a dynamic environment due to previous actions or decisions and events that are outside of the control of the decision-maker [7]. In addition to conventional one-time decisions, CPSS present dynamic decisions. The latter field is typically more complex than one-time decisions and much more relevant to the environments of Big Data and IoT. In addition, dynamic decisions occur in real time. In complex CPSS, if we wish to provide actionable knowledge, one must understand and explain the decision-making and action processes in such complex environments. The extraction of actionable knowledge will be consequently more demanding.

Two main influential streams [8, 9] are generally recognized to understand decision-making. The first stream refers to a rational approach that is based on formal analytic processes predicted by normative theories of probability and logic. The second stream, called naturalistic or intuitive theories, is based on informal procedures or heuristics to make decisions within the restrictions of available time, limited information, and limited cognitive processing. Bryant et al. [8] insist upon a continuum in decision strategy to adopt the approach that is best tailored to the situation and may use elements of the two approaches at the same time (Fig. 1.3).

More and more, IoT and Big data are perceived as two sides of the same coin where Big Data would be a subset of IoT [10–12]. As mentioned above, Big Data is evidently contextual to Cyber-Physical and Social Systems (CPSS) [13–17]. CPSS emerge from the interrelation of social, cognitive, information/cyber and physical worlds as pictured in Fig. 1.1b. Social and cognitive dimensions interface with the physical world through the cyber world.

1.2 Our World: Cyber-Physical and Social Systems (CPSS)

Our world is an interlocking collective of Socio-Technical Organizations (STO) recently referred to as Cyber-Physical Social Systems (CPSS) [13, 18, 19] in the literature. CPSS consist of inhomogeneous, interacting adaptive agents capable of learning: large number of groups of people hyperlinked by information channels and interacting with computer systems, and which themselves interact with a variety of physical systems in conditions of good functioning. Primary examples of CPSS include Command and Control Organizations such as 911/Emergency Response

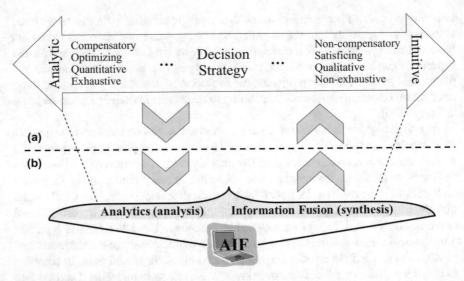

Fig. 1.3 Decision strategy: (**a**) the spectrum of decisions and (**b**) computer-based decision support. AIF, analysis and synthesis of information for decision-making

Systems and military organizations, as well as organizations that manage "critical infrastructures": transport, health, energy, defense, and security. Part (b) of Fig. 1.1 illustrates that CPSS emerge from the interrelation of social, cognitive, information/ cyber and physical worlds. The interrelation (cyber) is achieved by the means of what we call *information*.

Information overload and complexity are core problems to both military and civilian CPSS of today. Executives or commanders want better ways to communicate complex insights so they can quickly absorb the meaning of the data and act on it. That problem has also been referred to as Big Data in recent literature [20]. The advances in Information and Communications Technologies (ICT), in particular smart ICT, although providing a lot of benefits to improve dependability, efficiency and trustworthiness in systems, have also increased tremendously the networking capabilities so creating the conditions of complexity by enabling richer, real-time interactions between and among individuals, objects, systems and organizations. As a result, events that may once have had isolated consequences can now generate cascades of consequences, consequences that can quickly spin out of control (e.g., blackouts, catastrophes) and affect badly the system dependability or trustworthiness.

Dependability of a system (simple, complicated, or complex) reflects the user's degree of trust in that system. It reflects the extent of the user's confidence that it will operate as users expect and that it will not fail in normal use. The crucial dimensions of dependability are maintainability, availability, reliability, safety, and security. The high-level requirements for current and future CPSS are expected to be dependable, secure, safe, and efficient and operate in real time in addition to be scalable, cost-effective, and adaptive. Dependability can be achieved by effective high-quality *information*.

Cyber-Physical Systems (CPS) [21, 22] and the Internet of Things (IoT) [10, 11, 23] are big contributors to Big Data problems or opportunities. There is still a bit of confusion about the definition of Big Data whether it is best described by today's greater volume of data, the new types of data and analysis, social media analytics, next-generation data management capabilities, or the emerging requirements for more real-time information analysis. Whatever the label, organizations are starting to understand and explore how to process and analyze a vast array of information in new ways.

Cyber-Physical Systems (CPS) [24] are the integration of computation with physical processes. In CPS, embedded computers, and networks, monitor and control the physical processes, usually with feedback loops where physical processes affect computations and vice versa. That integration of physical processes and computing is not new as evidenced by predecessors to CPS called "embedded systems." However, embedded systems are rather "stand-alone" devices and they are usually not networked to the outside. What Internet and its evolutions are bringing is the networking of these devices. This is what we call Cyber-Physical Systems (CPS) [22] or Internet of Things (IoT) [25]. The following two definitions allow to see the distinction or the similarity between both CPS and IoT:

Helen Gill, NSF, USA [22]: *"Cyber-physical systems are physical, biological, and engineered systems whose operations are integrated, monitored, and/or controlled by a computational core. Components are networked at every scale. Computing is deeply embedded into every physical component, possibly even into materials. The computational core is an embedded system, usually demands real-time response, and is most often distributed."*

Benghosi et al. [25]: *". . . define the Internet of Things as a network of networks which enables the identification of digital entities and physical objects – whether inanimate (including plants) or animate (animals and human beings) – directly and without ambiguity,* via *standardized electronic identification systems and wireless mobile devices, and thus make it possible to retrieve, store, transfer and process data relating to them, with no discontinuity between the physical and virtual worlds."*

The generation of actionable knowledge to support decision-making in CPSS is challenged by the concurrent nature and laws governing the social, cognitive, cyber, and physical worlds. Figure 1.4 pictures crucial elements of that CPSS world and presents, at the same time, the global context of this book. The cybernetics functions of coordination, integration, monitoring, and control for dependable CPSS cannot be achieved efficiently without a profound understanding of the data-information-knowledge (DIK) processing chain that is practically realized through Analytics and Information Fusion (AIF) processes. The cybernetics functions summarize in some sort, the world of actions. System dependability in CPSS is an overall sine qua non-objective. Knowledge is always about the structure of a phenomenon rather than the essence of it. Relation is an ontological element of that structure. The contribution of this book is to discuss how relation and its calculus can be exploited

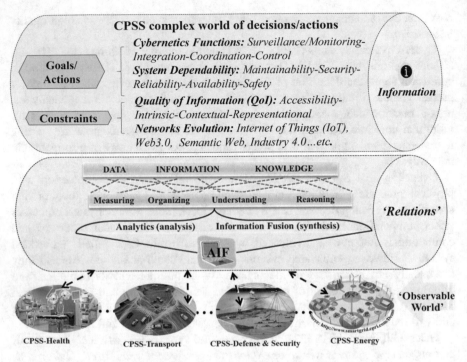

Fig. 1.4 Crucial elements of a CPSS complex world of decisions/actions

in the multifaceted process of creating actionable knowledge (Fig. 1.4) which is, by the way, the main goal of any AIF computer-based decision support systems.

1.3 Societal Behavior Face to Knowledge and Information

The massive pervasion of novel Information and Communication Technologies (ICTs) into contemporary society undoubtedly causes large-scale societal transformations and the awareness of major issues on the part of decision-makers. The new behaviors and strategies allowed by this pervasion allow us to open new fields of action whose outlines are already perceptible through [26, 27]:

- The tightening of the meshes of telematics networks.
- An optimized search for the organization and retention of knowledge in general.
- The lines of force of a new economy with the emergence of new types of commercial services (e.e-commerce), general exchange mechanisms, various services.
- The emergence of new relations between citizens and their administration (e.g., e-government).

- New relationships to knowledge, access, usage, and maintenance, and the establishment of new mechanisms of innovation.

The effects of this intrusion are concretely apparent from now on:

- By networking skills in various sectors of socioeconomic activity.
- By effectively capitalizing on different knowledge.
- By setting up and development of partnerships to better polarize the energies of production and innovation.

These transformations imply that the means of the nation, ever more numerous, are solicited to compete to a certain evolution of a society oriented by, if not fully dedicated to, knowledge. It aims to be more in line with the aspirations and needs of the citizen for knowledge. On the other hand, knowledge freely accessible to all, constitutes an incontestable democratization factor since eliminating all social discrimination.

The corollary of the information evolution of our society is the increase of its fragility: the systematized availability of knowledge, responding to a legitimate desire to make them accessible to all, increases the vulnerabilities. It is a safe bet that these will be unfortunately exploited by some pursuing criminal purposes. What happens every day on the networks is there to remind us. It is also to be feared that, in the event of a crisis, this vulnerability of the information society will become the subject of new forms of conflict. This can take the form of a war of meaning: the manipulation of deviant information about an adversary whom one seeks to destabilize psychologically.

It can also take the form of a war of potentials whose offensive nature is marked by the propagation of false information to weaken the adversary by targeting the sensitive elements (which may be of a diverse nature) constituting the essence of its potential. These are threats to information societies, which, despite their apparent harmlessness, must give rise to the same level of vigilance as is required by the usual forms of conflict. They are one of the many facets of the information war.

All nations have realized the importance of conserving on public networks all cognitive elements directly or indirectly involved in the socio-political and economic life. Huge or gigantic collective memories are worldwide created that require to be organized and managed for the benefit of all human communities. The access increasingly trivialized to servers, greatly democratized by the Internet, offers to all without social or cultural discrimination an access to contemporary knowledge. This is an important sociological fact that one must take the full measure because it will have a significant impact on social organizations, changing patterns of thought and action among its members, and, to some extent, on the modes of action of policy makers. Collective behavior will conform to new habits to learn to decide and act. This will also have an impact on citizens who now can directly perform administrative procedures using specialized servers.

Our relationship to knowledge is changing, becoming more direct and immediate, and furthermore showing changes in our learning techniques and in our modes of appropriation of knowledge. This will certainly impact the definition and implementation of future vocational training. The computer, networks, servers, and specialized software used together now possess all the attributes to become omniscient tutors and be most effective because they are available at any time. All ingredients and special-purpose nurturers are thus gathered to create a tremendous evolution of societies immerged in the world of information and communication. Already, a telematics infrastructure as the Internet, resting on finely capillaries-like networks, is both the foundation and nervous system of new informational situations.

The informational evolution concerns not only the major institutions of societies but also its basic cells: homes, where the computer becomes part of the family furniture. Connected to the Internet, the computer becomes a privileged instrument of dialogue for the entire family. The family becomes, ipso facto, by analogy with the concept of economic agent, an "*informational agent*" by consuming services and information products available on the World Wide Web. We must also expect to see some situations that generate their own "informational pollution" because no longer obedient to rational expressed needs. The outlook of a sustainable development will undoubtedly lead to eliminate superfluous inherent overabundant consumption and production in favor of the only treatments corresponding to a user 'strict needs to know'.

In information society, new possibilities for action are offered to actors that expand their intervention context and gear their knowledge. It follows a modern philosophy of action, which is observed in many places:

- Techniques, whatever their origin, must no longer be stationed in areas of elitist employment.
- Everything must be done to remove the technical point of contact between a server and a user: it should be kept away from technical arcana and generate maximum easements for implementation and operation.
- The repositories of information must be organized rigorously and practically. The emergence of new knowledge enables a rational exploitation and access facilitated by telematics.
- The information, from its initial creation, must be structured to be integrated into the memories of server computers.
- The coupling between a server supporting a domain of knowledge and a telematics infrastructure has a multiplier effect on a decision-maker. The decision-maker is a priori a non-specialist of the knowledge they seek to the extent that:
- The collective memory stores knowledge as diverse as specialized and from worldwide origin. The latter portends that we have access to knowledge derived from recent discoveries in the world since servers are timelessly updated.
- Access to information for all is almost instantaneous, whatever the extent of their geographical position is.

Boundaries become transparent to exchanged messages ipso facto making the transboundary information exchange mechanisms. This informational evolution announces changes for the usual way of thinking and acting hitherto maintained in a society often founding its modes of action on cultural bases. Novel forms of action appear that are more directed by rational domestication of a sort of "*informational energy*." For the new "*energy actors*" of the information age, involved in the major functions of management and finance, there is food for thought before designing and proceeding with entrepreneurship in organizations and various institutions.

The information also presents great interest to become a lever for saving energy, in the sense that its proper use in different operating environments allows both: reduce servo and improve regulations for instance in inventory management and control optimization of industrial processes. Along with these changes, there is no doubt that attitudes will change and help shape the contours of a new civilization, where the action will be led by a better mastery of knowledge and a more rational assimilation of knowledge.

1.4 Informational Situations

Wishing to know, having to know, or simply trying to stay aware represents a true informational challenge [26]. Are we witnessing an information revolution through a huge convergence of a multitude of techniques coming from various origins? It is more reasonable to substitute evolution to the term "revolution" since it is the combined effects of adaptation and maturation of several techniques. The integration of these techniques into advanced architectures of information systems materializes as an important factor of societal evolution, sparking changes in behavior and opening new fields of application. Evolution also generates a constant search for circumstantial adaptation of terms of employment or simultaneous implementation of those techniques as, for instance, the joint use of several modes of interaction: gesture designation with voice control, for instance.

Informational situations exemplify new contours drawn because of a systematic digitization of multimedia and of communication and data transmission networks as well as a large-scale miniaturization of information storage systems. This results in numerous consequences. Information contents are indeed the clear tendency to emancipate themselves from their traditional media, including paper, thereby focusing more on the "object of knowledge"or information content carried by the media or the message without much having to worry about the physical characteristics of the container or vehicle. The techniques related to the semantics of information have undoubtedly found a new tone and new concepts such as Information-Centric Networking (ICN)[3] [28] to network contents rather than devices that hold the contents. Internet usage has drastically shifted from host-centric end-to-end

---

[3]"*The ICN concept has born in the era that more and more users are shifting their interests to the content itself rather than the location or server where contents are stored. This content-centric behavior of user applications has rendered the point-to-point communication paradigm of IP networks inefficient.*"

communication to receiver-driven content retrieval. One important feature of ICN architectures is to improve the transmission efficiency of content dissemination. ICN has emerged as a promising candidate for the architecture of the Future Internet.

The informational development within the society has generated an ever increasing need to transport data and has imposed a big load on telematics networks, usually via existing telecommunications infrastructure and its evolutions. Careful observation of bandwidths in telecommunications infrastructure shows high availability to transport more than only voice signals. More means that a channel, in addition to the voice signal, would carry the various typical vectors of information: text, graphics, photos, videos. This objective is at the origin of multi-service networks: to propagate on dedicated tracks, the pictorial information (previously digitized) with or without animation, with the same apparent ease than with usual computer data (e.g., 5G). The research community is currently exploring advanced approaches to transform the Internet, as we know it today, into a system more capable of effective content distribution and sharing, according to today's (and tomorrow's) needs.

It is tempting to draw an analogy between information and raw material. From the perspective of raw material, information must undergo some processing steps before becoming a commodity because, despite an intangible nature, it must be transformed each time it must respond to new needs or get developed products to meet specific requirements. That differs little from the processing steps of the industrial revolution actions with respect to a mineral. Therefore, it can easily be likened to a commercial product undergoing the usual phases of transformation, (re) packaging, and distribution before the arrival at the market. Here the analogy stops because the information has a surprising substantive nature: when a transformation process is applied, then that does not cause a change to its intrinsic nature as in the case of a mineral.

Information can be consumed and replicated indefinitely under energy-free processing steps. Moreover, and this is not the slightest interest, it may, whatever the nature of the exchange mechanism used, be shared endlessly without the emitting source being altered. The information becomes a commodity in its mercantile sense, insofar as access to knowledge, deposited in a server, resulting in commercial transactions. The birth of this new market is not immune to the usual trading imperatives: production costs, profitability, competition. Note that the transmission of knowledge, through its various media, has never been completely free, its price is often dependent on its support implementation parameters. A book full of information can be sold at the same price as a book informatively devoid of interest, insofar as it represents the same production cost.

The current trend of information is continuously to be emancipated from its traditional media or vehicle. Data retrieved from a server and available immediately through telematics infrastructure account for a significantly higher quality of service compared with paper-based information geographically dispersed or difficult to access, especially by ordinary means of transmission (e.g., fax).

The Internet and its associated applications are, in this respect, a major factor of integration. The concept of information dissemination has not been more detrimental, as in the past. In this case, we buy only the information we need precisely without concern or the nature of the vector or its locality. The knowledge delivered by the server cannot be separated from its electromagnetic support. This, however, is not a handicap as in the case of print media, since the information generated by the server is no longer a frozen product. It can be enriched on demand using specialized software, depending upon the service requested. That will help to expand the informative field or restrict it by better focusing if necessary.

Raw information extracted from data servers can also give rise to successive transformations through the punctual use of specialized software. In doing so, the information is treated as a work in progress in a production cycle of a commercial product. The merchandise qualifier is fully justified within the context of raw material industry with the epiphenomenon of informational fields appearing more or less "open sky" with variable information concentration. Note that the information industry fits well into the context of sustainable development, since by its nature, it allows to go to manufactured products and consume less energy, while incorporating more "intelligence." This quality of smart product can, however, be achieved if we knew how to handle large amounts of information (e.g., IoT, Big Data) and assimilate as much knowledge to manufacture such products.

Millions of people now use online social network applications such as Twitter, Facebook, and Google+. This kind of application offers the freedom for end users to easily share contents on the Internet. For instance, there is a large consensus that the Internet of Things (IoT) will play a primary role in providing global access to services and information offered by billions of heterogeneous devices (or *things*), ranging from resource-constrained to powerful devices (and/or virtualized everyday life objects) in an interoperable way.

IoT aims to connect each device with the Internet, so that these devices can be accessed at any time, at any place, and from any network. Smart objects like smart washing machines, smart refrigerators, smart microwave ovens, smart-phones, smart meters, and smart vehicles connected via Internet enable applications like smart home, smart building, smart transport, digital health, smart power grid, and smart cities. When billions of these devices are connected to the Internet (IoT), combined with the data produced by multiple sources like Facebook, YouTube, etc., we end up in a situation that we call BigData. In fact, most of the IoT applications are *information-centric* in nature, since they target data regardless of the identity of the object that stores or originates them. For example, road traffic monitoring applications are ignorant to the specific car/sensor that provides the information.

Let us conclude this section by this citation from Burgin [29] that is still quite actual:

Creation of information technology of our time is, may be, the most important revolution in human evolution. Now information processing plays more and more important role in life of people and functioning of society. One of the most threatening features of the contemporary society is the common lack of ability to discriminate between useful, sound, and trustworthy information and a worthless, often harmful, and misleading stuff that has always been around but now with the advent of the Internet, e-mail systems and other communication tools, it multiplies with unimaginable rate. . . ., a lot of information carrying media are full of information that is of low quality and quite often imprecise and even false. The usage of such information to achieve a chosen objective may bring and often brings unwelcome results.

1.5   Mastering and Improving Knowledge

The context of better knowledge promotes a better perception, representation, and interpretation of reality. This should facilitate decision-making better suited to circumstances and actions that are wisely carried out on the world. In so doing, the world in which we are acting evolves. It follows that the state of it will change, by retroactive effect, resulting in a change of its cognitive state. Indeed, every time that ideas change as well as behaviors about the universe of action, this implies that a renewed vision inevitably has a direct impact on future actions. For all these reasons, it is necessary to have a global and permanent vision on the activities and processes generating information, as well as on the different information worlds, hence the idea to have an informational framework encompassing all. From there, the idea of an informational meta-space emerges, dedicated to better monitoring the informational flows and facilitating the coherence between knowledge and action and properly arranging knowledge to be actionable.

This singular or informational space, which some people call, is an *infosphere* [30–32]. This term has not yet been given a well-defined academic definition, giving rise to different connotations. The sphere invokes the concept of unifying volume, an informational integration between the physical, virtual, and cognitive worlds since everything that refers to it enters into the same plan of an informational integration. The notion of a well-accepted infosphere must also effectively facilitate the capitalization of knowledge by erasing any boundary between diverse fields of knowledge and proceeding, as it were, to a common denominator.

The boundaries, both those which exist naturally and those which have been created artificially (often justified by the emergence of difficulties in the perception of the universe), materializing subdivisions may prove to be later penalizing for an action. The infosphere has an undeniable interest in amalgamating all the knowledge, which favors the ability to "derive" knowledge useful to the conduct of an action, in particular by means of inference techniques. It also represents the ideal place to orient the acquisition of knowledge enabling its enrichment.

Finally, the infosphere constitutes a support adapted to the development of a knowledge strategy by better orienting the methods of collecting, organizing, and disseminating knowledge by improving the innervation of technical and technological fabrics by including them properly in objectives for policies or strategies and for the development of knowledge centers. At the level of political decision-makers, the infosphere can facilitate the functioning of crisis units, giving them early warning and decision-making skills, obtained by a mastery of the timely knowledge, constantly refreshed by a rational organization, the recognition of factual information about the crisis, its collection and treatment.

1.5.1 Toward a Better Mastery of Knowledge

Mastering knowledge depends first on a dilemma evoked by the question [26]:

- What is it necessary to know exactly or to what extent is it permissible to ignore when one intends to conduct an efficient action?

In fact, well beyond the relevant knowledge, which we need to have in order to respond to the concern for efficiency, there arises the question of the strict level of sufficiency of knowledge. It is necessary to stay conscious that to master the totality of knowledge remains illusory. An illusion that remains for many, in a simplistic vision, that the universal in its totality is completely accessible because of:

- An irresistible and continuous push from the web.
- A permanent tightening of the meshing of telematics infrastructures.
- A growing and apparent permeability between knowledge and all the worlds of action.
- The capitalization of knowledge, greatly facilitated by the fineness of networking.
- The federation of energies linked to innovation.
- The development of partnerships set up in a concerted information policy.

If you use a metaphor of a river that rushes out of its bed, the consequences of the flooding are hardly parable if the flood continues. This is the same situation with information overflow, or deluge of information, or Big Data, to which we quickly confront the incessant development of the web. The result is a certain disorder, which is difficult to control, since there is nothing to suggest a decrease in flows, on the contrary. This disorder, which is assimilated to a growing difficulty in the accessibility of knowledge, as soon as we consider it from the point of view of the universal, will modify our relationship to knowledge unexpectedly and probably in depth.

For the scholars of the eighteenth century, it was permissible and conceivable to enroll in a single framework all the knowledge of their time, an undertaking which, moreover, found its justification with Diderot in the realization of the Encyclopedia. Due to a weak evolution of sociological contexts and a very moderate progression of techniques, the availability of knowledge useful for any activity did not present difficulties, all seeming to proceed, it might be said, with a certain informational stability. It then became reasonable to assume that knowledge in general could comply with the requirements of a totalization, thus making it possible to establish quasi-exhaustive references.

With the concepts of infosphere and cyberspace or informational meta-space that concretize all environmental contexts and activities subordinate to the web, attitudes of the eighteenth century are no longer reproducible. These concepts, in fact, constitute a new way to "universal knowledge" by introducing, as it were:

- A *virtual access* so qualified insofar as it gives the impression of being able to access all knowledge, produced here and there, if all domains of the "*known*" are covered at a given moment.

- *Real access* because it cannot go beyond the framework of a subset of knowledge, in spite of the efforts deployed and the quality of the techniques implemented.
- *Effective and relevant access* which, in the case of real access, only achieves the desired useful knowledge.

Due to the capitalization of knowledge, the profusion of reticular means, and the sophistication of the search engines, it is now possible to gain access to all knowledge, whatever be its origin and nature. It should be noted that the "universal knowledge"referred to above may in fact be decomposed into different fields more truly accessible. Moreover, it would be futile to think that knowledge which is easily generalizable is nevertheless truly accessible. They are often the reflection of requirements corresponding to needs, responding to community interests which, as a result, are subject to rapid changes. The absence of proven durability removes the privilege of generality since they are of interest to their promoter at a given moment or a given period of time. It is therefore necessary to restrict the field which is really accessible to a useful subdomain which can make sense with regard to the actions envisaged.

1.5.2   Universality and Mastering Knowledge

To complete our discussion of the accessibility of "universal knowledge," it is useful to examine how the relations between knowledge and universality are established over time, according to the modes of communication, without forgetting their implications on semantics. This examination can be carried out in three stages [26]:

- In societies having only the mode of oral communication, exchanges took place in the same context, since the one who emitted a message could only do so in the presence of its recipient. The two communicating actors being "immersed" in the same situation, ipso facto (with a few rare exceptions) gave the same meaning to the objects of communication, thus achieving a certain semantic closure.
- The arrival of the printing press and the development of written materials (case1), initiated a new communication space in the sense that exchanges of information could be carried out without requiring a physical presence. The notion of common situation, the sharing of knowledge, and immediacy were blurred in favor of a circulation of messages distorting cultural differentiations and temporal shifts: a message that could be exploited much later than its production. The arrival of the textual supports (case2), did not take place without consequence. Linguistic conventions become more imperative than in case1 to allow exchanges, hence the rationale for: syntaxes, the birth of grammars, pragmatics, etc. All these efforts, when properly coordinated, contribute to a large extent to establishing the notion of universality. In case2 the relation to the sense changes with respect to case1 because every text carry within it, a signified, which must remain in spite of the diversity of users, contexts, temporal places and conditions of use: message

written yesterday must make sense as much in the present and future time and, of course, whatever the place of its writing.

- The new informational and communication space created by the generalization of electronic media brings to light a new form of universal which must be assimilated more to a global approach and to a planetary vision of the availability of the means, with the avowed objective, that all can communicate with all.

Surprisingly, this universal is non-totalizing insofar as a large number of the inhabitants of the planet remain excluded from the new agoras which materialize the infosphere and nothing further indicates whether updates remain to be done. The notion of totality singularizing the written supports is emptied of its meaning because the fields of knowledge are in perpetual evolution, just as the networks of hypertexts are constantly changing.

It should be noted, in opposition to what has been said in case2 that cyberspace becomes conducive to a certain return of contextualization with a difference in scale compared to case1 especially with the development of forums on the Internet. It is within these forums that represent the interests of certain communities that organizations are formed whose essence is to make sense in their rapprochement, when it is not simply a question of sharing it, because of the centers of common interest.

Without access to only limited and partial knowledge areas, ad hoc means are needed to gain access to the only areas of relevance, in other words, those that make sense in decisions and actions which it is desired to implement and ensure. This led us to suggest an approach that can help to achieve this objective by developing around the chain of knowledge a set of modalities that range from the way we grasp the universe of actions to its formalized representation.

1.6 Data, Information, and Knowledge

Burgin's books [29, 33] give a broad overview of the main mathematical directions to support general theories of information and knowledge. The general theory of information provides a unified context for existing directions in information studies, making it possible to elaborate on a comprehensive definition of information; explain relations between information, data, and knowledge; and demonstrate how different mathematical models of information and information processes are related. The material in this section is mainly from Burgin's books [29, 33].

1.6.1 What Is Data?

Data has experienced a variety of definitions, largely depending on the context of its use. Here are some expressions reflecting the usage of the term *data* that assign some meaning to this term:

- In information science, data is defined as unprocessed information.
- In other domains, data are treated as a representation of objective facts.
- In computer science, expressions such as a data stream and packets of data are commonly used.
- Having sources of data or working with raw data are other commonly encountered ways of talking about data.
- We can place data in storage, e.g., in files or in databases, or fill a repository with data.
- Data are viewed as discrete entities: they can pile-up, be recorded or stored and manipulated, or captured and retrieved.
- Data can be mined for useful information, or we can extract knowledge from data.
- Databases contain data.
- It is possible to separate different classes of data, such as operational data, product data, account data, planning data, input data, output data, and so on.
- Data are solid, physical, things with an objective existence that allow manipulation and transformation, such as rearrangement of data, conversion to a different form or sending data from one system to another.
- Data are symbols that represent properties of objects, events, and their environments.
- Data are products of observation by people or automatic instrument systems.
- Data is a set of values recorded in an information system, which are collected from the real world.

There are different kinds and types of data. As a substance, data can be measured. The most popular way to measure data is in bits where *bit* is derived from the term binary digit, 0 or 1. Namely, the vast majority of information processing and storage systems, such as computers, calculators, embedded devices, CD, DVD, flash memory storage devices, magneto-optical drives, and other electronic storage devices represent data in the form of sequences of binary digits.

1.6.2   What Is Information?

The question "What is information" is singled out as one of the most profound and pervasive problems for computer science. It reflects the kind of far-reaching issues that drive day-to-day research and researchers toward understanding and expanding the frontiers of computing. Few books concerning information systems clearly define the concept of information. One of the most common ways to define information[4] is to describe it as one or more statements or facts that are received by people and that have some form of worth to the recipient (Losee [34]).

---

[4]Etymologically the term information is a noun formed from the verb "*to inform*," which was borrowed in the fifteenth century from the Latin word "*informare*," which means "*to give form to*," "*to shape*," or "*to form.*"

Shannon himself never defined information and wrote only about the quantity of information and called it a theory of communication but not of information. Shannon [35] also was very cautious writing: *"It is hardly to be expected that a single concept of information would satisfactorily account for the numerous possible applications of this general field."* Burgin [29] reports several citations from researchers on the difficulty to define that term *"information."* Some of them are listed below:

- Belkin [36] argues (1978), *"The term information is used in so many different contexts that a single precise definition encompassing all of its aspects can in principle not be formulated."*
- Scarrott [37] writes (1989), *"During the last few years many of the more perceptive workers in the information systems field have become uneasily aware that, despite the triumphant progress of information technology, there is still no generally agreed answers to the simple questions—What is information? Has information natural properties? What are they?—so that their subject lacks trustworthy foundations."*
- Wilson [38] writes (1993), *"Information is such a widely used word, such a commonsensical word, that it may seem surprising that it has given 'information scientists' so much trouble over the years."*
- Flückiger [39] (1995) came to the conclusion: *"those working in areas directly related to information had apparently accepted that the problem with the definition would remain unsolved and considered Shannon's concept of information as the most appropriate."*
- Barwise and Seligman [40] write (1997), *"There are no completely safe ways of talking about information. The metaphor of information flowing is often misleading when applied to specific items of information, even as the general picture is usefully evocative of movement in space and time. The metaphor of information content is even worse, suggesting as it does that the information is somehow intrinsically contained in one source and so is equally informative to everyone and in every context."*
- Capuro and Hjorland [41] also stress (2003), *"for a science like information science (IS) it is of course important how fundamental terms are defined."* In addition, they assume that even *"discussions about the concept of information in other disciplines are very important for IS because many theories and approaches in IS have their origins elsewhere."*
- Burgin [29] remarks (2010), *"One more problem is that trying to tell what information is a necessary clear distinction is not made between a definition and an attribute or feature of this concept. Normally, describing or naming one or two attributes is not considered a definition. However, there are authors who stress even a single feature of information to the point that it appears to be a definition."*
- Opinions of other researchers [29] argue that: *". . . multifarious usage of the term information precludes the possibility of developing a rigorous and coherent definition."*

The situation with the difficulty of defining the term *information* is being illustrated by the famous ancient story of the blind men and an elephant[5] related in Burgin [29], p.39, and replicated as example 1.1.

Example 1.1 Ten Blind Men and An Elephant   Once upon a time there was a certain raja, who called his servant and said, "*Go and gather together near my palace ten men who were born blind... and show them an elephant.*" The servant did as the raja commanded him. When the blind men came, the raja said to them, "*Here is an elephant. Examine it and tell me what sort of thing the elephant is.*"

- **The first blind man** who was tall found the head and said, "*An elephant is like a big pot.*"
- **The second blind man** who was small observed (by touching) the foot and declared, "*An elephant is like a pillar.*"
- **The third blind man** who was always methodical heard and felt the air as it was pushed by the elephant's flapping ear. Then he grasped the ear itself and felt its thin roughness. He laughed with delight, saying "*This elephant is like a fan.*"
- **The fourth blind man** who was very humble observed (by touching) the tail and said, "*An elephant is like a frayed bit of rope.*"
- **The fifth blind man** who was daring walked into the elephant's tusk. He felt the hard, smooth ivory surface of the tusk and its pointed tip. "*The elephant is hard and sharp like a spear,*" he concluded.
- **The sixth blind man** who was small observed (by touching) the tuft of the tail and said, "*An elephant is like a brush.*"
- **The seventh blind man** who felt the trunk insisted the elephant was *like a tree branch*.
- **The eighth blind man** who was always in a hurry bumped into the back and reckoned the elephant was *like a mortar*.
- **The ninth blind man** was very tall. In his haste, he ran straight into the side of the elephant. He spread out his arms and felt the animal's broad, smooth side and said, "*This animal is like a wall.*"

Then these nine blind men began to quarrel, shouting, "*Yes it is!*," "*No, it is not!*," "*An elephant is not that!*," "*Yes, it's like that!*," and so on.

- **The tenth blind man** was very smart. He waited until all others made observations and told what they had found. He listened for a while how they quarreled. Then he walked all around the elephant, touching every part of it, smelling it, listening to all its sounds. Finally, he said, "*I do not know what an elephant is like. That is why I am going to write an Elephant Veda, proving that it is impossible to tell what sort of thing the elephant is.*"

---

[5] -- originating from India, having different versions, and being attributed to the Hindus, Buddhists, or Jainists.

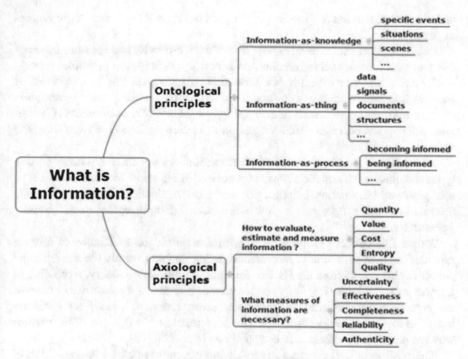

Fig. 1.5 Principles of information from Burgin's general theory of information [29]

Face to a huge diversity of meanings of the word *information*, elicited above, Burgin [29] develops the General Theory of Information (GTI). GTI makes it possible to unify these understandings and to explicate the essence of such a critical phenomenon. The parametric definition of information in GTI utilizes a system parameter called an *infological* system that plays the role of a parameter that discerns different kinds of information, e.g., social, personal, chemical, biological, genetic, or cognitive, and combines all existing kinds and types of information in one general concept of "*information.*" The GTI parametric type of comprehensive definition of information explains and determines what information is. Knowledge obtained in known directions of information theory such as statistical (Shannon 1948 [42], Fisher 1925 [43]), semantic (Bar-Hillel and Carnap 1953 [44]), algorithmic (Kolmogorov 1965 [45], Chaitin 1977 [46]), economic (Marshak 1971 [47]) may be treated inside GTI as its particular cases. Burgin's GTI is a system of principles as illustrated in Fig. 1.5. There are two groups of such principles: ontological and axiological.

Ontological principles reflect major features of information essence and behavior. The principles give an answer to the question "*What is information?*" In particular, information is defined as "*a phenomenon that exists in nature, society, mentality of people, virtual reality, and in the artificial world of machines and mechanisms created by people.*" The term information is used in different ways, including *"information-as-knowledge,"* *"information-as-thing,"* and *"information-as-process."*

Axiological principles explain how to evaluate, estimate, and measure information and what measures of information are necessary. Axiological principles provide a reference frame for a diversity of information measures, such as *". . ., information quality, information quantity, information value, information cost, information entropy, information uncertainty, average information score, information effectiveness, information completeness, information relevance, information reliability, and information authenticity."*

Information is a basic multifaceted phenomenon. As a result, it has many diverse types and classes. Information is better processed if we know to what class or type this piece of information belongs, for instance, distinctions between *"genuine information, false information, misinformation, disinformation, and pseudo-information."*

Werner Loewenstein [48] reveals that information is the foundation of life. He provides his own definition of information, being unable to apply the conventional definition that comes from the Hartley-Shannon's information theory. According to Loewenstein, *"information, in its connotation in physics, is a measure of order—a universal measure applicable to any structure, any system. It quantifies the instructions that are needed to produce a certain organization."* This is one of the reasons why the principle of information entropy is so important.

Finally, Burgin [29] provides the following general remark on the definition of information[6] as follows: *"Information is not merely a necessary adjunct to personal, social and organizational functioning, a body of facts and knowledge to be applied to solutions of problems or to support actions. Rather it is a central and defining characteristic of all life forms, manifested in genetic transfer, in stimulus response mechanisms, in the communication of signals and messages and, in the case of humans, in the intelligent acquisition of understanding and wisdom."*

To explicit his ontological principles, Burgin [29] introduces the notion of an "infological" system IF(R) of a system R. As an example of an IF(R) of an intelligent system R, is the system of knowledge of R. It is called in *cybernetics* the thesaurus[7] Th(R) of the system R. Identifying an infological system IF(R) of a system R allows to define different kinds and types of information relative to this system.

---

[6] Excerpts from Burgin's book as well: *". . . the term information has been used interchangeably with many other words, such as content, data, meaning, interpretation, significance, intentionality, semantics, knowledge, etc. In the field of knowledge acquisition and management, information is contrasted to knowledge. Some researchers assume that if Plato took knowledge to be "justified true belief", then information is what is left of knowledge when one takes away belief, justification, and truth."*

[7] A thesaurus can form part of an ontology. It is used in natural language processing for word-sense disambiguation and text simplification.

Ontological principles explain *"the essence of information as a natural and artificial phenomenon"*:

- *Substantial*: definition of information.

  - Separate information in general from information for a system R, i.e., in what context information is defined.
  - Information for a system R is a capacity to cause changes in the system $IF(R)$ of R; information is a capacity of *"things,"* both material and abstract, to change other *"things,"* thus information is intimately linked with time.
  - Information only exists in the form of *portions of information*.

- *Existential*: description of how information exists in the physical world.

  - For any portion of information, I, there is always a carrier C of this portion of information and, its representation, for a system R.

- *Dynamical*: how information functions.

  - A transaction/transition/transmission of information goes on only in some interaction of C with R.
  - One and the same carrier C can contain different portions of information for one and the same system R.

Axiological principles explain *"how to evaluate information and what measures of information"* are necessary.

- A measure of information I for a system R is some measure of changes caused by I in R.
- One carrier C can contain different portions of information for a given system R.
- There are three types of measures of information (note that *time* is involved): (1) potential or perspective; (2) existential or synchronic; (3) actual or retrospective.
- According to scale of measurement, there are two groups (qualitative, quantitative) which contain each three types of measures of information as illustrated in Fig. 1.6.
- According to spatial orientation, there are three types of measures of information: external, intermediate, and internal (Fig. 1.6).

Information in a broad sense can be defined as an ability of things, both material and abstract, to change other things. The Burgin's general theory[8] of information provides us tools for discerning information, measures of information, information

---

[8]The general theory of information does not exclude necessity of special theories of information, which can go deeper in their investigation of specific properties of information, information processes, and systems in various areas, such as Shannon's theory, Kolmogorov's complexity, semantic, economic, evolutionary, pragmatic, semiotic, and other special theories.

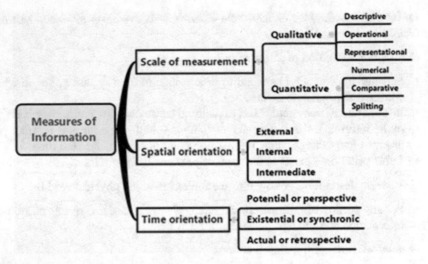

Fig. 1.6 Burgin's typology of measures of information

representations, and carriers of information. Example 1.2 presents a concrete illustration of these elements.

Example 1.2 Taking a letter written on a piece of paper, we see that the paper is the carrier of information, the text on it is the representation of the information contained in this text, and it is possible to measure the quantity of this information using Shannon entropy or algorithmic complexity (Kolmogorov).

Let us conclude this section by an important Burgin's recommendation: "*All this diversity of information measures confirms that it is necessary not only to invent new measures of information but also to elaborate methodology and theoretical foundations of information measurement.*"

1.6.3   What Is Knowledge?

Our civilization is based on knowledge and information processing. That is why it is so important to know what knowledge is. The principal problem for computer science and technology is to process not only data but also knowledge. Knowledge processing and management make problem-solving much more efficient if we are capable of distinguishing knowledge from knowledge representation; of knowing regularities of knowledge structure, functioning, and representation; and of developing software that is based on these regularities. Many intelligent systems are explicitly or implicitly predefined by the choice of knowledge representation.

Knowledge is intrinsically related to data and information. A lot of ideas, models, and several theories have been suggested treating various problems and studying different issues of knowledge that can be separated into the following directions:

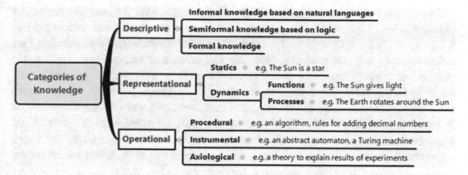

Fig. 1.7 Burgin's categories of knowledge

- **Structural analysis** of knowledge that strives to understand how knowledge is built; this is the main tool for the system theory of knowledge, knowledge bases, and artificial intelligence.
- **Axiological analysis** of knowledge that aims at explanation of those features that are primary for knowledge; this is the core instrument for philosophy of knowledge, psychology, and social sciences.
- **Functional analysis** of knowledge that tries to find how knowledge functions, is produced, and acquired; this is the key device for epistemology, knowledge engineering, and "cognitology."

In the structuration of knowledge, one can distinguish three categories of knowledge as shown in Fig. 1.7 with associated self-explained examples. Burgin [33] (Chap. 6) provides a rather detailed discussion about the structures and processes associated with Fig. 1.7.

The mathematical theory of knowledge developed by Burgin [33] puts the emphasis on structural and axiological analysis. As explained in his book, despite a millennium of effort by philosophers, there is no consensus on what knowledge is and, consequently, there are many definitions of knowledge. As explained in Example 1.2, knowledge is distinct from its representation. The representations are not knowledge itself as the same knowledge can have different representations. However, representations are a very important kind of structures in order to process knowledge.

Informal definitions of knowledge are not practical for computer processing of knowledge since computers can process only formalized information. Formal definitions of knowledge are often linked with some specific knowledge representation. There exist a great variety of formalized knowledge representation schemes and techniques: semantic and functional networks, frames, productions, formal scenarios, relational and logical structures, etc.

Example 1.3 Knowledge Versus Knowledge Representation Consider an event that is described in several articles written in different languages, for example, in English, French, Spanish, and Chinese, but by the same author. These articles convey the same semantic information and contain the same knowledge about the event, but the representation of this knowledge is different.

We always have knowledge about something, some object. However, to distinguish an object, we have to name it. A name may be a label, number, idea, text, and even another object of a relevant nature. Knowledge is acquired by an epistemic system E (see next chapter) under **the actions of information**, i.e., knowledge is the result of the information impact. Three basic stages of knowledge acquisition by a cognitive (intelligent) system are (1) information search and selection; (2) information extraction, acquisition, and accumulation; and (3) transformation of information into knowledge. That is why we treat knowledge in the context of epistemic structures. Examples of epistemic structures used by cognitive processes are concepts, notions, statements, ideas, images, opinions, texts, beliefs, knowledge, values, measures, problems, schemas, procedures, tasks, goals, etc. Note that an epistemic space is a set of epistemic structures or their representations with relations and operations. Examples are vector bundle, lattices, groups, or partially ordered sets.

1.7 Important Notions Related to Actionable Knowledge

Knowledge, experience, and information-processing are the primary foundations determining how goals of actions are established; how situations, opportunities, and risks are assessed; and how groups, cues, and patterns are interpreted. Knowledge, learning, and information-processing can be regarded as links between action and environment. Environment plays a key role in the acquisition of knowledge and the implementation of actions. The relations between knowledge, action, and environment are not simple and current models of decision-making may require refinements. The interdependencies of knowledge, action, and environment from different disciplinary perspectives, scales of analysis, time dimensions, and ontologies have not been sufficiently explored. Each scale of analysis bears specific insights that other scales cannot deliver. Time lags between knowledge acquisition (e.g., research) and successful action (e.g., innovations) can amount to many years or even decades.

The book of Meusburger et al. [1] "*Knowledge and Action*" addresses the issues discussed in the previous paragraph from business and social sciences perspectives. The book we are proposing takes the angle of data science and focuses on information and knowledge processing in dynamic environments. This section presents currently used high-level models of important notions related to actionable knowledge in dynamic environments: dynamic decision-making, situations, situation awareness (SAW), and analytics and information fusion (AIF).

1.7.1 Dynamic Decision-Making Models

Decision-making is involved in all aspects of our lives, and it is of particular importance for the critical CPSS (Fig. 1.8): health, transport, energy, and defense and security. With the advancement of Information and Communications

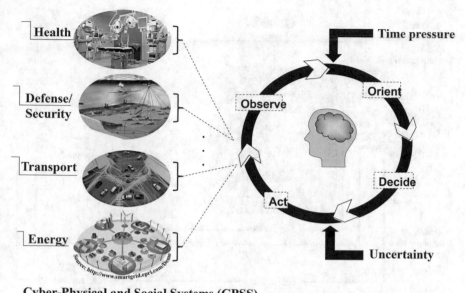

Cyber-Physical and Social Systems (CPSS)

Fig. 1.8 Illustrations of Boyd's OODA loop and four critical CPSS. (Source [50])

Technologies (ICT), these four environments become more and more complex and challenge decision-making. The oversimplified Boyd's Observe-Orient-Decide-Act (OODA) loop [49], illustrated in Fig. 1.8, is used to describe the decision process. Although the OODA loop might give the impression that activities are executed in a sequential way. In reality, the activities are concurrent and hierarchically structured. The processes of the loop are typically performed in a very dynamic and complex environment and are heavily influenced by factors such as uncertainty and temporal stress.

The four CPSS environments in Fig. 1.8 present cases where interdependent decision-making takes place in an environment that changes over time either due to the previous actions of the decision-maker or due to events that are outside of the control of the decision-maker [7]. In this sense, CPSS present, in addition to conventional one-time decisions, dynamic decisions. Dynamic decisions are typically more complex than one-time decisions. They occur in real time and involve observing the extent to which people can use their experience to control a particular complex system, including the types of experience that lead to better decisions over time.

In a problem-solving approach for complex environments such as CPSS, understanding and framing the problem are the most important steps. Over the years, multiple efforts have thus been deployed to better understand and explain decision-making. Fields like decision sciences, management sciences, administrative sciences, social choice, psychology, or naturalistic decision-making are examples of the growing effort in modeling and understanding the individual as well as organizational decision-making.

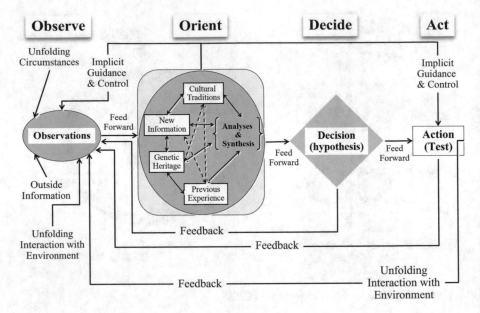

Fig. 1.9 The extended Boyd's OODA loop. (Source [50])

An organization should work toward identifying the demands and constraints (lack of information, lack of time) placed on decision-making in different aspects of its operations, then determine specific decision-making strategies best suited to these conditions. Tools and procedures that clarify the link between various (e.g., analytical and intuitive) decision strategies have the potential to enhance both plans and decisions made through all levels (e.g., strategic, operational, and tactical) of the organization. A decision-making process is defined as a set of activities and tasks that directly or indirectly support the decision-maker. The process and its activities could be well or poorly defined, rational, or irrational, or based on explicit assumptions or tacit assumptions.

The extended Boyd's OODA loop [49], that is shown in Fig. 1.9, presents what Boyd really proposed [51]. In the "*orient*" analysis, there is additional information that a user engages in analyzing and synthesizing a situation. The "*orient*" phase is the repository of our genetic heritage, cultural tradition, and previous experiences. Boyd's "*orient*" is the most important part of the OODA loop since it shapes the way we observe, the way we decide, and the way we act. The OODA loop model has been widely used to represent decision-making in military environments as well as in civilian domains. The OODA applications include information fusion [52–54], analytics [55], autonomic systems [56–59], cultural modeling [60], business intelligence [61, 62], cyber security [63], and semi-automated decision-making [54, 64].

Guitouni et al. [9] review a significant number of decision-making models and present a comprehensive view of the seemingly disparate approaches reported in the literature. They propose a unified decision-making framework that provides a setting to appreciate many contributions to the decision-making process modeling. It is developed around four domains of the decision-making activities: cognition, knowledge (information), organization, and observable effects. The framework could be mapped around four fundamental stages of the decision-making process: perception (observation), understanding (awareness), decision, and action. They assume that many feedback loops, control, and tasking protocols govern the framework. Chapter 2 of [65] presents a description of other known decision-making models: the expected utility theory [66], the prospect theory [67], the regret theory [68], bounded rationality [69–71], and multi-attribute heuristic [72, 73]. However, for the sake of discussion in this book, the extended Boyd's OODA model is sufficient.

1.7.2 Situation Awareness

Situation awareness (SAW) [74, 75] has emerged as an important concept around dynamic human decision-making in both military and civilian complex (e.g., public security, power networks, health networks, transport) environments. Situation analysis is defined as the process that provides and maintains a state of situation awareness for the decision-maker(s), and intelligent system (e.g., data mining, machine learning, information fusion, and analytics) is key enabler to meeting the demanding requirements of situation analysis in current and future complex environments. SAW is a state in the mind of a human. At the heart of this process is the provision of decision quality information to the decision-maker, thereby enabling timely situation awareness. This is on timeless requirement that is being immeasurably complicated by the Big Data 5Vs (variety-volume-velocity-veracity-value).

Blash in Chap. 2 of [76] entitled "*Situation Assessment and Situation Awareness*" provides more details on this topic. Figure 1.10 illustrates a theoretical model derived by Endsley of situation awareness (SAW) based on its role in dynamic human decision-making. SAW is defined [75] as the perception of the elements in the environment, within a volume of time and space, the comprehension of their meaning, and the projection of their status in the near future. The first level of SAW yields in the perception of the status, attributes, and dynamics of relevant elements in the environment. Endsley describes the comprehension process as follows: "*Comprehension of the situation is based on a synthesis of disjoint level 1 elements.*" Level 2 of SAW goes beyond simply being aware of the elements that are present to include an understanding of the significance of those elements in light of pertinent operator goals. Based on knowledge of level 1 elements, particularly when some elements are put together to form patterns with other elements, the decision-maker forms a holistic picture of the environment, comprehending the significance of objects and events. The third and last step in achieving situation awareness is the projection of the future actions of the elements in the environment. This is achieved through knowledge of the status and dynamics of the perceived and comprehended situation elements.

The situation awareness processes described by Endsley are initiated by the presence of an object in the perceiver's environment. However, processes related

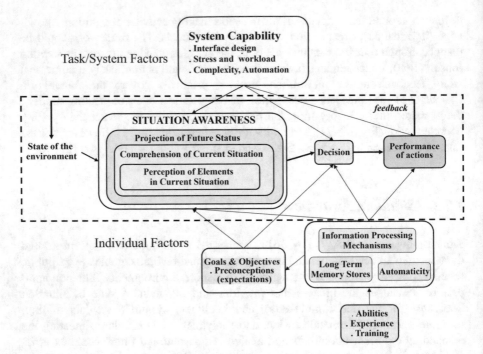

Fig. 1.10 Endsley's situation awareness model. (Source [50])

to situation awareness can also be triggered by a priori knowledge, feelings, or intuitions. In these situations, the picture is understandable, and projections in the future are possible, if any event, which has not been perceived at this time, can be found in the environment. Hence, hypotheses related to the possible presence of an object are formulated. The perceiver then initiates search processes in the environment that confirm or invalidate these hypotheses. Note that this type of SAW is possible only if mental models related to the possible objects are available.

If one compares the OODA loop with the SAW model of Endsley, one sees a close resemblance. In both models, one finds a decision-making part and an action part. In Endsley's model, SAW is one of the main inputs for decision-making. In the OODA loop, the processes Observe and Orient provide inputs for the decision-making process. One should recall, however, that situation awareness in Endsley's model is a state of knowledge and not a process. In her theory of SAW, Endsley clearly presumes patterns and higher-level elements to be present according to which the situation can be structured and expressed. SAW can be interpreted as the operator's mental model of all pertinent aspects of the environment (processes, states, and relationships).

There is a tight link between this mental model used to structure and express situation elements and the cognitive processes involved in achieving the levels of awareness. This link is known as the cognitive fit and requires an understanding of how the human perceives a task, what processes are involved, what are the human needs, and what part of the task can be automated or supported. This understanding is crucial and only achieved via a number of specialized human factor investigations known as cognitive engineering analyses [77–79].

1.7.3   Situations and Situation Analysis

The definitions of "*situation*" and "*situation analysis*" are based on the numerous works of J. Roy mainly from [65, 80]. He defined situation analysis (SA) as "*a process, the examination of a situation, its elements, and their relations, to provide and maintain a product, i.e., a state of situation awareness, for the decision maker*"; and situation as "*A specific combination of circumstances, i.e., conditions, facts, or states of affairs, at a certain moment.*"

The SA process is concerned with understanding the world. There is a real situation in the environment, and the SA process will create and maintain a mental representation in the mind of the decision-maker(s). The idea of "*awareness*" has to do with having knowledge of something. In addition to the cognition facet, awareness is also linked with the notions of perception and understanding/comprehension. The two basic elements involved in situation awareness are the situation and the person. The situation can be defined in terms of events, entities, systems, other persons, etc., and their mutual interactions. The person can be defined according to the cognitive processes involved in situation awareness (SAW), or simply by a mental or internal state representing the situation.

Roy [80] states that the main two basic situation elements are "*entity*" and "*event.*" He pursues the description of elements of a situation, illustrated in Fig. 1.11, the following way: "*. . .an entity is an existing thing (as contrasted with its attributes), i.e., something that has independent, separate, self-contained, and/or distinct existence and objective or conceptual reality. An event is something that happens (especially a noteworthy happening). Hence, entities exist, while events occur. A scenario is defined as a sequence of events. A group represents a number of individuals (entities and/or events) assembled together or having some unifying relationship, i.e., an assemblage of objects/events regarded as a unit. The term activity refers to the notions of action, movement, and motion. It is appropriate when something has the quality or state of being active, i.e., when something is characterized by action or expressing action as distinct from mere existence or state. Finally, one may think of a global situation as being composed of a set of local situations. The term scene, i.e., a single situation in this set, could be used to refer to a local situation.*"

Roy [80] finishes the description of basic elements of a situation (Fig. 1.11): "*A fact is defined as something that has actual existence, or an actual occurrence. It is a piece of information presented as having objective reality. Clearly, a fact has to do with the notions of truth, verity, and reality. Data is indeed factual information used as a basis for reasoning, discussion or calculation. Cues, i.e., features indicating the nature of something perceived, are generated and received from the various data/*

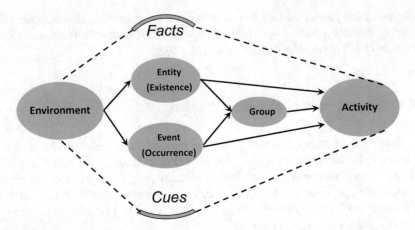

Fig. 1.11 Basic elements of a situation

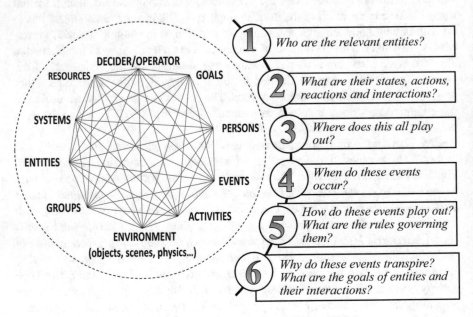

Fig. 1.12 Six epistemic questions for situation assessment. (Source [50])

information sources. Hence, the SA process manipulates data and information that include cues, about facts, obtained from external sources, and inferences drawn by the system itself."

If we are talking about situation analysis support systems, we refer to systems (e.g., AIF based) that can provides answers, in whole or in part, to the six basic epistemic questions excerpted from Sulis [81] that are listed in Fig. 1.12. Phrased in a different way, similar questions appeared in Nicholson [82] with respect to situation awareness: *"What are the objects of interest? Where are they? How are they moving? Where have they been? Where might they be going?"* Objects could refer to either physical objects, such as vehicles, or symbolic objects such as terrorist plans.

1.7.4   Analytics and Information Fusion (AIF)

Situation analysis (SA) is defined as the process that provides and maintains a state of situation awareness for the decision-maker(s). Managing the complexity of situation understanding and of the decision space is a crucial challenge. This challenge is compounded by the wide spectrum and diversity of data that must be analyzed, processed, fused, and eventually transformed into actionable knowledge. The data explosion comes with the advent of advanced sensing and the diversity and volume of data from multiple sources and forms (unstructured and open sources, voice records, photos, video sequences, social networks, etc.). The decision-makers/analysts cannot cope with the flow of material, with potentially severe consequences on the quality of decisions and a negative impact on the performance of operational processes.

Consequently, to support the military decision-makers/analysts in making sense out of the data, a field of research called information fusion (IF) begun about 25 years ago and has been slowly maturing in terms of the establishment of an underlying science and in terms of standardized engineering methodology. The overall objectives of IF technologies are then (1) to support decision-makers/analysts in making sense out of an ever-increasing volume of data and complexity; (2) to improve the quality of information, reduce uncertainty, and support real-time decisions and actions; and (3) to ensure and improve dependability face to an ever-increasing complexity (networked aspect and heterogeneity).

In both defense and civilian businesses, situation analysis provides the actionable knowledge for planning and acting. In the military domain, information fusion (IF) systems, built from smart technologies, have emerged as a key enabler to support SA for complex environments (hostile or cooperative). In the general business and industry domain, the term "analytics" [83, 84] is used to pursue roughly same objectives considering different focus and constraints. Constraints seem getting closer with the advancement of network technologies, so there are some foreseen benefits for the convergence of these two application communities. Both information fusion and analytics are the application of computer sciences and technologies, operational research, cognitive engineering, and mathematical approaches and methods to support situations understanding.

The data fusion model maintained by the Joint Directors of Laboratories' Data and Information Fusion Group (JDL DIFG) is the most widely used approach for categorizing data fusion-related functions [85]. The JDL distinction, among fusion "levels" in Fig. 1.13, provides a valuable way of differentiating between data fusion processes that relate to the refinement of "objects," "situations," "threats," and

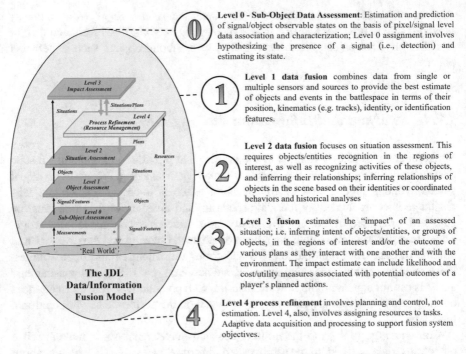

Level 0 - Sub-Object Data Assessment: Estimation and prediction of signal/object observable states on the basis of pixel/signal level data association and characterization; Level 0 assignment involves hypothesizing the presence of a signal (i.e., detection) and estimating its state.

Level 1 data fusion combines data from single or multiple sensors and sources to provide the best estimate of objects and events in the battlespace in terms of their position, kinematics (e.g. tracks), identity, or identification features.

Level 2 data fusion focuses on situation assessment. This requires objects/entities recognition in the regions of interest, as well as recognizing activities of these objects, and inferring their relationships; inferring relationships of objects in the scene based on their identities or coordinated behaviors and historical analyses

Level 3 fusion estimates the "impact" of an assessed situation; i.e. inferring intent of objects/entities, or groups of objects, in the regions of interest and/or the outcome of various plans as they interact with one another and with the environment. The impact estimate can include likelihood and cost/utility measures associated with potential outcomes of a player's planned actions.

Level 4 process refinement involves planning and control, not estimation. Level 4, also, involves assigning resources to tasks. Adaptive data acquisition and processing to support fusion system objectives.

Fig. 1.13 Description of the processing levels of the JDL Data/Information Fusion Model

"processes." In addition to the descriptions provided in Fig. 1.13, the following remarks bring more clarification.

- *Level 1—Object Assessment*: Level 1 involves tracking and identification that includes reliable location, tracking, combat ID, and targeting information. Such information may be of high quality particularly when it leverages multiple sensors to provide robustness to spoofing or deception, reliability in case of sensor malfunctions, and extended space time coverage due to the diversity of observations.
- *Level 2—Situation Assessment*: Level 2 focuses on relations. Inferring relationships of objects in the scene based on their identities or coordinated behaviors and historical analyses; the capability for the automated system to estimate certainties about object identities and activities; and the capability to request human assistance or additional information from sensors or databases to resolve ambiguities.
- *Level 3—Impact Assessment*: Level 3 addresses methods for constructing and learning a wide variety of models of threat behavior; methods for reasoning with uncertain and incomplete information for assessing threats from object activities and relationships; methods for efficient analytics.
- *Level 4—Process Refinement*: Level 4 processing involves planning and control. This is the level where smart technologies can be used at profit for attaining the overall fusion system objectives.

In most defense applications, data/information fusion processing tends to be hierarchical in nature due to the inherent hierarchies built into defense organizations and operations. As a result, the fusion process also progresses through a hierarchical series of inferences at varying levels of abstraction (data-information-knowledge). The exploitation of contexts in the inference processes supports an increase in semantics that is only obtained by *"what one can do with the information?"*, i.e., *"actionable"* knowledge. There are numerous books available that present more definitions, explain concepts in detail, and develop mathematical techniques and models given in Fig. 1.13. For more detailed insights, the following books on data and information fusion are recommended [50, 65, 76, 86–89].

Eckerson [90] defines **analytics** as follows: *"everything involved in turning data into insights into action."* This is a quite broad definition that could include "data and information fusion" but does not help understand analytics from an applied point of view. Unlike information fusion community, the analytics community did not benefit from a well-structured organization like JDL to fix the terminology to ease communication among communities. The definition of analytics associated with Big Data becomes more and more confusing with various vendors, consultants, and trade publications defining and offering new technologies. As Sheikh points out in his book [91], analytics is one of the hot topics on today's technology landscape (also referred to as Big Data). Analytics is not new and originates from business intelligence. It has been rejuvenated with Big Data. Figure 1.14 illustrates Sheikh's proposal [91] to define analytics based upon both business and technical implementation perspectives.

The business value perspective looks at data in motion as it is generated through normal conduct of business. For this data, there are three variations of value: the

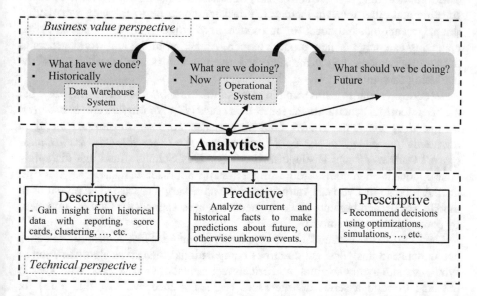

Fig. 1.14 Sheikh's proposal definition of analytics. (Adapted from [50])

present, the **past,** and the **future,** in the exact order as represented in Fig. 1.14. When data is created, referenced, modified, and deleted during normal business activities, it lives in an operational system. The operational system at any given time can tell us what we are doing now. The data from day-to-day operations (e.g., selling merchandise, reviewing applications, etc.) is accumulated for recordkeeping and starts to build history in a data warehouse. Reporting can then be done to help understand how a business did (e.g., total sales) in the last month, quarter, or year. These analysis reports provide managers the tools to understand the performance of their departments. This leads into the question that "Analytics" should help to answer, *"what should we be doing within departments and business units to improve business performance?"* Any tools, technologies, or systems that help with this question can qualify to be in the analytics space.

The technical implementation perspective described in Sheikh [91] is also the tangent adopted by Das [55] in his book on computational business analytics. Das' book describes the characteristics of analytics in terms of the techniques used to implement analytics solutions. In the analytics literature, three general types of analytics make consensus even though terminology may differ: descriptive, predictive, and prescriptive. A lot of good reference books are available to describe these techniques particularly under data mining and machine learning fields [92–96].

Descriptive and predictive analytics link with the Orient step in the OODA loop (Fig. 1.9), whereas the Decide step corresponds to prescriptive analytics. While originally developed for military applications, the OODA loop has been applied frequently in the private industrial sector because of its commonsense approach to making decisions. Das [55] instantiates the OODA loop in the business domain as follows: (1) observation is declining revenue figures; (2) orientation is to identify causes for declining revenue and to fully understand the company's overall financial situation and other relevant factors; (3) decision could be to enhance a marketing campaign, upgrade products, or introduce new products; and (4) action is the marketing campaign or new product launch. An action in the real world generates further observations such as the increased revenue or customer base because of the marketing campaign.

Of course, business teams stand in place of the *"solo pilots in this metaphor,"* but the behaviors and benefits are the same. Deitz says [97]: *"In warfare and in business, the speed at which the OODA Loop is executed becomes the largest factor in disrupting the enemy in the battle space, and the competition in the business space."* Gartner VP and Distinguished Analyst Roy Schulte offers that disrupting your opponent often comes from using a systems approach that shortens your own OODA loop so that you can "turn inside" the adversary's process. The pressures of mobile, social, and big data make the need urgent to support the OODA loop steps by powerful analytics software.

Das [55] discusses a model-based approach to analytics that contributes to define and understand analytics. He describes computational models as a combination of symbolic, sub-symbolic, and numerical representations of a problem. This is obtained through a series of inferences (e.g., inference cycle of Fig. 1.15) for description, prediction, and prescription: the tree kinds of analytics. Structured

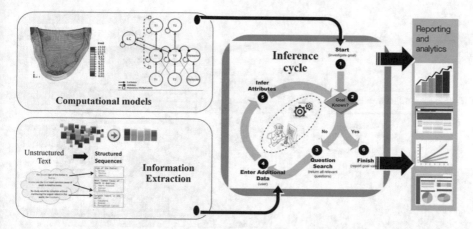

Fig. 1.15 Das' model-based analytics. (Source [50])

input in the form of transactional data is fed into an inference cycle for the model to produce analytical results. If the input is unstructured text, structured sequences need to be extracted as illustrated in Fig. 1.15. The knowledge base on which implicit facts can be inferred is composed of computational models in the form of rules and structure relational data in the form of facts. Figure 1.15 illustrates this approach.

Das [55] explains how to build these computational models by mimicking human reasoning. One of the steps is to represent by expressive graphical constructs and linguistics variables, the internal mental models of business analysts about "things" they observe and with which they interact. Computational models can also be viewed as patterns that are embedded within huge volumes of transactional data continuously generated by many business processing systems. Such models can therefore be extracted or learned via automated learning methods.

1.8   Positioning Relational Calculus with Respect to Actionable Knowledge

As explained in the previous sections, the concept of structure is of prime importance for understanding the phenomenon of information and knowledge. Relation is an ontological element of a structure. A structure has been defined as a representation of a complex entity that consists of components in relations to each other. The perspective adopted in this book for actionable knowledge is that of information processing where data are being transformed into knowledge that can be used to accomplish actions. The processes to perform this transformation are grouped under two main appellations: analytics and information fusion (AIF).

This is through AIF processes that relational calculus [27] can bring a contribution in the creation of actionable knowledge. Relational calculus here is loosely seen

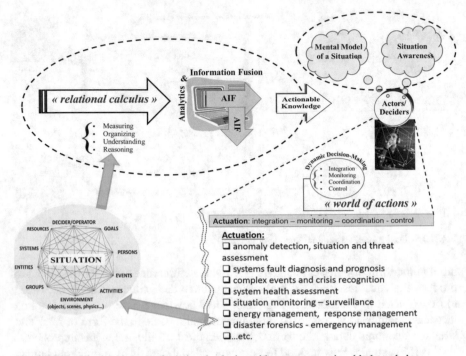

Fig. 1.16 Positioning the use of relational calculus with respect to actionable knowledge

as methods of computation for relations and its associated theories. Computations for relations are often embedded into various and numerous techniques used to implement an AIF system for a specific actuation. Figure 1.1a can now be extended to illustrate the interrelations among the main concepts related to actionable knowledge described in this introduction. This is illustrated in Fig. 1.16. Relations and relational calculus can support the AIF processes for measuring-organizing-understanding-reasoning with data, information, and knowledge. For instance, relational calculus is quite useful for high-level information fusion which is fundamentally reasoning about objects and their relations to each other.

1.9 Structure of the Book

The aim of this book is to present advanced knowledge concepts and its formalization to support the analytics and information fusion (AIF) processes that aim at delivering actionable knowledge. The book offers three major contributions: (1) the concept of "relation" and its exploitation (relational calculus) for the AIF processes, (2) the formalization of certain dimensions of knowledge to make possible a semantic growth along the AIF processes, and finally (3) the modeling of the interrelations within the couple (knowledge, action) to gain sense along a processing chain (data to information to knowledge to decision-action). These contributions are necessary for the transition of a concept "(knowledge, action)" from business and social sciences to data science, computer science, and engineering.

The book is organized into seven chapters. Chapter 1 gives an introduction on actionable knowledge, its related notions, and to what general context actions are going to take effect? What is actionable knowledge? From what angle, this book is approaching it? Where and how do we position relational calculus with respect to actionable knowledge? The context of Cyber-Physical and Social Systems is briefly described. Important related notions of knowledge, dynamic decision-making, situations and situation awareness, and analytics and information fusion are being introduced. These notions are necessary to position relational calculus in the processes of creating actionable knowledge.

Chapter 2 studies properties and dimensions of knowledge. We treat knowledge in the context of epistemic structures and from a semiotic basis. Symbols and signs and their interpretations are necessary to acquire meaning. Knowledge representation and acquisition are key issues. To understand and process knowledge, it is important to know that there are various types, sorts, dimensions, and kinds of knowledge. This chapter emphasizes the multidimensionality of knowledge and discusses dimensions such as ontological, semantic, temporal, and reference. Formalization and quantification of knowledge are discussed in the last two sections.

Chapter 3 focuses on the information-processing aspect of a knowledge chain. We discuss the relationship between data, information, and knowledge with their associated imperfections (i.e., imperfect world). Distinctions in terms of semantic status, role, and processing are important since they have a significant impact on their conceptual and implementation approaches. The notion of "quality" is of prime importance to any information processing, and particularly for a knowledge processing chain, since the objective is to improve, through processing, data, information, and knowledge quality. Various aspects of Quality of Information (QoI) are then discussed with emphasis on uncertainty-based information.

Chapter 4 presents preliminaries of crisp and fuzzy relational calculus to support the discussion in the subsequent chapters. Under the terms crisp and relational calculi, we group here the basic elements allowing to perform different operations on the calculation of the mathematical relations and also examine the interest of their properties according to their respective contexts of use.

Chapter 5 looks at what facilitates the relevant decision-making and the modalities that can make the action (effect) more efficient. The aim is to provide a better understanding of the couple (knowledge, action). There is a strong dependency between the notion of knowing about a given world and the decisions that can be made and consecutively the potential actions that can be undertaken. In Chap. 5, we bring the notion of "*mastering knowledge*" for efficient actions. Mastering knowledge amounts to having a coherent set of means to represent the most useful knowledge in the context of the action and to know how to resort as necessary to the appropriate formalizations to model the situations.

Chapter 6 proposes a discussion on the usage of relational calculus when applied to the analytics and information fusion (AIF) core processes that support the generation of actionable knowledge. This chapter discusses how relations and its calculus make AIF processes more capable, technologically speaking, of supporting the processing chain of transforming data into actionable knowledge.

Chapter 7 provides general conclusions.

References

1. P. Meusburger, B. Werlen, and L. Suarsana, *Knowledge and action*: Springer Nature, 2017.
2. N. Stehr, "Knowledge societies," *The Wiley-Blackwell encyclopedia of globalization,* 2012.
3. P. Barnaghi, A. Sheth, and C. Henson, "From data to actionable knowledge: Big data challenges in the web of things [Guest Editors' Introduction]," *IEEE Intelligent Systems,* vol. 28, pp. 6-11, 2013.
4. K. De Smedt, D. Koureas, and P. Wittenburg, "FAIR digital objects for science: from data pieces to actionable knowledge units," *Publications,* vol. 8, p. 21, 2020.
5. L. Cao, "Actionable knowledge discovery and delivery," *Wiley Interdisciplinary Reviews: Data Mining and Knowledge Discovery,* vol. 2, pp. 149-163, 2012.
6. R. Batra and M. A. Rehman, "Actionable Knowledge Discovery for Increasing Enterprise Profit, Using Domain Driven-Data Mining," *IEEE Access,* vol. 7, pp. 182924-182936, 2019.
7. S. Ghosh, *Algorithm design for networked information technology systems*: Springer, 2004.
8. D. J. Bryant, R. D. Webb, and C. McCann, "Synthesizing two approaches to decision making in command and control," *Canadian Military Journal,* vol. 4, pp. 29-34, 2003.
9. A. Guitouni, K. Wheaton, and D. Wood, "An Essay to Characterize Models of the Military Decision-Making Process," in *11th ICCRT Symposium, Cambridge UK,* 2006.
10. K. Pawar and V. Attar, "A survey on Data Analytic Platforms for Internet of Things," in *Computing, Analytics and Security Trends (CAST), International Conference on,* 2016, pp. 605-610.
11. T. Dull, "Big data and the Internet of Things: Two sides of the same coin," *SAS Best Practices,* 2015.
12. N. Jesse, "Internet of Things and Big Data–The Disruption of the Value Chain and the Rise of New Software Ecosystems," *IFAC-PapersOnLine,* vol. 49, pp. 275-282, 2016.
13. Z. Liu, D.-S. Yang, D. Wen, W.-M. Zhang, and W. Mao, "Cyber-physical-social systems for command and control," *IEEE Intelligent Systems,* vol. 26, pp. 92-96, 2011.
14. H. Zhuge, "Cyber-Physical Society-The science and engineering for future society," *Future Generation Computer Systems,* vol. 32, pp. 180-186, 2014.
15. G. Xiong, F. Zhu, X. Liu, X. Dong, W. Huang, S. Chen, *et al.,* "Cyber-physical-social system in intelligent transportation," *IEEE/CAA Journal of Automatica Sinica,* vol. 2, pp. 320-333, 2015.
16. J. Zeng, L. T. Yang, M. Lin, H. Ning, and J. Ma, "A survey: Cyber-physical-social systems and their system-level design methodology," *Future Generation Computer Systems,* 2016.
17. P. Jiang, K. Ding, and J. Leng, "Towards a cyber-physical-social-connected and service-oriented manufacturing paradigm: Social Manufacturing," *Manufacturing Letters,* vol. 7, pp. 15-21, 2016.
18. J. Zeng, L. T. Yang, M. Lin, H. Ning, and J. Ma, "A survey: Cyber-physical-social systems and their system-level design methodology," *Future Generation Computer Systems,* vol. 105, pp. 1028-1042, 2020.
19. J. J. Zhang, F.-Y. Wang, X. Wang, G. Xiong, F. Zhu, Y. Lv, *et al.,* "Cyber-physical-social systems: The state of the art and perspectives," *IEEE Transactions on Computational Social Systems,* vol. 5, pp. 829-840, 2018.
20. M. Chen, S. Mao, and Y. Liu, "Big data: A survey," *Mobile networks and applications,* vol. 19, pp. 171-209, 2014.

21. R. Baheti and H. Gill, "Cyber-physical systems," *The Impact of Control Technology,* pp. 161-166, 2011.
22. H. Gill, "From vision to reality: cyber-physical systems," in *Presentation, HCSS National Workshop on New Research Directions for High Confidence Transportation CPS: Automotive, Aviation and Rail,* 2008.
23. A. Al-Fuqaha, M. Guizani, M. Mohammadi, M. Aledhari, and M. Ayyash, "Internet of things: A survey on enabling technologies, protocols, and applications," *IEEE Communications Surveys & Tutorials,* vol. 17, pp. 2347-2376, 2015.
24. L. D. Xu and L. Duan, "Big data for cyber physical systems in industry 4.0: a survey," *Enterprise Information Systems,* vol. 13, pp. 148-169, 2019.
25. P.-J. Benghozi, S. Bureau, and F. Massit-Folléa, *L'internet des objets: quels enjeux pour l'Europe*: Maison des Sciences de l'homme, 2009.
26. M. Barès, *Maîtrise du savoir et efficience de l'action*: Editions L'Harmattan, 2007.
27. M. Barès, *Pratique du calcul relationnel*. Paris: Edilivre, 2016.
28. K. Yu, S. Eum, T. Kurita, Q. Hua, T. Sato, H. Nakazato, *et al.*, "Information-centric networking: Research and standardization status," *IEEE Access,* vol. 7, pp. 126164-126176, 2019.
29. M. Burgin, *Theory of information: fundamentality, diversity and unification* vol. 1: World Scientific, 2010.
30. L. Floridi, *The fourth revolution: How the infosphere is reshaping human reality*: OUP Oxford, 2014.
31. A. Biem, E. Bouillet, H. Feng, A. Ranganathan, A. Riabov, O. Verscheure, *et al.*, "IBM infosphere streams for scalable, real-time, intelligent transportation services," in *Proceedings of the 2010 ACM SIGMOD International Conference on Management of data,* 2010, pp. 1093-1104.
32. L. Floridi, "On the intrinsic value of information objects and the infosphere," *Ethics and information technology,* vol. 4, pp. 287-304, 2002.
33. M. Burgin, *Theory of Knowledge: Structures and Processes* vol. 5: World scientific, 2016.
34. R. M. Losee, "A discipline independent definition of information," *Journal of the American Society for information Science,* vol. 48, pp. 254-269, 1997.
35. C. Shannon, "Shannon Collected Papers, Sloane, Wyner (eds), Prediction and Entropy of Printed English," *Hoboken, NJ: John Wiley & Sons,* pp. 194-208, 1993.
36. N. J. Belkin, "Information concepts for information science," *Journal of documentation,* 1978.
37. G. Scarrott, "The nature of information," *The Computer Journal,* vol. 32, pp. 262-266, 1989.
38. T. Wilson, "Trends and issues in information science–a general survey," *Media, Knowledge and Power,* pp. 407-422, 1993.
39. D. F. FLÜCKIGER, "Contributions towards a unified concept of information. 1995," Tese (Doutorado)–Faculty of Science, University of Berne, 2004.
40. J. Barwise and J. Seligman, *Information flow: the logic of distributed systems*: Cambridge University Press, 1997.
41. R. Capurro and B. Hjørland, "The concept of information," *Annual review of information science and technology,* vol. 37, pp. 343-411, 2003.
42. C. E. Shannon, "A mathematical theory of communication," *Bell system technical journal,* vol. 27, pp. 379-423, 1948.
43. R. A. Fisher, "Theory of statistical estimation," in *Mathematical Proceedings of the Cambridge Philosophical Society,* 1925, pp. 700-725.
44. Y. Bar-Hillel and R. Carnap, "Semantic information," *The British Journal for the Philosophy of Science,* vol. 4, pp. 147-157, 1953.
45. A. N. Kolmogorov, "Three approaches to the quantitative definition ofinformation'," *Problems of information transmission,* vol. 1, pp. 1-7, 1965.
46. G. J. Chaitin, "Algorithmic information theory," *IBM journal of research and development,* vol. 21, pp. 350-359, 1977.
47. J. Marshak, "Economics of Information Systems (1971)," in *Economic Information, Decision, and Prediction,* ed: Springer, 1974, pp. 270-341.

48. W. R. Loewenstein, *The touchstone of life: Molecular information, cell communication, and the foundations of life*: Oxford University Press, 1999.

49. J. R. Boyd. (1987). *A discourse on winning and losing* [Unpublished set of briefing slides available at Air University Library, Maxwell AFB, Alabama]. Available: http://www.ausairpower.net/APA-Boyd-Papers.html

50. É. Bossé and B. Solaiman, *Fusion of Information and Analytics for Big Data and IoT*: Artech House, Inc., 2016.

51. C. Richards, "Boyd's OODA loop," *Slideshow. URL: http://www.dni.net/fcs/ppt/boyds_ooda_loop.ppt [Online]*, 2001.

52. P. Valin, E. Bossé, A. Guitouni, H. Wehn, and J. Happe, "Testbed for Distributed High–Level Information Fusion and Dynamic Resource Management," in *Int. Conf. on Info Fusion*, 2010.

53. E. Shahbazian, D. E. Blodgett, and P. Labbé, "The extended OODA model for data fusion systems," in *Proceedings of the 4th International Conference on Information Fusion (FUSION2001)*, Montreal, 2001.

54. E. Blasch, R. Breton, and É. Bossé, "User Information Fusion Decision Making Analysis with the C-OODA Model," in *High-Level Information Fusion Management and Systems Design*, E. Blasch, É. Bossé, and D. A. Lambert, Eds., ed: Artech House, 2012, pp. 215-232.

55. S. Das, *Computational Business Analytics*: Taylor & Francis, 2013.

56. J. Strassner, J. Betser, R. Ewart, and F. Belz, "A Semantic Architecture for Enhanced Cyber Situational Awareness," in *Secure and Resilient Cyber Architectures Conference*, 2010.

57. J. Strassner, J.-K. Hong, and S. van der Meer, "The design of an autonomic element for managing emerging networks and services," in *Ultra Modern Telecommunications & Workshops, 2009. ICUMT'09. International Conference on*, 2009, pp. 1-8.

58. J. Strassner, S.-S. Kim, and J. W.-K. Hong, "The design of an autonomic communication element to manage future internet services," in *Management Enabling the Future Internet for Changing Business and New Computing Services*, ed: Springer, 2009, pp. 122-132.

59. J. Strassner, "Knowledge Representation, Processing, and Governance in the FOCALE Autonomic Architecture," *Autonomic Network Management Principles: From Concepts to Applications*, p. 253, 2010.

60. E. Blasch, P. Valin, E. Bosse, M. Nilsson, J. van Laere, and E. Shahbazian, "Implication of culture: user roles in information fusion for enhanced situational understanding," in *Information Fusion, 2009. FUSION'09. 12th International Conference on*, 2009, pp. 1272-1279.

61. J. Taylor, *Decision Management Systems: A Practical Guide to Using Business Rules and Predictive Analytics*: Pearson Education, 2011.

62. M. Minelli, M. Chambers, and A. Dhiraj, *Big data, big analytics: emerging business intelligence and analytic trends for today's businesses*: John Wiley & Sons, 2012.

63. D. Dittrich, A. Center, and M. P. Haselkorn, "Visual Analytics in Support of Secure Cyber-Physical Systems."

64. R. Breton, "The modelling of three levels of cognitive controls with the Cognitive-OODA loop framework," *Def. Res. & Dev. CA-Valcartier, DRDC TR*, vol. 111, 2008.

65. É. Bossé, J. Roy, and S. Wark, *Concepts, models, and tools for information fusion*: Artech House, Inc., 2007.

66. D. Von Winterfeldt and W. Edwards, *Decision analysis and behavioral research* vol. 604: Cambridge University Press Cambridge, 1986.

67. D. Kahneman and A. Tversky, "Prospect theory: An analysis of decision under risk," *Econometrica: Journal of the Econometric Society,* pp. 263-291, 1979.

68. G. Loomes and R. Sugden, "Regret theory: An alternative theory of rational choice under uncertainty," *The economic journal,* pp. 805-824, 1982.

69. G. Gigerenzer, "Bounded rationality models of fast and frugal inference," *REVUE SUISSE D ECONOMIE POLITIQUE ET DE STATISTIQUE*, vol. 133, pp. 201-218, 1997.

70. H. A. Simon, "Rational choice and the structure of the environment," *Psychological review,* vol. 63, p. 129, 1956.

71. P. M. Todd and G. Gigerenzer, "Bounding rationality to the world," *Journal of Economic Psychology,* vol. 24, pp. 143-165, 2003.

72. J. Rieskamp and U. Hoffrage, "Inferences under time pressure: How opportunity costs affect strategy selection," *Acta psychologica,* vol. 127, pp. 258-276, 2008.
73. U. Hoffrage, "When do people use simple heuristics, and how can we tell?," ed: New York: Oxford University Press, 1999.
74. M. R. Endsley and D. J. Garland, *Situation awareness analysis and measurement*: CRC Press, 2000.
75. M. R. Endsley, "Toward a theory of situation awareness in dynamic systems," *Human Factors: The Journal of the Human Factors and Ergonomics Society,* vol. 37, pp. 32-64, 1995.
76. E. Blasch, E. Bosse, and D. A. Lambert, *High-level Information Fusion Management and Systems Design.* Boston & London: Artech House, 2012.
77. J. W. Gualtieri, S. Szymczak, and W. C. Elm, "Cognitive system engineering-based design: Alchemy or engineering," in *Proceedings of the Human Factors and Ergonomics Society Annual Meeting,* 2005, pp. 254-258.
78. D. D. Woods, "Cognitive technologies: The design of joint human-machine cognitive systems," *AI magazine,* vol. 6, pp. 86-86, 1985.
79. K. J. Vicente, *Cognitive work analysis: Toward safe, productive, and healthy computer-based work*: CRC press, 1999.
80. J. Roy, "From data fusion to situation analysis," in *Information Fusion (FUSION), 2001 4th Conference on,* 2001, pp. 1-8.
81. W. Sulis, "Archetypal Dynamics," in *Formal descriptions of developing systems.* vol. 121, J. Nation, I. Trofimova, J. Rand, and W. Sulis, Eds., ed: Kluwer Academic Publishers, 2003, pp. 180-227.
82. D. Nicholson, "Defence Applications of Agent-Based Information Fusion," *The Computer Journal,* vol. 54, pp. 263–273, 2011.
83. C. V. Apte, S. J. Hong, R. Natarajan, E. P. D. Pednault, F. A. Tipu, and S. M. Weiss, "Data-intensive analytics for predictive modeling," *IBM Journal of Research and Development,* vol. 47, pp. 17-23, 2003.
84. G. H. N. Laursen and J. Thorlund, *Business Analytics for Managers: Taking Business Intelligence Beyond Reporting*: Wiley, 2010.
85. A. N. Steinberg, C. L. Bowman, and F. E. White, "Revisions to the JDL data fusion model," in *The Joint NATO/IRIS Conference,* Quebec City, 1998.
86. D. L. Hall and J. M. Jordan, *Human-Centered Information Fusion*: Artech House, Incorporated, 2010.
87. D. L. Hall and S. A. H. McMullen, *Mathematical Techniques in Multisensor Data Fusion*: Artech House, 2004.
88. M. Liggins, D. Hall, and J. Llinas, *Handbook of Multisensor Data Fusion: Theory and Practice, Second Edition*: Taylor & Francis, 2008.
89. S. Das, *High-Level Data Fusion*: Artech House, 2008.
90. W. W. Eckerson, *Secrets of Analytical Leaders*: Technics Publications, 2012.
91. N. Sheikh, *Implementing analytics: A blueprint for design, development, and adoption*: Newnes, 2013.
92. Z. Chen, *Data mining and uncertain reasoning: an integrated approach*: Wiley New York, 2001.
93. P.-N. Tan, M. Steinbach, and V. Kumar, *Introduction to data mining* vol. 1: Pearson Addison Wesley Boston, 2006.
94. I. H. Witten and E. Frank, *Data Mining: Practical machine learning tools and techniques*: Morgan Kaufmann, 2005.
95. C. M. Bishop, *Pattern recognition and machine learning* vol. 4: springer New York, 2006.
96. J. Han, M. Kamber, and J. Pei, *Data mining, southeast asia edition: Concepts and techniques*: Morgan kaufmann, 2006.
97. F. Burstein and C. Holsapple, *Handbook on decision support systems 2: variations* vol. 2: Springer Science & Business Media, 2008.

Chapter 2
Knowledge and Its Dimensions

Abstract This chapter studies the properties and dimensions of knowledge. We treat knowledge in the context of epistemic structures and from a semiotic basis. Symbols and signs and their interpretations are necessary to acquire meaning. Knowledge representation and acquisition are key issues. To understand and process knowledge, it is important to know that there are various types, sorts, dimensions, and kinds of knowledge. This chapter emphasizes the multidimensionality of knowledge and discusses dimensions such as ontological, semantic, temporal, and reference. Formalization and quantification of knowledge are discussed in the last two sections.

2.1 Introduction

Knowledge remains the domain by excellence of philosophers since the nature of the issues, and questions challenge the human mind so deeply [1]. Questions like the following can arise: Under what mechanisms an individual can memorize knowledge and transmit it? What would be a useful ratio between implicit and explicit knowledge to act under reasonable terms? It is beyond the scope of our discussion to address these questions. The issues focused in this chapter are apprehension and knowledge representation prior to defining a suitable formalism that remains, in effects, the required essential for any subsequent artificial reasoning.

Any automaton[1] geared to support human decision must be indeed endowed with reasoning ability or interpretation depending on the circumstances and the context of its employment. Most often, that capacity will be achieved through an inference mechanism. Inference cannot operate without a priori properly formalized knowledge. Knowing remains one of the basic mental activities of a human being that can amount to 20% of his human lifetime. In the same line of thought, the successive stages of his education and professional training will represent a major part of his

---

[1] Automaton: "*a machine that performs a function according to a predetermined set of coded instructions, especially one capable of a range of programmed responses to different circumstances*"

© The Author(s), under exclusive license to Springer Nature Switzerland AG 2022
M. Barès, É. Bossé, *Relational Calculus for Actionable Knowledge*, Information Fusion and Data Science, https://doi.org/10.1007/978-3-030-92430-0_2

time until the age of 20–25 years. All his subsequent actions are in part conditioned by two aspects:

- A sort of permanent knowledge, originally initialized and maintained over time
- Timely knowledge to be acquired continuously, depending on needs

Humans must not only learn but also understand to act. It requires not only to possess a thinking capacity but also to have the following skills:

- To attach a meaning to shapes and relations connecting them
- To make the necessary bridges between data, information, and knowledge
- To take advantage of doing in the process of knowledge enrichment

These are features already outlined in his time by G. Vico [2] called the "Ingenium."[2]

This chapter discusses the properties and dimensions of knowledge. We treat knowledge in the context of epistemic structures and from a semiotic basis. Symbols and signs and their interpretations are necessary to acquire meaning. Knowledge items are epistemic structures. Beliefs are epistemic structures as well but associated with descriptive knowledge. Knowledge representation and acquisition are key issues. To understand and process knowledge, it is important to know that there are various types, sorts, dimensions, and kinds of knowledge. This chapter emphasizes the multidimensionality of knowledge and discusses dimensions such as ontological, semantic, temporal, and reference. Formalization and quantification of knowledge are discussed in the last two sections.

2.2   Structures and Knowledge Structures

Everything has its structure, and according to the contemporary approach in methodology of science, scientists gain and publish information obtained from the structure under investigation rather than the essence and nature of the studied phenomena. The concept of structures is of prime importance for understanding the phenomenon of information. For instance, data structures are essential for algorithms and programming. It is impossible to build a database without an explicit construction of data structures. In the Platonic tradition, the global world structure has the form of three interconnected worlds (Fig. 2.1a). Modern sciences (C. Pierce, Popper, Burgin, Piaget, and others) have evolved the Plato's triad[3] toward a more understandable triad known as the "*Existential Triad*" of the world (Fig. 2.1b).

---

[2] In Vico et al.: the ingenio is defined as the ability to "*express the relationships between the objects.*" . . . "*the virtue and active force with which the intellect together, unites and found similarities, relationships and reasons for things.*"

[3] Note that the term triad used here is different from the notion of a triplet. A triad is a system that consists of three parts (elements or components), while a triplet is any three objects. Thus, any triad is a triplet, but not any triplet is a triad. In a triad, there are ties and/or relations between all three parts (objects from the triad), while for a triplet, this is not necessary.

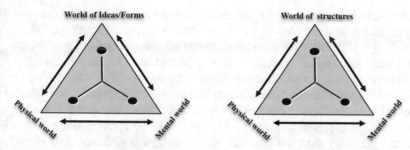

a) The Plato triad of the world   b) The modern sciences existential triad of the world

Fig. 2.1 Triad of the world: (**a**) Plato's triad and (**b**) modern science triad

In Fig. 2.1b, the *"Physical World"* is interpreted as the physical reality studied by natural sciences, while ideas or forms might be associated with structures, and the *"Mental World"* encompasses both individual conscience and social conscience. The world of structures includes knowledge, such as the intellectual contents of books, documents, and scientific theories. It is a kind of structures that are represented in people's mentality [3]. Note that no material thing exists without a structure and that the world of structures does exist.

In the mental world, there are real "things" and "phenomena." For example, there exist happiness and pain, smell and color, love and understanding, impressions and images of stars, tables, chairs, etc. In the physical world, there are the real tables and chairs, sun, stars, stones, flowers, butterflies, space and time, molecules and atoms, electrons, and photons. Knowledge, per se, forms a dynamic component of the world of structures. Each of these three worlds is hierarchically organized, comprising several strata. For instance, the hierarchy of the physical world goes from subatomic particles to atoms to molecules to bodies to cells to living beings and so on— similarly, from the first level of the mental world with individual mentality to a second level of group and social mentalities, which include collective sub-conscience and collective intelligence [3].

Example 2.1 Existential triad A computer has its hardware as the physical component, everything that is in its memory is the mental component, and all structures and systems of the physical and mental components are the structural component.

What is a structure? The notion of a structure is used in a variety of fields: mathematics, physics, chemistry, biology, etc. Scientists study theoretical structures corresponding to natural phenomena that are called models of studied phenomena. A structure in psychology refers to an organization of components that perform related psychological functions. Family, culture, religion, science, education, law, economy, and politics, all are social structures. Social relations are structures. Social structure is constituted by rules and resources governing and available to social agents. Information structures play an important role in decision theory and information economics. Mathematics is to be thought of as nothing other than the study of pure structures, and formal and informal structures are studied by all sciences, applied mathematics, and philosophy. In modern science and philosophy, structuralism is a general approach based on intrinsic structures invariant relative to transformations (Piaget [4]). The main assumption of structuralism is that different things and phenomena have similar essence if their structures are identical and, consequently, to investigate a phenomenon means to find its structure.[4] According to Burgin [3] and in spite of many achievements in a variety of disciplines, structuralism was not able to develop a general definition of structure as it is demonstrated in the book of Piaget [4]. Thus, there is not a generally accepted definition of a structure, and the word structure has different meanings in different disciplines. We adopt here the definitions of Burgin [3]:

Definition 2.1 A *structure* is a representation of a complex entity that consists of components in relations to each other. This definition implies that two structures are identical if and only if (a) their nonrelational components are the same; (b) their relation components are the same; and (c) corresponding components stand in corresponding relations.

There are three main types of structural representations: (1) Material in the *Physical World*, (2) Symbolic in the *Mental World*, and (3) Intrinsic in the *Structural World*.

Definition 2.2 A *mathematical structure* is an abstract (symbolic) system $A = \langle M; R_1, R_2, \cdots, R_n \rangle$ where M is a set and $R_1, R_2, \cdots, R_n$ are relations in this set. A *mathematical structure*[5] is a set with relations in this set and axioms that these relations satisfy. A more extended definition considers relations between elements, relations between elements and relations, and relations between relations (higher orders).

Definition 2.3 A *structure* is a *schema*[6] with variables of two types: object variables and connection variables. Any *complex system* has, as a rule, several structures and a *structure* can be static or dynamic. In Burgin [3], schemas and variables are formally defined to make Definition 1.3 exact.

Definition 2.4 A *system* consists of components and relations between components.

---

[4] At first, structuralism emerged in linguistics. Although Ferdinand de Saussure (1857–1913) did not use the term structure preferring the term form, his understanding in many respects coincides with the "*structuralistic*" approach.

[5] Structure gives a scientific explication of both concepts: ideas of Plato and forms of Aristotle. Aristotle (structural realism) asserts that the realization of a form of an object in one's mind is as real the instantiation of the corresponding form in external reality and both forms exhibit the same powers and properties and the same necessary relations to other forms. As forms are special kinds of structures, this means that mental structures correctly represent structures in nature and society.

[6] A mathematical definition of a *schema* is given in Burgin M. (2006). Mathematical Schema Theory for Modeling in Business and Industry. Simulation Series, 38(2), 229.

Example 2.2 The *structure*[7] of the *system* (a cloud, rain, the ground) is the fundamental triad (Essence 1, connection, Essence 2). Here Essence 1 is interpreted as a cloud, Essence 2 is interpreted as a ground, and connection is interpreted as the rain.

Example 2.3 In logics, there is a large diversity of various *structures* studied by different authors such as *syllogistics* (Aristotle), *classical propositional* logic and *classical predicate* logic, *algebraic* logic, *autoepistemic* logic, and *belief* logic. (See Burgin [5], p.444, for a discussion on the main structures of thinking: *concepts*, *judgments* and *conclusions*, and other examples of *structures*.)

2.2.1   Symbols and Signs: Knowledge Semiotic Basis

Structures are represented by *signs* and *symbols*, being themselves symbols of things, systems, and processes. To understand this, let us consider theoretical understanding implied by models of the concepts symbol and sign. Knowledge exists only insofar as we know beforehand how to identify and collect signs from the real world. Our ability to organize them into coherent groups then constitutes an indispensable basis. Signs are assimilated to manifestations of the real or observable world and beyond any ontological dissertation about their nature and characteristics. Measuring and sensing systems, human or artificial, can capture signs: light signal, graphic sign on a written document, acoustic wave, bit sequence number, etc.

In his influential study of linguistic signs developed by Ferdinand de Saussure, the sign relation is dyadic (Fig. 2.2), consisting only of a form of the sign (the signifier) and its meaning (the signified). Language is made up of signs, and every sign has two inseparable sides: The first is the signifier (French *signifiant*)—the "*shape*" of a word, its phonic component, i.e., the sequence of graphemes (letters), e.g., "c," "a," "t," or phonemes (speech sounds), e.g., /kæt/. The second is the signified (French *signifié*)—the ideational component, the concept, or object that appears in our minds when we hear or read the signifier, e.g., a small domesticated feline. (The signified is not to be confused with the "*referent*." The former is a "*mental concept*," while the latter the "*actual object*" in the world.) In the case of

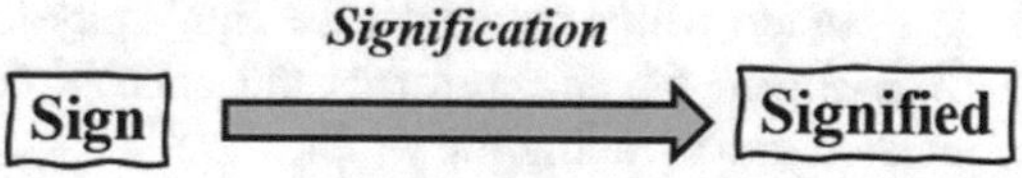

Fig. 2.2 The dyadic sign triad of "de Saussure"

Signification

Sign ⟹ Signified

---

[7]Structures exist not only in languages, society, or human personality, but everywhere. Consequently, it is necessary to study structures not only in linguistics, anthropology, psychology, or sociology, but in all sciences: in natural sciences such as physics, social sciences and humanities such as sociology and psychology, and technical sciences such as computer science.

Fig. 2.3 The Ogden and
Richards' semiotic triangle

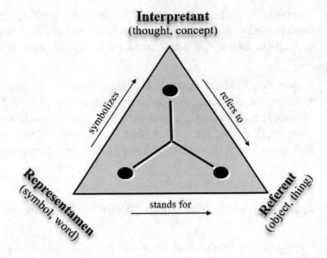

C. S. Peirce's theory, the sign is defined as a triadic relation as "something that stands for something, to someone in some capacity." This yields a definition for the word sign which is a triadic relationship between a *representamen* [6], an *interpretant* (purely conceptual), and the *referent* (the actual object referred to by the *interpretant* and to which the *representamen* stands for). This trichotomy was depicted by Ogden and Richards [7] into what is known today as the "Semiotic Triangle" (Fig. 2.3).

It is remarkable that despite the abundance of the literature on the matter, there is no authoritative definition for the notion of sign. We will therefore, safely turn to that defined in linguistics by de Saussure [8], which, rather than emphasize the temporal evolution of elements of language (diachronic paradigm) focuses on the basic elements of language that allow it to exist independently from its evolving context (synchronic paradigm). According to de Saussure, a language is at any given moment a consistent whole, separate from its evolution; similarly, a decision-maker can act based on a current state without knowing a previous state.

In the synchronic paradigm, the linguistic sign, which is defined as the relationship between two notions, is an entity that exists only in this relationship. This entity, known as a dyad, contains the following:

- A *signifier*, which represents the sign's physical aspect, e.g., the words on a page
- A *signified*, which represents the concept conveyed, e.g., the meaning that is created upon reading the words

There is an obvious dependency between signifier and signified since the signified cannot exist without the signifier. However, the signified can be arbitrarily generated. The word that comprises letter-signifiers /c/h/i/e/n/ evokes a concept related to the canine sphere in the mind of a French reader, while /h/u/n/d/ evokes the same for a German reader. The same sequence of letter-signifiers can instead, for a member of a breeders' association, evoke a particular breed. The sign can also be

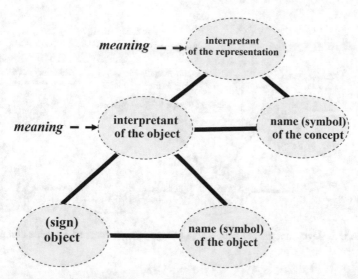

Fig. 2.4 A two-level *fractality* of Peirce's semiotic model

defined analytically as an agent of meaning: a phrase conveys meaning with respect to the text in which it appears, a word does the same in relation to the phrase in which it appears. F. de Saussure performed a systematic study of signs in linguistics at the beginning of the century. The theory was dubbed semiology and came to be known over time as semiotics.

Data are produced from sets of signs. Identifying them must be done cautiously since they have a big impact on how to represent and process knowledge. The nature of the sign also determines the choice of encoding that can be applied to them and through which they can acquire a first level of meaning. There are two major theories about the way in which signs acquire the ability to transfer information: the concept introduced into linguistics by F. de Saussure [9, 10] and the other major semiotic theory developed by C. S. Peirce [6, 11]. Both F. de Saussure and Pierce understand the defining property of the sign as being a relation between a few elements.

Peirce implied that signs establish meaning through recursive relationships that arise in sets of the three main semiotic elements (*a fundamental triad*): sign object, sign name, sign meaning. It means that each component of a sign can acquire the role of another component as represented in Fig. 2.4. This defines the *fractality* property, which tells that the structure of the whole is repeated in the structure of its parts. Note that if the meaning of a sign is knowledge of the first level, then the meaning of the representation of that sign is knowledge of the second level, i.e., its metaknowledge (see chapter 2 of Burgin [5] and Fig. 2.5 below).

U. Eco [12] explains that any given (material) sign is required to be an element of the expression-plane and must therefore be conventionally correlated to one or more elements of the content-plane as illustrated in Fig. 2.6 (the Eco-Hjelmslev's model). With this model, it is possible for the expression of a sign to have more than one

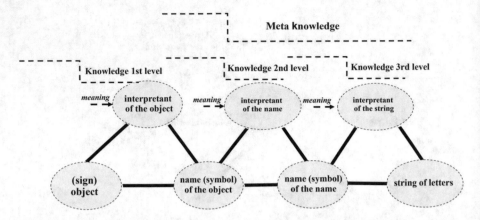

Fig. 2.5   A three-level sign triad of Peirce: sign of sign of sign, name of the name of the name

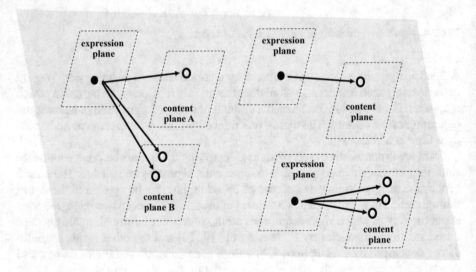

Fig. 2.6   The Eco-Hjelmslev's model of sign

content, while the content of a sign may have more than one expression as in example 2.4 extracted from Burgin [5].

Example 2.4 Such symbol (material sign) as 1 can denote a digit of a decimal numerical system (Content 1), a digit of a binary numerical system (Content 2), or the number one (Content 3), while its expressions can be such material symbols as "1," "1," "1," or "1."

Signs are the first relation between two types of materiality, one from the world perceived or in apprehension, the other from a system of representation required to transform and manipulate elements of signification generated by signs. A sign is

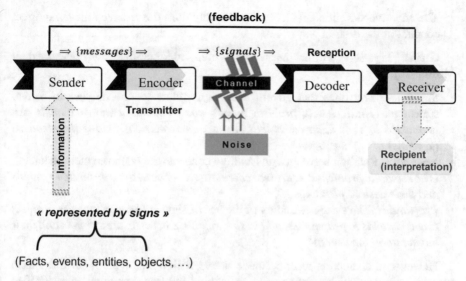

Fig. 2.7 Signs and symbols in the Shannon's model of information process

only relevant in relation to the signified that it conveys, which is relevant within the operational information system in which it is used. In the case of a voice recognition system operating within a noisy environment, only certain signs assimilated to certain formant characteristics are considered relevant, while others are ignored as they do not make sense in the context of the recognition algorithms.

In an information process, the sign is the generating event of an activity evolving over time through different processors. For instance, it can be described according to Shannon's model [13]. In Fig. 2.7, the code relates the signifier, represented by the signals, to the signified, that is, the message. The transmission channel generates emissions, e.g., noise, which, if unprocessed, may alter the signified attributed to the sign. While the unicity of the channel processor in Shannon's model simplifies description with physical parameters, it makes difficult generalization in information theory.

However, let's recall here the brief discussion presented in chapter 3 of [14] with respect to Shannon and information theory. Shannon's theory has had an immense impact on the technology of information storage, data compression, information transmission, and coding. Shannon's theory was readily remarked as a "syntactic" theory of information, but some authors claim that it may address "semantic" issues as well [15]. Graben [15] recalls from *"The Mathematical Theory of Communication"* [13], three levels of communication problems (A-B-C):

- The technical problem A: How accurately can the symbols of communication be transmitted?
- The semantic problem B: How precisely do the transmitted symbols convey the desired meaning?

- The effectiveness problem C: How effectively does the received meaning affect conduct in the desired way?

Graben [15] makes a direct link with the three semiotic dimensions discussed by Morris [16]:

- *Syntactics:* The rules that govern how signs relate to one another in formal structures. (*The technical problem A with noisy communication channels can be tackled by redundant codes that introduce correlations among the elements (symbols) of the messages.*)
- *Semantics:* Relation between signs and the concepts they denote (their denotata). (*The semantic problem B addresses correlations between the transmitted symbols and their desired meanings.*)
- *Pragmatics:* The systematic study of the use of signs. (*The effectiveness problem C corresponds to pragmatics, addressing relations between the symbols and their impact upon the users.*)

However, a distinction must be made between sign and symbol since the term symbolic representation is often used. A symbol is often perceived as a signified referring to another signified, e.g., the lion refers to the notion of courage. The symbol differs from the sign because it relates two terms that are similar, while the sign relates a container to a unit of meaning. Symbol systems are designed to be interpretable, that is, to receive external interpretation. Unlike the sign, the symbol does not create a dependency between its two terms. The choice of symbol can be arbitrary. For example, why is courage symbolized by a lion rather than by a tiger? Recall that the word symbol comes from the Greek word "symbolon" (σύμβολον),[8] a clay piece consisting of two parts, which fit together in the manner of a wire cut [17]. Each part of a "symbolon" is being owned by one of the branches of the same family. In ancient Greece, when one member of a branch visited the other, he/she could be recognized by the adjustment of the parts of the symbolon. Beyond the anecdote, it is interesting that the symbol of family membership derived its meaning from the union of the two physical parts of the symbolon.

One can note other characteristics attached to the notion of a symbol as:

- Its close relationship with the signifier: the element of materiality acting as a medium for the evocation "family membership" in the case of the symbolon.
- The compositional aspect: the union of several salient, generally material, features grounding the status of symbol, which incidentally may be granted to any object, including formal objects.
- The non-arbitrary aspect: the interlocking of the symbolon's parts, which, where possible, ascribes meaning to the notion of family membership.

---

[8] σύμβολον: (Ancient Greek) Originally, an object cut into two, two hosts (φίλος and ξένος) each retained a half; two close parts used to recognize holders and to prove the previously contracted hospitality relations (φιλότης)

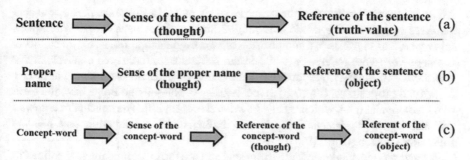

Fig. 2.8 *"Fractality"* in the conceptual structures of Frege that are like those of Russell

2.2.2 Concepts, Names, and Objects

According to Burgin [5], concepts used in science are important types of symbols. A concept is defined as a general idea derived from specific instances, i.e., a concept is a symbolic linguistic representation of these instances. Concepts/notions are signs/symbols of ideas and have structures like the general structure of a symbol. Often a concept is treated as a *unit of knowledge* and considered as a constituent of a proposition. Concepts are ways of thinking of objects, properties, and relations. We refer the reader to Chapter 5 of Burgin [5], for a comprehensive description of various models of concepts brought by philosophers and other scientists (e.g., concept triangle of Russell, concept triangle of Frege, others). Figure 2.8 illustrates some of these concepts.

For instance, Russell states that a concept has a name and two more constituents: objects and their exemplifications. Russell calls the relation between concepts and their particular "exemplifications" *denotation*. This gives us the following structure of each component of a concept: *name, denotation*, and *meaning*. As a name is itself an object, it has a name[9] and a denotation and sense (or meaning), as any other object. It means that each component of a concept can acquire the role of another component. As a result, the structure of a concept has the property called *fractality* (illustrated in Fig. 2.8), which tells that the structure of the whole is repeated/reflected in the structure of its parts.

As a name is itself an object, it has a name[10] and a denotation and sense (or meaning), as any other object. It means that each component of a concept can acquire the role of another component. As a result, the structure of a concept has the property called *fractality* (illustrated in Fig. 2.8), which tells that the structure of the

---

[9] A name is defined as an expression of a natural language that denotes a separate object, collection of similar objects, properties, relations, etc. Exact names of concepts are called terms. Concepts do not exist without names.

[10] A name is defined as an expression of a natural language that denotes a separate object, collection of similar objects, properties, relations, etc. Exact names of concepts are called terms. Concepts do not exist without names.

whole is repeated/reflected in the structure of its parts. In Fig. 2.8a, *Sense* connects *Sentence* with *Reference*, while in the Fig. 2.8b, *Sense* connects *Proper Name* with *Reference*, and Fig. 2.8c is a composition of two fundamental triads: *Sense* connects *Concept-word* with *Reference*, while in the second one, *Reference* connects *Sense* with *Referent*. Burgin [5] reports that Tatievskaia [18] has demonstrated that proper names and propositions (Fig. 2.8) in the theories of Russell and Frege have the form of a fundamental triad, which is called a *named set* [19, 20]. The concept of a *named set*, also called *fundamental triad*, is a mathematical construction that has the following structure (Fig. 2.9):

Named sets exist independently from sets and are the most fundamental entities in the whole mathematics. There are different kinds of *named sets*: set theoretical, categorical, mereological, etc. (see examples 2.5 and 2.6). The most customary are *set theoretical named sets* visually presented in Fig. 2.10.

Fig. 2.9 Structures of two named sets

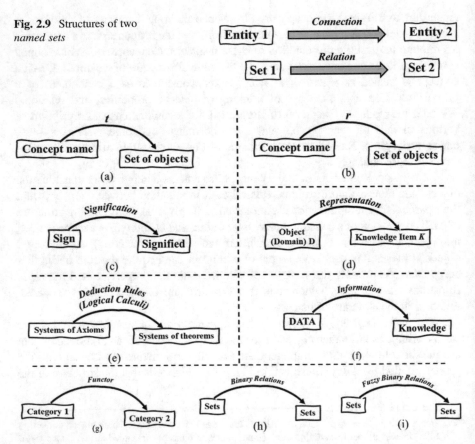

Fig. 2.10 Illustrations of various named sets: (**a**) set theoretical model of a concept, (**b**) fuzzy set-theoretical model of a concept, (**c**) de Saussure triad, (**d**) knowledge structure, (**e**) knowledge system named set, (**f**) the DIK cognitive pyramid, (**g**) categorical named sets, (**h**) set-theoretical named sets, and (**i**) fuzzy set-theoretical named sets

Example 2.5 Examples of *named sets* other than being *set theoretical*. Categorical *named sets* are different from set theoretical named sets since an arrow in a category is a *named set* but does not include sets as components. *Named sets* with physical components, such as (a woman and her name), (an article and its title), (a book and its title), and many others, are not set theoretical. Cars and their owners, books and their authors, and the traditional scheme of communication (sender-message-receiver) constitute other named sets.

Example 2.6 People can even see some *named sets*. When it is raining: a cloud (s) (Entity 1), the Earth where we stand (Entity 2), and flows of water (the correspondence). When we see a lightning: a cloud(s) (Entity 1), the Earth where we stand (Entity 2), and the lightning (the correspondence). There are many named sets in which Entity 1 is some set, Entity 2 consists of the names of the elements from the Entity 1, and elements are connected with their names by the naming relation.

A *named set* consists of three components (Fig. 2.10) and each component plays its unique role and has a specific name. In a *set theoretical named set X*, X and I are some sets, collections, or classes and f is a binary relation between X and I. The set (collection or class) X is called the *support*, the set (collection or class) I is called the *set of names*, and f is called the *naming relation* of the *named set X*. Many mathematical systems are particular cases of *named sets* or *fundamental triads*. The most important of such systems are fuzzy sets, multisets, graphs and hypergraphs, and topological and fiber bundles.

Numerous types of concept models are detailed in Chapter 5 of Burgin [5] and illustrated in Fig. 2.10. The *set-theoretical* model (Fig. 2.10a and Fig. 2.10h) is based on set theory and represents a concept as a set of objects that belong to the concept. The *fuzzy set-theoretical* model (Fig. 2.10b and Fig. 2.10i) is based on fuzzy set theory and represents a concept as a fuzzy set of objects that to some extent belong to the concept. Both structures are similar but in contrast to the relation t, which simply assigns each object to the concept name, the relation r shows to what extent each object belongs to the concept with the given name.

Objects can be ideal entities, such as concepts, ideas, algorithms, or names. Categorical named sets (Fig. 2.10g) are different from set theoretical named sets. A functor is a mapping between categories. It does not include sets as components. The third example (Fig. 2.10c) of a named set is the concept of *sign* in works of Ferdinand de Saussure,[11] where a basic property of a sign is that it points to something different from itself. He introduced a structure of sign in the form of a triad (corresponding to a *named set* – Fig. 2.10c). The knowledge structure of Fig. 2.10d is a specific case of named sets.

Each element or system of knowledge refers to some object domain because knowledge is always knowledge about something. It means that for any knowledge system (element) K, there is a domain D of real or abstract objects and K describes the whole D or its part. Any calculus is a named set {Axioms, deduction rules,

---

[11] He is sometimes called the father of theoretical linguistics.

theorems}, where expressions accepted without proofs (axioms) are transformed into expressions that are proved (theorems) via rules used in the process of deducing these theorems. Logical calculi (Fig. 2.10e) combine axioms, deduction rules, and theorems into unified systems.

The knowledge pyramid (data-information-knowledge) (Fig. 2.10f), discussed in the next chapter, is another example of a named set or chains of named sets where data are reflected (connected) to knowledge via information. This is realized through three essential components of information processing: data mining, information retrieval, and knowledge discovery (acquisition and production). The information retrieval process via a chain of named sets takes the raw results from the process of extracting trends or patterns from data (data mining) and transforms them into useful and understandable information. The retrieved information is transformed into knowledge through the knowledge discovery process, which places information into a definite knowledge system (Fig. 2.10e), which is coordinated with the knowledge system of a user. It is worth noting that there are distinctions in working with data, processing information, and acquiring or producing knowledge. Below are some definitions and examples to clarify terms used in Fig. 2.10.

Definition 2.5 An *object* is anything that is considered as a whole, i.e., in its entirety, and has a name. There are physical objects, mental objects, and structural objects.

Example 2.7 *Knowledge is always related to some real or abstract object or domain.* Object to which knowledge is related may be a galaxy, planet, point, person, bird, letter, novel, action, operation, event, process, love, surprise, picture, song, sound, light, square, triangle, etc. We call such an object *knowledge domain* or *knowledge object*.

Definition 2.6 A *concept* is defined as a general idea derived from specific instances, i.e., a concept is a symbolic linguistic representation of these instances.

Definition 2.7 A *name* is defined as an expression of a natural language that denotes a separate object, collection of similar objects, properties, relations, etc. Exact names of concepts are called terms.

2.2.3 Epistemic Structures and Spaces

Knowledge can be treated in the context of epistemic structures[12] because knowledge items are epistemic structures. Beliefs are epistemic structures associated with descriptive knowledge. However, beliefs are related only to declarative or

---

[12]Epistemology is the theory of knowledge, which primarily studies the nature and structure of knowledge and how knowledge is or should it be acquired, validated, preserved, revised, updated, and retrieved. It means that epistemology contains metaknowledge (*knowledge about knowledge*).

descriptive knowledge while there are also other epistemic structures, to which knowledge is intrinsically attached such as operational knowledge and representational knowledge. "Blocks" of knowledge are identified with structured quantum knowledge items, and we consider such quantum knowledge items as signs and symbols. Knowledge structures are of a crucial importance and various methodologies have been developed to utilize them for various purposes, such as learning, knowledge management, knowledge representation, and knowledge processing.

Definition 2.8 An *epistemic structure* (ES) is a structure that represents or contains information about some domain. This is a basic structure of cognition.

Example 2.8 Examples of structures—*Explicit structures or forms* of a physical macroscopic object: e.g., geometric forms or shapes, differential forms. *Symbolic form* is a formalized structure: E.g., formulas in chemistry, syllogisms in formal logic. *Pattern* is a stable structure. *Idea* is an informal inexact structure. Three main groups of mathematical structures: order structures, algebraic structures, and topological structures.

Knowledge does not exist by itself but belongs to an agent A (a knowledgeable system or a knower), that has knowledge K about the domain (object) D. Knowledge is in fact a cognitive representation of knowledge objects, D, which is intrinsically linked to their symbolic representations. For instance, a symbolic representation of an object D may be a sentence in natural language, a logical formula, a mathematical expression, and so on. According to Burgin's general theory of information [3], data, knowledge, and information all belong to the world of structures having carriers in both mental and physical worlds.

Sets with relations construct epistemic spaces. A symbolic *epistemic space* is a set of cognitive epistemic units or their representations with relations and operations. Note that not all epistemic structure (knowledge) representations are symbolic. For instance, algorithms embodied in computer chips are hardware representations of operational knowledge that is not symbolic. Elementary knowledge units are often expressed by logical propositions such as:

"*An object O has the property P*"

or

"*The value of a property P for an object O is equal to b.*"

It is important to discern knowledge and its representations, as well as epistemic structures and their representations. Epistemic structures are usually represented by symbolic systems in general and symbols in particular. As a rule, a knowledge unit has several representations as shown in Table 2.1.

What would be an epistemic structure for knowledge in a mathematical form? Consider the existence of a universal knowledge set, a knowledge multiset,[13] W of knowledge units or knowledge items and consider a class A of dynamic systems, e.g., intelligent knowledgeable agents. Note that a traditional approach to

---

[13] In mathematics, a multiset is a modification of the concept of a set that, unlike a set, allows for multiple instances for each of its elements.

Table 2.1 Knowledge representations of "*a person A is seven feet tall*"

The proposition	"*A is seven feet tall*"
The proposition	"*The height of A is seven feet*"
The equality	$H(A) = 7$ ft where $H(X)$ is the property height of a person
The truth of the predicate	$H(A,7)$ where $H(X,h)$ is the predicate with X being a person and h being a number of feet
The element	$(A; 7)$ of a relational database

Fig. 2.11 A *named set*
X = the fundamental triad
($S(\mathbf{X})$, n($\mathbf{X}$), $N(\mathbf{X})$)

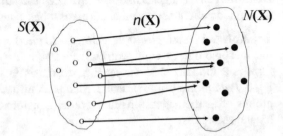

knowledge, which is called the representational *atomism*[14] according to which knowledge is built from some basic elementary units by combining them into a more complex structure.

The first approximation to real epistemic/knowledge systems and information processes is to use the classical mathematical setting, i.e., *sets*. However, applications of mathematical structures to real-life phenomena have shown limitations of sets and several generalizations of sets have been suggested. The most popular of these generalizations are fuzzy sets and multisets. A *multiset* [21] is similar to a set but can contain different copies of the same elements. Multiset is an interesting notion since many real cognitive systems contain several copies of the same element of knowledge stored in different parts of the computer memory or of the brain.

Named sets, as said in Burgin [22], is the "*most fundamental mathematical construction encompassing all generalizations of ordinary sets and provide unified foundations for the whole mathematics.*" A *named set* is also referred to as a fundamental triad. A triad is different than a triplet.[15] Examples of various types of *named sets* are provided in Fig. 2.10. Figure 2.11 below illustrates the named set X which is the fundamental triad ($S(\mathbf{X})$, $n(\mathbf{X})$, $N(\mathbf{X})$) where:

---

[14] A number of important theorists in ancient Greek natural philosophy held that the universe is composed of physical "atoms," literally "uncuttable."

[15] Using the term triad, it is necessary to distinguish it from the notion of a triplet. A triad is a system that consists of three parts (elements or components), while a triplet is any three objects. Thus, any triad is a triplet, but not any triplet is a triad. In a triad, there are ties and/or relations between all three parts (objects from the triad), while for a triplet, this is not necessary.

- $S(\mathbf{X})$ is the *support* of $\mathbf{X}$
- $n(\mathbf{X})$ is the *naming correspondence,* also called *reflection*
- $N(\mathbf{X})$ is the *set of names* or *reflector*

A popular type of named sets is a named set $\mathbf{X} = (S(\mathbf{X}), n(\mathbf{X}), N(\mathbf{X}))$ in which $S(\mathbf{X})$ and $N(\mathbf{X})$ are sets. The *reflection*, $n(\mathbf{X})$, consists of connections between their elements. When each connection is represented by a pair (x, a), where x is an element from $S(\mathbf{X})$ and a is its name from $N(\mathbf{X})$, we have a *set theoretical named set*, which is a *binary relation*. There are many *named sets* that are not *set theoretical* such as an *algorithmic named set* $\mathbf{A} = (X, A, Y)$ that consists of an algorithm A, the set X of inputs, and the set Y of outputs. Another example is the *Categorical named sets* that are different from set theoretical named sets since an arrow in a category [23] is a fundamental triad but does not include sets as components. *Named sets* with physical components, such as (people and their names), (books and their authors), and many others, are not set theoretical. A different example of a *named set* (fundamental triad) is given by the traditional scheme of communication (*sender, message, receiver*) represented by the Shannon model of communication (Fig. 2.7).

Figure 2.12 presents other examples of named sets. A *fuzzy set A* in a set U is the triad (Fig. 2.12a) $(U, \mu_A, [0, 1])$, where $[0, 1]$ is an interval of real numbers, $\mu_A : U \to [0, 1]$ is a membership function of A, and $\mu_A(x)$ is the degree of membership in A of $x \in U$ [24, 25]. A mathematical category (Fig. 2.12b) is also a triad (named set) where relationships between categories are called functors and relationships between relationships are called natural transformations [23]. Finally, a vector bundle (Fig. 2.12c) is also a fundamental triad [26]. In this book, the discussion will be limited to ordinary and fuzzy sets.

Burgin [5] considers information as the knowledge content rather than treating information in the same category as data and knowledge. Data are used not only for generating knowledge but also for deriving other epistemic structures such as beliefs and fantasies as shown in Fig. 2.13. Distinctions between knowledge, beliefs, and fantasies depend not on the object domain but are estimated by comparison to another knowledge system. People process and refine information and form beliefs, knowledge, ideas, hypotheses, etc. People's beliefs are formed by the impact that some information has on people's mind. Beliefs and fantasies are structured in the same way as knowledge but, for beliefs, they are not sufficiently justified and, for

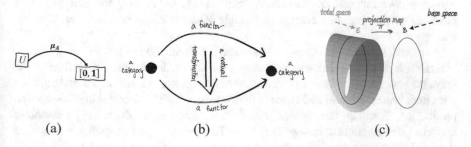

| | | |
| (a) | (b) | (c) |

Fig. 2.12 Fundamental triads as *named sets*: (**a**) a fuzzy set, (**b**) a category, (**c**) a vector bundle

Fig. 2.13 Epistemic
structures: knowledge,
beliefs, and fantasy

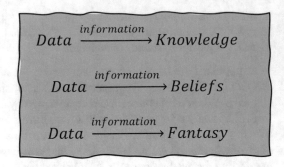

fantasies, they are not justified at all. Thus, the concept of information and the schema presented by Burgin integrate beliefs and fantasies into the system data, information, and knowledge proposed by his general theory of knowledge [27].

Example 2.9 (From Burgin [5]) *"For instance, looking at the Moon in the sky, we* **know** *that we see the Moon. We can* **believe** *that we will be able to see the Moon tomorrow at the same time and we can* **fantasize** *how we will walk on the Moon next year."*

2.3   Knowledge Systems

Knowledge can be seen as the culmination of a chain of entities of different nature [1]. The essence of their nature is dependent upon the perception that we give signs or signals of the observable world involved in the conduct of actions. It also depends on a semantic status defined as the degree of meaning that those signs are likely to convey. According to philosophers, knowledge can be present even in the absence of a close relationship between the knowing agent and the knowledge object. As soon as the informative content of an information allows some pairings with other elements of cognition (e.g., those in permanent memory), it becomes possible to reveal an even stronger sense. The information content is enriched in the knowledge processing chain when information is being transformed to knowledge with a corresponding increase of semantics. Incidentally, amplifying the emergence of meaning quickly becomes normative attitude of any decision-maker for an efficient acting.

   A global knowledge system comprises a domain consisting of an immense number of objects, relations, processes, transformations, and interactions. The knowledge universe comprises a diversity of knowledge systems and items together with their representations and carriers that we can consider as cognition, knowledge production, learning, knowledge acquisition, knowledge discovery, reasoning, knowledge management, and applications. These processes and applications utilize data and knowledge representations, e.g., texts, schemas, and formulas.

Burgin [5], p.311, succinctly describes a knowledge system this way: *"Knowledge by its essence is knowledge of something, namely, about the knowledge object or knowledge domain, which forms the external structure of knowledge per se. Consequently, any knowledge system has two parts: cognitive, that is, knowledge per se, and substantial, which consists of the knowledge object (knowledge domain) with its structure (internal or external). These two parts are connected by a relation (correspondence), which is conveyed by the word "about" in English. Cognitive part of a knowledge system is also called symbolic because, as a rule, it is represented by a system of symbols. Cognitive/symbolic parts of knowledge form knowledge systems per se as abstract structures, while addition of substantial parts to them forms extended knowledge systems."*

Two meanings can be attached to the term knowledge system:

1. Knowledge system as a concise representation of knowledge.
2. Knowledge system as a structure consisting of knowledge elements and relations between them.

Below, we provide informal definitions of various kinds of knowledge systems that will be useful for the discussion in the rest of this chapter: knowledge item, knowledge unit, knowledge quantum, and knowledge element.

2.3.1   Knowledge Item, Unit, Quantum, and Element

Knowledge can intuitively resemble a meta set of aggregated informational (knowledge) items of a particular field according to an explicit finality. Knowledge results in a subtle construction made from multiple levels of aggregation:

1. An aggregation of signs from an observable world to the most benefit of decision-makers
2. An aggregation of data that will be used to support any act of informing
3. An aggregation of knowledge items where structure and depth depend upon the level of culture or expertise of the actor

Informal definitions are provided below for four kinds of knowledge systems:

- *Knowledge item* is a knowledge system that is considered separately of other knowledge systems.
- *Knowledge unit* is a knowledge item that is used for constructing other knowledge systems and treated as a unified entity.
- *Knowledge quantum* is a minimal, in some sense, knowledge unit.
- *Knowledge element* is an element of a knowledge system (structure).

The quantum[16] level of knowledge consists of "knowledge bricks" elements and "knowledge blocks" of knowledge that are being used to build other knowledge systems. For instance, propositions and predicates knowledge are quantum elements of macrosystems, such as logical calculi and formal theories in logic and mathematics. These "knowledge bricks" are cemented together via relations to construct some larger blocks and then knowledge systems are built from such blocks.

A knowledge *unit* is a knowledge *item* that is used for constructing other knowledge systems and treated as a unified entity. A knowledge *item* is a more general concept than a knowledge *unit* since a knowledge *item* can be a part of a knowledge *unit* or consists of several knowledge *units*. Any knowledge *unit* is also a knowledge *item*, but the opposite is not always true. The following examples are provided to make the distinction between a knowledge *unit* and a knowledge *item*.

Example 2.10 (From Burgin [5]) An example of a knowledge item: *"This book is about knowledge. Its title is "Theory of Knowledge: Structures and Processes." It has many pages. It has nine chapters."* All knowledge from this textbook is a knowledge item. Knowledge in several unrelated stories (novels) is a knowledge item. Knowledge about several unrelated persons is a knowledge item.

Example 2.11 (From Burgin [5]) An example of a knowledge unit: "This book is about knowledge." This statement is also *a knowledge quantum*. Knowledge in one story (novel) is a knowledge unit. Knowledge about one person is a knowledge unit. Knowledge about members of one family is a knowledge unit.

Example 2.12 (From Burgin [5]) An example of a knowledge element: *"a book."* At the same time, there are also composite knowledge elements, e.g., *"an interesting book."*

Example 2.13 A knowledge base: all knowledge from this knowledge base is one knowledge item or it can be considered as an organized system of knowledge items (units).

2.3.2 Knowledge and Course of Actions

To *act* upon, or in the world, implies for each human to acquire knowledge of several informational items about someone or something via various ways: perception, discovery, or learning. Knowledge is awareness or facts, information, descriptions, or skills acquired through experience, education, and culture. Human knowledge proves to be an indispensable prerequisite for any rational action or for heading a company according to sound rules.

---

[16]*Quantum* is the Latin word for amount and, in modern understanding, means the smallest possible discrete unit of any physical property, such as energy or matter.

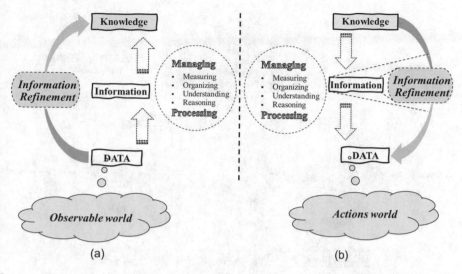

Fig. 2.14 Data-information-knowledge: (**a**) transforming data into knowledge and (**b**) transforming knowledge into data

There are differences between knowledge, information, and data. The relationships between data, information, and knowledge will be discussed in the next chapter. For now, let express the differences this way:

- Knowledge is a compressed, goal-oriented structural depository of refined information.
- Data is a raw symbolic depository of information.
- Conversion of data into knowledge is a refinement of information as illustrated in Fig. 2.14.

Conversion of data into knowledge can be thought as a refinement of information to provide knowledge of a specific event or situation. Data, under the influence (action) of additional information, are transformed into knowledge as illustrated in Fig. 2.14a. The upward arrows in Fig. 2.14a suggest that information creates knowledge. The downward arrows in Fig. 2.14b indicate that it is possible to create data from knowledge. For instance, when an e-mail is sent, the text, which is usually a knowledge representation, is converted to data packages, using a definite type of information processes (refinement) prior to be transmitted (action) to the recipient. Another example for Fig. 2.14b is the management of sensors using a priori knowledge. Obviously, the information processes to convert data into knowledge in Fig. 2.14a are different of those to convert knowledge to data of Fig. 2.14b.

Fundamental to knowledge is the notion of informational enrichment (or semantic growth) that starts from the information and is obtained via human intervention or supported by a computer-based intelligent process (Fig. 2.15). Enriched knowledge will, of course, have a stronger impact on decisions. Note that knowledge brings to existence a *world of action*, in the sense that it takes a form other than those more or

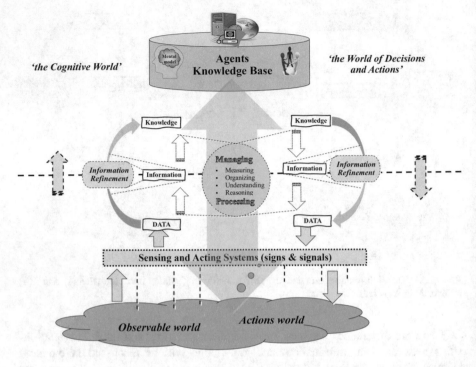

Fig. 2.15 Course of actions

less conscious of the human mind. A course of action, illustrated in Fig. 2.15, normally involves a few phases:

1. A perception phase expressed by identifying the most appropriate signs to represent the observable world. This is the first level of formal representation dictated by a cyber world where there is a need to define data.
2. A data processing phase where diverse information processing schemes and refinements are applied to data along its transformation chain toward knowledge; here, a second level of formal representation that uses different logics to support machine interpretation is required (e.g., knowledge-based systems).
3. A passage in the cognitive world for pattern matching and reasoning phases preceding a decision.
4. A return to the observable world, specifically in the world of decisions-actions where the effect of this decision will make sense.

Figure 2.15 suggests that data, or more precisely data representations, are physical (rooted in the world's physical properties) but not information. On the other hand, knowledge is rooted in the cognitive world of the individuals. Information is what a knower can use for creation and extraction of knowledge from data, given his/her capacity. Physical extraction essentially depends on the types of knowledge

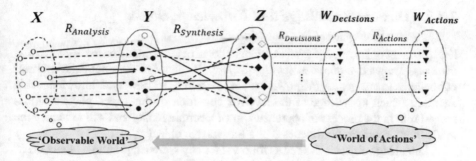

Fig. 2.16 The epistemic space structure of a course of action composed of a chain of *named sets*

recipient: algorithms and logic, models and language, operations and procedures, goals, and problems.

It is important here to realize the difference between knowledge and cognitive information. For instance, in a teaching process, the teacher does not give knowledge itself but provide cognitive information causing changes that may result in the growth of knowledge for a recipient. Note that information transactions may also result in the decrease of knowledge. An obvious case for the decrease of knowledge is misinformation and disinformation that is often used to corrupt opponent's knowledge in information warfare. "Fake news" [28, 29] are the flavor of the day for disinformation in civilian affairs.

The notion of epistemic structures can be brought here to represent the course of action illustrated in Fig. 2.15. Since named sets are fundamental triads, they can, by a composition of these (sets + their relations), be the foundation for the design of a computer-based system to support the execution of the generic course of action discussed in Fig. 2.15. This means that we must define the named sets and their associated information processes with respect to decisions and actions as pictured in Fig. 2.16. For instance, we may choose fuzzy sets or a hybridization of different kind of sets as named sets to design a computer-based support system depending on the specific application. In fact, the overall design objective is to get a computer-based support system that is conceptually and theoretically correct (epistemic structures), capable of addressing large problems (computability and tractability, managing and processing DIK), and useful (dependability) to users (domain of applications).

The course of action shown in Fig. 2.15 involves necessary phases that are described in the language of named sets below and illustrated in Fig. 2.16.

1. A perception and data processing phases (*analysis and synthesis*) expressed by fundamental triads $(X, R_{\text{Analysis}}, Y)$, $(Y, R_{\text{Synthesis}}, Z)$ to represent the observable world and the diverse information processing refinements to transform data into knowledge
2. A passage in the cognitive world for pattern matching and reasoning phases $(Z, R_{\text{Decisions}}, W_{\text{Decisions}})$ to define decisions and actions
3. A return to the observable world with actions $(W_{\text{Decisions}}, R_{\text{Actions}}, W_{\text{Actions}})$

2.3.3 Domain Knowledge and Knowledge Object

Each element or knowledge system refers to some object domain *because knowledge is always knowledge about something (object)*. Knowledge does not exist by itself but belongs to an agent A (a knowledgeable system or a knower). A knowledgeable agent, in which knowledge or the level of cognition of this agent, plays a fundamental role in the process of apprehension of a certain reality and will influence the mode of appropriation (awareness) of a knowledge object.

A knowledge object is defined as follows: for any knowledge system (element) K, there is a domain D of real or abstract objects and K describes the whole D or its part. A domain could be considered as one object or several objects. In such a way that we can derive the following diagram (Fig. 2.17), which is a specific case of *named sets*.

The relationship between objects, domains, and knowledge items is illustrated in Fig. 2.17 using named sets concept. It says that knowledge is a cognitive representation of knowledge objects and knowledge domains, which is intrinsically related to their representations that are often symbolic. Recall that any decision-making or action in (or on) a universe is conditioned by the prior acquisition and nature of a certain knowledge resulting itself from an aggregation of knowledge whose quality is even largely dependent on the way in which we first perceive the things of this world. We will consider here that every phase of apprehension of the real world depends on several criteria of perception:

- The intrinsic qualities of the perceptual system: the noisy operation of the sensors
- The psychological, possibly physiological, dispositions of the knowing agent (the knower): sensitivity and stress
- The means of interpretation in the cognitive world from the mental representations that the knower made of the perceived world

This form of perception allows an object, essentially physical in its world of belonging, to acquire via the knower a new mode of existence: a new entity in an immaterial world. If the latter has not received an ad hoc symbolic representation and has not acquired a sufficient sense in relation to its potential user predisposing it to be assimilated by subsequent computer processing, then it will not be operative in the sensitive world as illustrated if Fig. 2.18. Note that a symbolic representation of a knowledge object is a knowledge item.

The intercession of the knowing agent is fundamental in the process of knowing, as can be illustrated by the case of a person who says that he knows Africa without ever having stayed there, but simply because she made a note of information by reading documentation. This is a level of knowing that may represent a totally

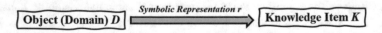

Fig. 2.17 Illustration of a knowledge object as a specific named set (D, r, K)

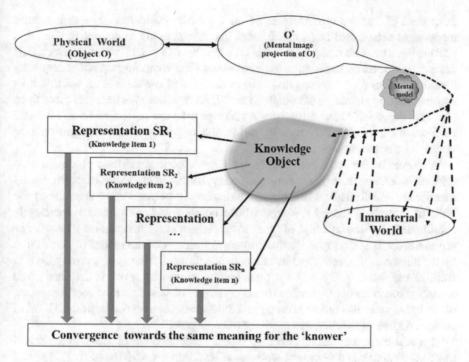

Fig. 2.18 Knowledge acquisition process

imperfect knowledge. The knowing agent perceived a certain reality of the sensitive world influenced by:

1. The points of view treated by the magazine
2. The quality and authoritative authority of the author on the subject considered
3. His/her cultural level

To refer to a reading or cartographic knowledge corresponds to two distinguishable modes of understanding the reality: "knowledge of Africa." In fact, this means to acquire a partial knowledge of the real world. These two modes of acquisition reflect different relationships between the knowing agent and the known (physical) object. In other words, it is a differentiated appropriation of a knowledge object by the knower, which will be translated into different forms of representation: first mental, material, then possibly symbolic with advanced information treatments. Figure 2.18 illustrates the process of knowledge acquisition. The "interpretation" at the level of understanding by the knowing agent must also consider the nature of reference, which is dependent on the material vector of representation (literal, documentary, pictorial).

By using this notion of knowledge object, one tries to consider at best these different attitudes and modality which can account for them. To better situate this notion, we will give the meaning of three terms which, in our opinion, play a key role in the process of knowledge acquisition: (1) a *fact*: an entity which manifests itself from a set of features characteristic of the universe considered, (2) a *reference*: a mapping to established facts, in the sense that it is possible to memorize them in a sustainable way. This mapping is based on similarities or neighborhood characters. It depends of course on experience and know-how. (3) *Perception*: the act of acquiring data through an ad hoc perceptual system on the real world. It is an act that may require an implementation of complex mental mechanisms. We note that these three elements are conditioned differently with respect to the notion of *time*. *Fact* and *perception* can evolve very rapidly, unlike the *reference* whose changes obey to longer time constants.

Whatever be the considered universe, the knowledge acquisition process stays a difficult exercise since it depends on many factors with some of them hardly identifiable and formalizable. This has undoubtedly become more acute since the birth of knowledge-based systems, whose essence and performance are largely conditioned by the quality of the basic ingredient: information (knowledge represented). This quality is itself obviously a function of the modes of acquisition by the human of the reality and its universe of actions. Different observers placed in front of the same reality will not perceive the same way, even if they are given unambiguous rules of observation. In addition, one observes that the comprehension of the universe is also linked to certain qualities of the human operator: their level of culture, degree of training, and capacity of conceptualization.

It seems more reasonable, outside the philosophical point of view, to say that if one fails to apprehend the world, one can at least express a vision of it. This would prevent computer systems designers, who represent the world as they see it and rely on formalisms that have little connection with the encountered reality, to design applications that correspond imperfectly to the customer needs.

The notion of "knowledge object" must thus enable the object (physical) or entity acquired in the physical world to be substituted for a notion of "vision" of that object. We will now consider that the knowledge acquisition about an object amounts to acquiring the vision of a "perception" of that object. The object observed in the world remains essentially physical, whereas its acquisition by the human takes it into an immaterial world where, according to the knowing agent, its *signified* will carry a greater or less expressiveness. The notion of knowledge object is independent of any connotation of "instrumentalization," for example of a structured object, which is, moreover, only a possibility of representation among many others. However, to process information about that knowledge object with an instrument like a computer, we do need to represent it using the notion of structures (knowledge item) of the previous section.

2.3.4   Representation of Knowledge

The acquisition of a larger part of knowledge about the world is realized through conditions concerning the situational state of the knowing agent and his/her modes of perception for a given time and space. It is indeed difficult to dissociate the facts from their perceptions relative to a space-time representation. The notion of knowledge object carries the need to have several levels of representation, according to the fineness (stratification) level of interpretation.

First level of representation   A knowledge object makes somehow allusion to a window that is moved on an observation space, identified by a time axis and axes x, y, noting that the window is positioned more or less precisely in x, y as well as in time. Some facts belong to the window with certainty. Other elements have a modulo membership criterion in the range of certainty (0–1). Some imperfections that are attached to the perception of all information coming from the real world appear in this window and need to be represented.

Second level of representation   Any knowledge object can eventually give rise to a finer interpretation and be represented in a *trihedron* whose axes are the reference, the fact, the perception as illustrated in Fig. 2.19. The three planes of this trihedron express (1) perception-reference axes: for the explanation of the behavior, (2) axes reference-fact: for the expression of judgment, and (3) axes perception-fact: for the formal expression of the facts and their perceptions.

Time is of course omnipresent in these levels of representation. Facts depend on time. Perception depends on facts. References depend on both perception and facts. Any representation of a knowledge object expressed in the trihedron (Fig. 2.19) is very strongly dependent on time as:

$$\text{Knowledge\_object}(t) = \varphi\left[f_{\text{fact}}(t), f_{\text{perception}}(t), f_{\text{reference}}(t)\right].$$

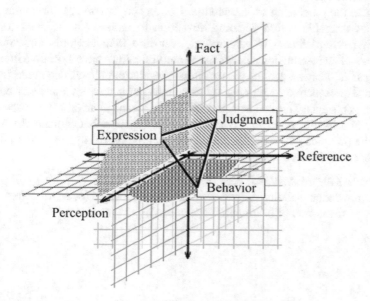

Fig. 2.19 Trihedron to represent a "knowledge object"

2.4 The Multidimensionality of Knowledge

A knowledge chain, organized and implemented rationally, must allow its user to better conduct his/her action insofar since it can offer several services such as the following [1]:

- The use of necessary *referents* to point to the best location of information to acquire before acting
- The need to know if the requested knowledge, inherently having a *signified*, always make sense in the context in which it is expected to act
- The knowledge structuring or the relational mapping of concepts, and explanations on the history of their formation: the "why and how" about their relationship, linkage, and nesting or filiation
- The issuance of the required knowledge in time or in the time windows where it remains potentially valid

The four previous remarks highlight several dimensions of interest that lead us to reveal a multidimensional aspect of knowledge. Note first that any decision often requires relying on reference knowledge that can present the following three properties:

- *Universality*: (entire sets) access in full knowledge
- *Existentiality*: (partial sets) access only to a class of specific knowledge
- *Uniqueness:* a reference knowledge, specific or unique.

Somehow knowledge can also be seen in terms of a "*meaning*" vector. It is therefore necessary to consider the place and role of the semantic dimension of knowing in the processing of knowledge. Concerning knowledge structuring, it is sometimes desirable before deciding an action to understand the need for such knowledge, its enrichment, its parenthood, over time. In other words, understanding the genesis of an informational background can sometimes have a positive impact on the process of decision-action. We can thus consider the knowledge chain from an ontological perspective to be able to somehow get a trace of the links between successive concepts. This can be achieved by the means of an ad hoc representative structure of a relational network combined with a system that explains the reasons behind the presence of some notional elements. Finally, time plays also a predominant role for action in two aspects:

1. The access to "*constrained*" time (temporal constraints).
2. The knowledge validity gained has not been negated over time (e.g., time window, the notion of information *timeliness*).

2.4.1   Dimensions and Characteristics of Knowledge

To be used profitably, It may be advantageous to place a knowledge object (defined above) in an immaterial space of several dimensions, to assess its sensitivity to certain criteria that can be defined on the considered dimensions. Dimensions are basic descriptive characteristics of epistemic structures in general and knowledge structures in particular. Each dimension has different gradations and/or modalities. Some of these gradations are discrete, while others can be considered as continuous. Table 2.2 presents a list of dimensions, properties, and modalities for epistemic knowledge structures. This is the result of a sort of a brainstorming analysis based of Chap. 2 of Burgin [5]. Note that each dimension integrates specific knowledge properties usually containing these properties as its components. For the interested reader, Chap. 2 of Burgin [5] contains a comprehensive treatment of Table 2.2 items with definitions and examples.

Table 2.2 is not an exhaustive list of knowledge dimensions and characteristics. There are also other characteristics of knowledge mentioned in [5]. Novelty, specificity, and amount of knowledge are three of them. Novelty shows the extent to which the content is original (e.g., repetition, duplication, contradiction). Knowledge domain is what this knowledge is about (e.g., contextual information). Specificity refers to degree of detail of the knowledge in a message. Amount of knowledge has many different meanings and measures. This is also intimately related to measures of information, e.g., Hartley–Shannon entropy, algorithmic complexity, number of known facts or ideas.

2.4.2   Knowledge Correctness

To conduct a high-level analysis of the dimensions and characteristics of knowledge, we need to frame it with the categories and aspects of knowledge as pictured in Fig. 2.20. This is what Burgin [5] did with Table 2.2 items. For instance, consider the first row of Table 2.2, *knowledge correctness*, with the series of informal definitions and examples presented below.

Correctness, relevance, and *consistency* are the main properties that reflect relation of knowledge to some system C of correctness conditions. For instance, consider operational knowledge, i.e., a computer program. This knowledge informs computer how to perform computations. Correctness becomes a critical issue in software and hardware production and utilization because of the increased social and individual dependence upon computers and computer networks.

Definition 2.9   A knowledge system (unit) K is correct with respect to a system C of conditions if it satisfies all conditions from C.

Table 2.2 Brainstorming blackboard—knowledge dimensions and characteristics

Dimensions – Characteristics	Properties	Modalities/Components

Correctness:
 - syntactic
 - semantic
 - pragmatic
	- relevance
	  - (domain, problem, goal) - related
	 - consistency
	 - truth
	 - completeness
		- description
		  - accuracy
		  - exactness
		  - precision/completeness
		 - attribution
		  - domain interpretability, representability, applicability

Confidence:
 - syntactic
 - semantic
 - pragmatic
	- psychological confidence
	  - internal, external
	  - exterior
	 - certainty
	  - knowledge item is certain for the knower.
	  - source certainty: knowledge item is coming from some source.
		- assurance
		 - validation
		 - groundedness
		 - confidence level, interval, coefficient
		 - degrees of confidence
		 - degrees of certainty

Complexity:
 - syntactic
 - semantic
 - pragmatic
'is always complexity of doing something'
	- knowledge acquisition, transmission, integration
	 - knowledge-to-action
	  - **actionable knowledge**
	 - learning
	 - resources management
	  - used resource.
	 - diversity of processes
	 - utilization, transformation, problem, cognitive complexity
		- degrees of complexity
		 - algorithmic
		 - tractability
		 - relations
		 - networking
		 - efficiency
		 - robustness
		 - clarity, accessibility
		 - time and space
		 - dynamics

Significance: - syntactic - semantic - pragmatic	- graduality - importance - value	- selection criteria (quantity of knowledge) - representations - numerical, ordered, nominal scales.
Efficiency: - syntactic - semantic - pragmatic	- goals - problem-solving - resources	- tractability - quality of solutions - reliability, exactness, relevance
Reliability - syntactic - semantic - pragmatic	- knowledge content - knowledge source - knowledge transmission and production	- accuracy, veracity, credibility, correctness and validity. - true/false knowledge
Meaning	- sense-making, sense giving - meaning of a knowledge *item* - contexts	- denotation/connotation - domain of reference - representation of sense
Temporal	- events and agents - processes - time semantics - relevance	- time representation - timeliness - time algebra - time validity
Reference	- quantification - contexts - domain knowledge	- order of a referent - relational predicate

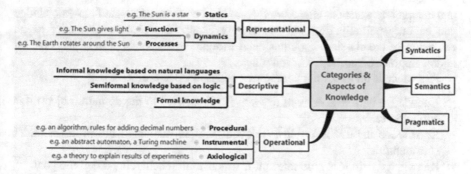

Fig. 2.20 Categories and aspects of knowledge

Example 2.14 *Correctness*: A knowledge system is usually represented by a propositional calculus. Such propositional knowledge is correct if this calculus is consistent. Existence of various correctness conditions results in a variety of correctness types: *truth, correlation, consistency*. Defining correctness of propositions, it is possible to take into consideration three aspects of propositions—syntactic, semantic, and pragmatic aspects. For each of these aspects, we can attribute one criterion of knowledge correctness:

1. The syntactic criterion has the form of a syntactically correct English sentence
2. The semantic criterion is that the proposition is true in the sense of classical logic
3. The pragmatic criterion is that the proposition has a model in the real world

Correctness is a gradual property because in many cases, conditions can be satisfied only partially. We talk about degrees of correctness so that correctness is a fuzzy property. It is possible to introduce different measures of correctness functions on knowledge items that can take numerical values, vector values, or even values in a partially ordered set.[17] Knowledge correctness has description and attribution modalities that are relational properties of knowledge since they depend on relations between knowledge and some domains (objects). For instance, accuracy, exactness, and precision are components of the description modality of knowledge correctness. Probability represents only one aspect of knowledge vagueness and inexactness. Other theories, such as imprecise probabilities and fuzzy set theory, have been developed to represent better these properties of knowledge.

Example 2.15 *Exactness:* This is how a descriptive knowledge item represents a larger or a smaller domain in comparison with its assigned domain and the extent of the existing difference. For a long time, one assumed that knowledge is always true

---

[17] A *partially ordered set* (also poset) formalizes and generalizes the intuitive concept of an ordering, sequencing, or arrangement of the elements of a set. A poset consists of a set together with a binary relation indicating that, for certain pairs of elements in the set, one of the elements precedes the other in the ordering. The relation itself is called a "partial order." https://en.wikipedia.org/wiki/Partially_ordered_set

and completely exact but after a while, one has begun to understand that knowledge can be only partially true and moderately exact. The probability theory as an extension of the classical logic has been brought by researchers to express partial and/or imprecise knowledge or ignorance.

Relevance is an important component of knowledge correctness.

- Knowledge item K is *domain* relevant if it is related to the domain (object) it is attributed to.
- Knowledge item K is *problem* relevant if it is related to the problem under consideration.
- Knowledge item K is *goal* relevant if it is useful in achieving a definite goal.

Definition 2.10 The domain relevance of a knowledge item shows the extent to which this knowledge item is related to some issue (*domain, problem, goal*) of a considered domain or how does that bear on the issue.

As relations can be stronger or weaker, the degree of relevance may be higher or lower. To represent this degree in a quantitative form, it is possible to introduce different measures of (*domain, problem, goal*) relevance, the scale of which can be either the two-element set {0, 1} or the interval [0, 1] or the set of all non-negative real numbers. Treating relevance as a fuzzy property allowing different degrees of relevance is an appropriate way to represent this property.

Example 2.16 *Relevance*: A knowledge item represented by a statement can be clear, accurate, and precise, but not relevant to the question at issue (*domain, problem, goal*). For instance, knowledge about elementary particles is irrelevant to music or art. Knowledge of geometry is irrelevant to moral issues.

Further details, definitions, and examples of dimensions, characteristics, properties, and components of Table 2.2 can be found in Burgin [5], according to the three semiotic dimensions (Fig. 2.20). For instance, *consistency* is an important relational characteristic of a knowledge system. The traditional approach to knowledge consistency implies eliminating elements of knowledge that are called contradictions. The expression "It is a car, and it is not a car" is contradictory by the rules of classical logic. Within this logic, *consistency* is an absolute property quality, but in general for various systems, this property is relative, including for logical systems.

Definition 2.11 A system R from K is *consistent* (inconsistent) with respect to C if it satisfies (does not satisfy) all conditions from C (e.g., propositional knowledge – absence of contractions).

An important type of knowledge correctness is truthfulness in three different types: domain-oriented, reference-oriented, and attitude-oriented.

Definition 2.12 Consider an agent A that has knowledge K about the domain (object) D, then the truthfulness K means that (condition from C) the description that K gives for D is true.

Thus, if we name a truthfulness function, *truth* (K, D) of the knowledge K about the domain D, then that function can take two values—true and false. That function, *truth*, gives conditions for differentiating knowledge from similar structures, such as beliefs or fantasies. Truthfulness is closely related to accuracy of knowledge but are different properties.

Knowledge accuracy says how close the knowledge K is to the absolutely exact knowledge. For instance, statements "π is approximately equal to 3.14" and "π is approximately equal to 3.14159" are both true, but the second statement is more accurate than the first one. We see that conventional truthfulness can indicate only two possibilities: complete truth and complete falsehood. It is possible to treat *truthfulness* and *correctness* as linguistic variables in the sense of Zadeh [30]. For instance, we can separate such classes as highly correct/true knowledge, sufficiently correct/true knowledge, weakly correct/true knowledge, weakly false knowledge, and highly false knowledge.

We refer the reader to Chap. 2 of Burgin [5] for a more detailed discussion on the rest of Table 2.2 such as confidence in and certainty of knowledge, complexity and clarity of knowledge, significance, importance and value of knowledge, efficiency and reliability of knowledge, precision and completeness of knowledge.

2.5   Meaning of Knowledge: The Semantic Dimension

Knowledge exists in different forms and shapes. Speaking about meaning of knowledge, it is necessary to consider knowledge carriers and representations. Various images can be carriers of knowledge. The most popular form is symbolic expressions since it is much easier to assign meaning to expressions than to other forms and shapes of knowledge representations, e.g., for emotions or feelings. A discipline called *semantics* has been developed and applied in semiotics, linguistics, computer science, and logic to study the meaning of expressions (forms and shapes).

Any information or knowledge possesses some *signified*. As such, it represents a feature that directly impacts how to drive an action or make a decision. History attests unfortunate consequences induced by the attitude of decision-makers who have not been able to interpret or represent, a *signified* attached to a series of events, or, worse, a community of actors who did not agree or wish to share a common understanding of those events. This underlines the importance that we must now give to the meaning dimension of knowledge: a semantic dimension tailored to operational information systems purposes and usable in artificial reasoning schemes.

The semantic dimension of knowledge requires special considerations. What specific elements can give sense to informational elements (represented knowledge item) that are required prior to execute an action? How can we characterize their modes within the knowledge chain? The semantic dimension in question here is not simply looking for a grammar giving advanced syntax rules whenever they exist. It cannot thus be achieved, for example, through an approach called generative grammars which, under the influence of Chomsky [31, 32], have been widely used in linguistics and also in computer sciences (compilers). These grammars cannot be

Fig. 2.21 (a) Sulis' archetypal dynamics and (b) the semiotics triangle

applied to knowledge since they are too focused on structural and grammatical recognition required for combinations of sets, ordered according to strict rules. Anyway, they are unable to take into account the semantics, including that of a text.

The semantic dimension must be considered from the perspective of facilitating the interpretation of the knowledge required for acting. In fact, it means to fetch the most relevant *signified* with respect to the execution of a given action. This involves being able to identify and strengthen the semantic status of the different levels of a knowledge processing chain. This also implies to facilitate the reduction of the implicit, as well as the ambiguity, as it can occur in statements like "*Claude stopped to snorkel*": implicit because this assumes that Claude dived before and ambiguous since nothing indicates if this judgment is temporary or permanent.

The perspective expressed in the previous paragraph so that meaning is tight with actions is the viewpoint of archetypal dynamics as well. Sulis [33] introduced Archetypal Dynamics (AD) as follows: "*Archetypal dynamics is a formal framework for dealing with the study of meaning laden information flows within complex systems.*" This is a formal framework for dealing with the study of the relationships between systems, frames, and their representations and the flow of information among these different entities. The framework consists of a triad: semantic frame (representation), realizations (system), and interpretation (agent/user). Real systems relate to semantic frames through one of the dimensions of that triad represented in Fig. 2.21 [14].

The underlying dynamics is based upon the idea of a combinatorial game [33]. A semantic frame is an organizing principle that ascribes meaning in a coherent and consistent manner to phenomena that have been parsed into distinct entities, mode of being, modes of behaving, and modes of acting and interacting. The following excerpt from Sulis [33] clarifies the distinction between the triad dimensions interpretation-representation-realization: "*A representation is understood to be a particular situation, involving a representational frame which serves as a kind of universal object, whose realizations may be associated in some manner with the realizations and interpretations of some other semantic frame. The representation is not itself a realization or interpretation. Realizations and interpretations are ontological constructs, whereas representations are epistemological constructs like symbols and signs.*"

The semantic dimension must therefore accommodate certain norms to reduce the effects of ambiguity and help limit the influence of an interpretive context. The semantics, usually considered as the prerogative of linguistic, cannot be completely avoided in the design of intelligent automatic systems (including embedded systems) or to optimize execution of actions and decision-making. It is, therefore, important for a chain of knowledge to clarify the nature and define, as far as possible, the role of its semantic dimension.

A common definition (and summary) of semantics is to say that it is the study of meaning attached to languages, but as noted by P. Vaillant in his semiotics glossary [34], such a definition is very general as it indistinctly can apply to formal systems as in the Tarski's theory of models [35] as well as to human language. "Here we use the term semantic that designates a discipline descriptive of senses of a given language, and as such, may therefore, also be applied to an image than a linguistic text. Semantics is then a subordinate to semiotics: the first being rather technical and the second rather theoretical." Semantics is to remain the elected discipline of meaning. It must be clearly distinguished from both phonology (only concerned with sounds) and semiotics. Regarding the latter, the distinction cannot be made easily because according to F. de Saussure [36], semiotics should address all signs without exception and, therefore, consider semantics as its part included.

2.5.1 Production of Sense

The problem of meaning is old, recurrent, and difficult as it addresses various questions concerning elements of a knowledge processing chain:

- Does the required information or knowledge within the context of an action have really the sense given to it by the knowledge being or the sense assigned to it within a process?
- Is it possible to produce or enhance the significance of a piece of knowledge from other sources, for example, using deduction mechanisms?
- Is it possible to "produce" sense formerly?
- Under what modalities can meaning be represented and attached to the various elements of a chain of knowledge?
- How can we represent and store elements of meaning? In other words, is it possible to arrange a semantic memory in perspective to end up with a facilitator for actions?

Seeking to answer these questions develops the semantic dimension of a knowledge chain. That, in a first step, can be considered concretely in two aspects:

1. Consider the implications of a theory to "produce" sense
2. Propose modalities of representation, and to a lesser extent, production of a signified, and seek ways to instrument those modalities

A binary word has no semantic status by itself; it cannot therefore be used without the addition of technical features, conventions. For instance, conventions to specify the usage are necessary, which can be coding, transmission techniques, morphological rules, etc. Situating an action in a world, and specifically acting on its most representative elements of reality, shows an emergence of a certain materiality in which the projected action acquires a sense. This is not only meant to describe a *signified* but, beyond its required modes of representation, to acquire the means for that *signified* that it used in a context of action, to adopt or bear an unequivocal meaning. Any lack of uniqueness does not allow a knowledge item to make sense in relation to an action carried out in a specific context, especially for an action resulting from a collaborative decision. Producing sense [37, 38], sense giving [39, 40], and making sense [41, 42] are all very related and sometime use interchangeably.

Consider now a set defining a semantic category. Here, category means an intuitive designation of a criterion for belonging to a more global set; for instance, man and woman belong to mankind. The semantic categories are designated by $\{C_1, C_2\}$: C_1 representing gender "masculine" and C_2 the gender "feminine." Intuitively, we see that C_1 and C_2 make sense with respect to a relation of association that connects them, which is also based on presuppositions. Masculine only makes sense in relation to feminine and vice versa. Making it, is to lay down a principle of opposition. In addition, define two logical operations described as assertion sign "$\vdash$" and negation sign "$\neg$." The following expressions can be written: $\neg C_1$, contrary to C_1; $\neg C_2$, contrary to C_2; then $\vdash(\neg C_1) \Rightarrow \vdash (C_2)$.

So C_2 maintains a relation of contradiction with respect to C_1. Using these different expressions, we can construct the *semiotic square*, also known as the *Greimas' square* [43]. This is a tool used in structural analysis for the relationships between semiotic signs through the opposition of concepts, such as feminine-masculine or beautiful-ugly. Greimas considered the semiotic square to be the elementary structure of meaning. Figure 2.22 illustrates the semantic square for the concept "beautiful-ugly."

The structure of Greimas in linguistics, despite its simplicity, is effective to model the conditions to produce sense. In practice, two steps are proposed to make Greimas' square operational:

1. Set up the vertices with the contents on which negation and assertion operations can be applied
2. Apply negation and assertion operations on the vertices that contain tangible sense related to the specific case under stud.

2.5.2   The Semantic Traits or Semes

Linguists are used to introduce a basic unit of meaning, the semantic trait, usually designated by the term "*semes*" to serve as elements of discrimination between two

Fig. 2.22 Greimas' semiotic square (C_1, C_2) represents the axis of contraries: beautiful $\leftrightarrow$ ugly $(\neg C_1, \neg C_2)$ represents the axis of sub contraries: not-beautiful $\leftrightarrow$ not-ugly $(C_1, \neg C_1)$, $(C_2, \neg C_2)$ are the axes of contradictions

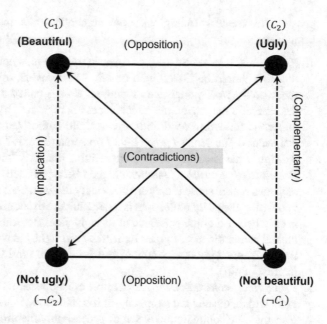

neighboring *signified*. The notion of seme,[18] closely related to studies on semantics, was introduced from the field of cognitive psychology. The meaning of a same word should be distinguishable in statements that differ in their informative content. For example, the meanings of "warm" can be distinguished in the following two statements:

1. "A warm summer night" with a semantic trait denoted as physics
2. "A warm welcome" with the trait connoted as moral

As just seen, a seme is not defined rigorously since it depends upon a system of signs. The message is supposed to convey the signified from all semes (multiple semes) present or under development. Polysemy[19] is the capacity for a sign (such as a word, phrase, or symbol) to have multiple meanings (that is, multiple semes or sememes and thus multiple senses), usually related by contiguity of meaning within a semantic field. The semanticists have proposed a theory to show how it was

---

[18] *Seme*, the smallest unit of meaning recognized in semantics, refers to a single characteristic of a sememe. These characteristics are defined according to the differences between sememes. The term was introduced by Eric Buyssens in the 1930s and developed by Bernard Pottier in the 1960s. It is the result produced when determining the minimal elements of meaning, which enables one to describe words multilingually. Such elements provide a bridge to componential analysis and the initial work of ontologies. (Source: https://en.wikipedia.org/wiki/Seme_(semantics))

[19] Polysemy is a pivotal concept within disciplines such as media studies and linguistics. The analysis of polysemy, synonymy, and hyponymy and hypernymy is vital to taxonomy and ontology in the information-science senses of those terms. It has applications in pedagogy and machine learning, because they rely on word-sense disambiguation and schemas. Source: https://en.wikipedia.org/wiki/Polysemy

possible to create meaning with the semantic traits mainly based upon three principles.

1. *Principle of a semantic axis*: a semantic trait alone is not sufficient to engender a sense, it should be linked with others. The network of selected traits produces sense on the reality analyzed or studied. This is called a semantic axis. Consider the case of a narrow tunnel which access is regulated by a traffic light operating under the following conditions: green light: authorized access, red light: forbidden access. The semes / green light / on semantic axis / green light / vs (versus) / red light / has meaning in the interpretative context: "tunnel access."

2. *Principle of opposition:* A semantic axis being defined, the presence of a couple of semes makes sense if they are obviously distinct. This reiterates the principle of opposition posed by philosophers to say that the existence of the left side of a road is only because of the existence of its right side. Returning to the process "access tunnel" semantic axis, / green light / vs / red light / makes sense, insofar, / green light / differs without any doubt from / red light/ (and fortunately it is so, in the interest of all!).

3. *Principle of contradiction:* Any notion presupposes its opposite. This is a principle which, despite the simplicity of his statement, is much more important than it seems. The contradiction of seme / green light / mentioned by / ~~green light~~ / (the seme contrary is indicated by its strikethrough) means that not / green light / can generate anything else than its opposite / red light /.

In the interpretative context of "tunnel access," and assuming the existence of a third light / blue light / rendered by the seme: / blue light / could mean to break. As shown in the previous example, a seme becomes relevant with respect to a semantic axis and the axis of the opposites. The presence of these three principles is usually represented on a diagram similar to the semiotic square to model the impacts of meaning in a given application.

2.5.3 Sense and Action

An initial observation must be made in the context of a process to control the elements of sense: they are only accessible by successive indirections, and they defy any simple causal explanation. Specialists usually classify them with respect to two types of actions: a dyadic type that roughly corresponding to what is called stimulus-response and a triadic kind related to the concept of learning.

For the dyadic type, the act is automatic without the presence of a signified, for example, putting hands on your ears when you hear a loud noise. In the triadic case, the knowledge acquired by learning or by training affects the action. The signifier, a "red light" signal, must be taken into account: it is mandatory. In the example of the process "tunnel access," the signifier does not act directly as in the case of a stimulus-response (dyadic). Note that a triadic act has several characteristics opposed to those of Shannon's model.

There is a reversibility in a process of meaning as a reciprocal coupling is automatically created between the seme / red light / and the action: stop. This reciprocity is opposed to the one-way link "transmitter-receiver" of the Shannon model. Similarly, the signified obtained by decoding the signal "red light" does not necessarily determine the action "stop," while in Shannon, transmitting a signal through a channel is necessarily deterministic.

2.5.4 Connotation and Denotation

The apprehension of a semantic dimension would not be complete if we do not also clarify the roles of denotation and connotation. Connotation and denotation are two principal methods of describing the meanings of words. Connotation refers to the wide array of positive and negative associations that most words naturally carry with them, whereas denotation is the precise, literal definition of a word that might be found in a dictionary as represented in Fig. 2.23.

The connotation helps to enrich meaning by adding new attributed interpretive fields. The semantic axis, / green / vs / red /, represents a common meaning and shared by the community of the tunnel users. All messages sent from the traffic light occurrences, which need to be consistent with the structure of the semiotic square, make up what experts call a denotative code. Denotation involves representing and attaching an unambiguous (unequivocal) meaning to the system of signs to avoid confusion. Any user in the example of this tunnel must have denotative code as a reference. Denotation seen as such opposes to connotation because it can be likened to an interpretation of the message meaning, which suggests that it may be as many connotations as interpreting agents. The connotation will add meaning but in return induces a serious difficulty since the notion of a common reference, so be shared by a community of agents, no longer exists. In the example of tunnel access, each configuration can thus be connoted via commands such as stop, go, slow down, and shift.

Fig. 2.23 Informal definitions of connotation and denotation

2.5.5 Representations and Manipulations of Sense

After evoking elements of theory to produce a sense, we will look for practical ways in which we can consider representing and manipulating a sense. The first difficulty is that there are no specific representation tools, leading to seek and adapt systems used in artificial intelligence for knowledge representation. All tools discussed from now on are all based on graph theory and its applications [44, 45]. Graph theory is well formalized [44] and it is possible to take advantage of some of its interesting properties in the different representations used to bear sense: connectivity, partial graph (tree), etc. Using graphs has two advantages:

1. Can provide open representations (e.g., tree, arborescence)
2. Can provide closed representations with connected graphs (existence of either chains or cycles in the structure)

Open Representations

Two types of representations are used in the world of literature: the thesaurus and the semantic network. The thesaurus is a very ancient attempt to represent meaning by special links between keywords. The goal of any thesaurus is, firstly, to improve indexation to better cover the meaning conveyed by source documents, and secondly, by increasing, at the end of the documentary chain, the relevance of a search and consequently, increasing the meaning of the responses. A thesaurus is a reference work that lists words grouped together according to similarity of meaning (containing synonyms and sometimes antonyms). A thesaurus is generally in the form of an arborescence where nodes can designate concepts, represented by a grouping of keywords, and the leaves in an arborescence can be simple keywords.

A semantic network is a network that represents semantic relations between concepts. This form of knowledge representation is a directed or undirected graph consisting of vertices, which represent concepts, and edges, which represent semantic relations between concepts. Most semantic networks [46] are cognitively based. They also consist of arcs and nodes which can be organized into a taxonomic hierarchy. It is also possible to represent logical descriptions such as the existential graphs of Peirce [47] or the related conceptual graphs of Sowa [48].

Closed Representations

Frames are closed representations since they capture a certain picture of the world through a rigid structure and secondly there are connected graphs. They are an extension of semantic networks by strengthening semantic conferred on common-sense knowledge, in particular, taking into account their relationship and procedural aspects. The notion of frame was originally introduced by M. Minsky [49] to represent structures, which can roughly equate to prototypes. A prototype is a kind of typical element of a class or a category that it belongs and from which generalizations are possible. A well-designed framework allows the aggregation of knowledge about an object as well as its associated reasoning elements. The description of an object or a concept is possible by clarifying the meaning given to its properties and attributes. A semantic frame[20] can be thought of as a conceptual structure describing an event, relation, or object and the participants in it.

The fundamental idea of a frame system is rather simple: a frame represents an object or a concept. Attached to the frame is a collection of attributes (slots), potentially filled initially with values. When a frame is being used, the values of slots can be altered to make the frame correspond to the particular situation at hand. According to an interpretation by Minsky [49], the slots of a frame might represent questions most likely to arise in a hypothetical situation represented by the frame. Frames are stored as ontologies of sets and subsets of the frame concepts. They are similar to class hierarchies in object-oriented languages, although their fundamental design goals are different. Frame-based knowledge representation has been exploited in a non-exhaustive list of several applications such as semantic web [50, 51], object-oriented and markup languages [52, 53], ontologies [54–57], and the FrameNet[21] project and in information fusion [14, 58, 59].

2.5.6   Contexts

Data acquire meaning through context or contextual information and knowledge. Context establishes the basis for discerning meaning of its subjects and may occur at many levels. Exploitation of contextual knowledge is necessary for situation awareness and decision support [14]. One of the most widely accepted definitions of context is from Dey [60] phrased as follows: "*Context is any information that can be used to characterise the situation of an entity. An entity is a person, place, or object that is considered relevant to the interaction between the user and the application, including the user and the applications themselves.*"

This general definition of context has been extended by Zimmermann et al. [61] who introduce a definition comprising three canonical parts: a definition per se in general terms, a formal definition describing the appearance of context, and an operational definition (pragmatics) characterizing the use of context and its dynamic behavior. Figure 2.24 shows the five categories of the formal definition as well as the elements of the operational extension. Zimmermann *et al.* [61] extend the definition by the description of the following five categories of elements: *individuality, activity, location, time, and relations.* These five fundamental context categories determine the design space of context models. The description of the five categories of elements is presented in Table 2.3. We refer the reader to Zimmermann *et al.* [61]

---

[20]*Frame semantics* is a theory developed by Charles J. Fillmore that extends his earlier case grammar. It relates linguistic semantics to encyclopaedic knowledge. The basic idea is that one cannot understand the meaning of a single word without access to all the essential knowledge that relates to that word. (Source: https://en.wikipedia.org/wiki/Frame_semantics_(linguistics))

[21]FrameNet project—https://framenet.icsi.berkeley.edu/fndrupal/. FrameNet maps meaning to form in contemporary English through the theory of Frame Semantics.

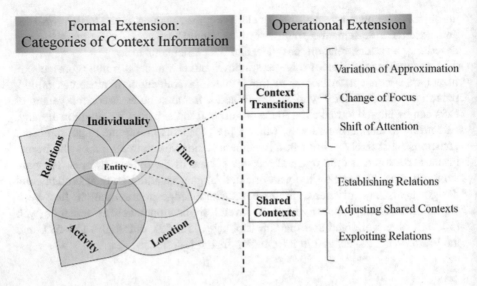

Fig. 2.24 Definition of context according to Zimmermann et al. [61]

for more comprehensive details. Their underlying motivation is to provide a structure that bridges the user-developer gap from a general concept (e.g., Dey's definition) easily understandable by the user to its engineering by software developers.

Dey's definition clearly states that context is always bound to an "*entity*" and that information that describes the situation of an entity is context. The activity predominantly determines the relevancy of context elements in specific situations, and the location and time primarily drive the creation of relations between entities and enable the exchange of context information among entities. The task itself is also part of the context as it "characterizes" the situation of the user.

Context awareness must be distinguished from *situation awareness* discussed in [14, 62]. In part of the literature, the term has been restraint to a property of mobile devices that is defined complementarily to location awareness (e.g., users of smart phones). From "Wikipedia" [63], context awareness has been related to different fields: "*Context awareness originated as a term from ubiquitous computing or as so-called pervasive computing which sought to deal with linking changes in the environment with computer systems, which are otherwise static. The term has also been applied to business theory in relation to contextual application design and business process management issues.*"

2.6 The Temporal Dimension of Knowledge

For purpose of discussion here, consider the temporal dimension unfolds into two usage modes: (1) a *batch mode* (non-*real-time*) where knowledge has been previously acquired and stored and (2) *real-time* where knowledge is solicited

Table 2.3 Description of the five categories of context elements [61]

Individuality: comprises anything that can be observed about an entity, typically its state	*Natural Entity Context:* comprises the characteristics of all living and non-living things that occur naturally and are not the result of any human activity or intervention
	Human Entity Context: covers the characteristics of human beings (e.g., the user's necessities, the user behavior, …, preferences in language, color schemes, modality of interaction, menu options or security properties, and numberless other personal favorites)
	Artificial Entity Context: denotes products or phenomena that result from human actions or technical processes. In a broad sense, this category covers descriptions for any human-built things like buildings, computers, vehicles, books, and many more
	Group Entity Context: A group is a collection of entities, which share certain characteristics, interact with one another, or have established certain relations between each other. The primary purpose of using groups is to structure sets of entities and to capture characteristics that only emerge if entities are grouped together
Time	Subsumes time information like the time zone of the client, the current time, or any virtual time. Intervals of time also constitute a fundamental requirement on the context model. To capture and express recurring events, intervals are a significant feature for modeling user characteristics
Location	Describes location models that classify the physical or virtual (e.g., the IP address as a position within a computer network) residence of an entity, as well as other related spatial information like speed and orientation
Activity	Covers the activities the entity is currently and in future involved in and answers the question "What does the entity want to achieve and how?" It can be described by means of explicit goals, tasks, and actions
Relations	*Social Relations:* describes the social aspects of the current entity context. Usually, interpersonal relations are social associations, connections, or affiliations between two or more people. For instance, social relations can contain information about friends, neutrals, enemies, neighbors, co-workers, and relatives
	Functional Relations: indicates that one entity makes use of the other entity for a certain purpose and with a certain effect, e.g., transferring a specific input into a specific output
	Compositional Relations: is the relation between a whole and its parts. In the aggregation, the parts will not exist anymore if the containing object is destroyed. For example, the human body owns arms, legs, etc. The association is a weaker form of the composition

immediately (e.g., to respond to an emergency). However, in view of efficient action, a notion of opportunity must be considered (e.g., to access critical knowledge). In what follows, timely knowledge is assimilated to what is made accessible (by technical means) on opportune time. The question is whether the information required to conduct an action is available at the needed moment? This moment can be both associated to a point or with a finite interval on a time axis (e.g., a window of opportunity).

The temporal dimension is omnipresent in decision-making. Actions are often subject to temporal constraints. The decision-maker, whatever be the actions, acts in a dynamic information world. The various transformations of information from data to action (the actionable knowledge processing chain) to satisfy the needs of an application are evolving over time and impose time delays. These delays are nearly quantifiable and even more difficult to formalize: the dissemination of guidelines, sending orders, and receiving reports require time more or less long. Modeling and reasoning over time apply to information and knowledge about phenomena or events that are highly dynamic and marked by a great variability.

Temporal knowledge must be treated in its integrality as any other knowledge in an information processing chain for actions. Thus, it presents the same general problem of representation as discussed in the context of symbolic computing. The formalisms to represent time must be determined with great care to allow different connotations (from actor's point of view) of temporal knowledge, and above all, to lay the foundations of temporal reasoning.

Any acting entity (agent) in the world (e.g., smart controller or automaton) must adjust its/his/her reasoning scheme over time in order to reach an assigned goal with its/his/her current state of knowledge (awareness): awareness dynamically gained through interpretation mechanisms. A necessary formalization of temporal knowledge is sought to be either to represent the phenomena observed or evolving. The formalization of temporal reasoning falls into two approaches: (1) the first being based on numerical models and (2) the second falling more under a symbolic representation of time. Regarding the numerical approach, temporal events are somehow defined on a time reference and with time constraints such as start and end dates, time windows, and calculation of critical paths.

Temporal reasoning is being addressed through "extensions" of classical formal logic: modal and temporal logics with a definition of operators to handle knowledge in the past and the future, reified logic (quasi-logical) [64] where every time element has a timeless attribute both a description: in terms of instants (logic of McDermott [65]), and in terms of intervals (logic of Allen [66]).

To construct the basis of a temporal dimension, the first idea that comes to mind is to relate the occurrence of events with the real axis. Each occurrence is thus associated with a point in time that can be interpreted: this event occurred from a certain time can be expressed in some metric. This simple tracking, whether it is sufficient in many physical systems, however, cannot be generalized to informational systems for supporting decision-making since they often need to identify specifically a temporal knowledge with respect to several possible time scales. Consider the following information: an explosion occurred between 15:00 and 15:

30 on February 20, 1978. This example shows the need to introduce several situations of tracking: referring to calendar time (month-day-year), referring to a schedular hour interval, etc. This last point will cause problems since it will require to handle time intervals in different granularity, which intervals can be expressed vaguely or imprecisely.

2.6.1   Temporal Dimension and "Signified"

Any information that expresses time, partially or not, does not play the same role for an agent and the various processes that must use it but do not carry the same *signified*. Temporal knowledge is properly interpreted unless it has its own semantics. Before trying to establish a basis, you must first identify the temporal "interpretation needs" to satisfy:

- The nature of the different temporal entities: date, time, event occurrence
- The structural links between entities, causality, sequentiality
- The standardization of time scales and data: dates and times within intervals
- The time references (time basis) to facilitate the resetting of clocks, management of cooperative processes, the expression of certain parallelisms.

In addition, some information may be considered as a more or less an implicit reference to the time without being temporal entities as such. This type of information is important to possess a semantic to allow, at need, its importation in a temporal reasoning scheme. Several examples can be given information that is referring to a comment made in the past or the present: and assumptions applicable to the present, the future, and the past. For instance, the phases of a decision support process, including the conduct of an action, may face heavy temporal constraints. These constraints can surface when tasks have different priorities, with tasks presenting multiple interruptions, in the presence of parallel processes, and finally, when a shareability of common resources is required.

The temporal dimension also imposes to consider some obligations related to specific situations. For instance, consistency must be maintained after an update, which is mandatory to preserve rationality of decisions. Meeting deadlines is preferable to an endless quest for rigorous results. Coping with approximate results is essential if it turns out that a refinement would be too time-consuming.

2.6.2   Temporal Dimension Validity

When dealing with symbolic representations of the world and not with the actual objects that compose them, be aware that Aristotle's Law of Identity (A is A: each thing is identical with itself) may prove defective insofar since the representation must reflect the change in the context of evolution of A during the reasoning process.

In other words, we need to express formally that A was true at a certain instant and may well not be true in the next moment. Under the conventional or classical logic framework, it means that the world in which deductions are made (as in the case of a production rules system) remains stable in a certain way. The mode of operation is said to be monotone. However, classical logic presents an important inadequacy in representing knowledge belonging to an unstable or dynamic world.

This inaptitude will result in the emergence of new technical issues. To check for the validity of knowledge, assertion must be done on a time window. To do that assertion, a so-called logic of intervals shows some promises. In this logic, the timeless components associated with intervals are of three types: (1) properties, (2) events, and (3) processes.

This categorization is derived from work carried out in semantics on what is referred to as "aspect," that is, the intrinsic temporal structure of the predicate, as distinguished from grammatical tense. These properties describe the static aspect of a process occurring over an interval I. Thus, to indicate the property:

"*Machine A is stopped*" is true over all of time I, we use the rule: $\text{HOLD}(A, I)$. This supposes of course that $\text{HOLD}(A, I)$ is true if and only if:

$$\forall J \subset I \Rightarrow \text{HOLD}(A, J) :: \text{true}$$

Events and processes represent the dynamic aspect of things. If we want, for instance, to express the event E: "*Machine A has stopped*" in the interval I, we write: $\text{OCCUR}(E, I)$. Unlike properties, an event is not temporally decomposable. Therefore $\text{OCCUR}(E, I)$ is verified if and only if for all point J we have $(J \text{ in } I)$ which is not verified. We note by $\text{OCCUR}(P, I)$ the fact that the process P occurring during I is true, $\text{OCCUR}(P, I)$ is verified if and only if $J \subset I$ is not verified, then $\text{OCCUR}(P, I) :: \text{false}$.

Based on the above elements, J.F. Allen [66] defined a theory of causality and action. The predicate ECAUSE enables thus to say that an event E is the cause of an event E_j: $\text{ECAUSE}(E_i, I_i, E_j, I_j)$ that we interpret by: E_i, occurring in the interval I_i is the cause of E_j which occurred in interval I_j. Allen defines an action as an event or process caused by an agent. The predicate ACAUSE allows stating that agent A is the cause of event E by $\text{ACAUSE}(A, E)$.

If presently the time window is reduced to a point, it is preferable to apply a logic that can represent points on an axis of time. The logic of instants defined by Mc Dermott [65] is recommended here. It is characterized by its discrete nature, and the temporal basic object is considered a representative state of a moment in the universe or of an instant. The following two relations indicate the position of events in relation to others: before, same moment, noted $\leq$, a relation of precedence having all of the characteristics of an order relation (see next chapter): reflexivity, anti-symmetry, transitivity.

We specify that for Mc Dermott:

- There is no single past even if it is unknown.
- The future is undetermined.

We observe that contrarily to the formalism of J.F. Allen, the logic of instants allows manipulating the metrical aspect of time in that it becomes possible to manage numerical constraints over durations. According to Mc Dermott, facts and events are atemporal components.

- *Fact*: a fact is defined by the set of states in which it is true. The notation (TSP) allows indicating that the fact P is true in the state S, hence $S \in P$.
- *Event*: an event is defined based on intervals in order to express duration and process events. An interval is defined then as a totally ordered and connected set of states. Similarly to what was said for facts, an event is defined as a set of intervals.

Moreover, Mc Dermott considers that an event can be the cause either of another event or the cause of a new fact (a certain delay can occur between cause and effect). This causality is indicated by the predicate ECAUSE.

Example 2.17 ECAUSE(A, E_1, E_2, I) signifies that event E_1 is always followed by event E_2; the effect follows the cause in the interval of time I unless fact A becomes false in the meantime.

2.7 The Ontological Dimension of Knowledge

In some circumstances, the objective of getting efficient actions may raise questions regarding whether the requested information is appropriate and to what degree? It brings out some questions on the impact of that information have had on the knowledge creation process.

- How well that information represents "useful" knowledge and to what extent it can, ontologically speaking, represent concepts?
- What nature, at the ontological level, these knowledge items do originate?
- What are the terms and conditions of their existence in the knowledge processing chain?

In the context of specific actions, it may be desired to understand lineage over time of certain knowledge items. Understanding their sequence can be useful for undertaking judicious action. Why did a certain notion trigger another, over time and under which conditions did this occur? To begin answering such a question, a certain number of ad hoc means should be first defined and implemented within the knowledge chain, enabling "traceability" (to use the common term) between constituting elements of knowledge. This implies examining the possibility of giving the knowledge chain specific modalities allowing the existence and relational environment established over time between different constituting elements of knowledge to be retraced.

The reason for attaching an ontological dimension to the processing knowledge chain is to provide a traceability of notions. We will postulate, first, that it is possible to organize knowledge in domains and that the ontological approach is fully justified when one wants to be able to make inferences in an organization of concepts or notions (necessarily arborescent), a "sine qua non" condition, for obtaining a notional trace. To establish the traceability between notions, we will rely on the techniques of an ontology.

Ontology, although part of the vocabulary of researchers in artificial intelligence, is a borrowing from philosophy. For philosophers, ontology means "*sensu stricto,*" the study of being, the universal aspect of what exists. In philosophy, ontology is the study of the being at the third level of abstraction, e.g., abstraction of the quantitative to retain only the universal. Any ontological approach thus presupposes abstracting concrete individuals (living or not), including formal entities such as mathematical ones, to that which is general or universal to them.

At this level of abstraction, characteristics that are identified are referred to with terminology such as having an essence, existing, subsisting, and being, e.g., any verb that can represent the "being as being." This means performing first-order philosophy, according to Aristotle. Artificial intelligence researchers reuse the notion of ontology by adapting it to the usage that they make of it in the scope of knowledge representation and agree on a commonly accepted terminological basis. The use of this philosophical notion strictly pertaining to the being as being is justified because artificial intelligence researchers consider that only what exists can be represented.

According to T. Gruber [67], an ontology is based upon a conceptualization, an abstraction, or simplified view of the world that is represented for a certain purpose. It must be understood as a formal and explicit representation of how to conceptualize a particular knowledge domain. It allows us to represent, more or less formally, notions and their relations simply through definitions of properties, attributes, constraints, axioms, possibilities. A well-designed ontology represents the adequate support for shareability within a distributed organization. It offers a formal, shared interpretation on a given domain of knowledge and ultimately facilitates its exploitation by both human actors and processors. This is one of the reasons why some members of the military community see this as a means of addressing the thorny problem of interoperability of command systems. The first objective of an ontology is to clarify the nature of things and to facilitate their order in the universe in which one wants to be able to act effectively. This implies that its promoters put a particular concern to combining certain qualities within their conceptual effort. Qualities required for an ontology are intelligibility and *clarity, consistency, extensibility,* and a *minimum core encoding.* The first goal of ontology is to specify the nature of things and to facilitate their ordering in the environment in which we want to act efficiently. This implies, during conceptualization, a concerted effort to combine the following properties:

- *Clarity:* Any definition must be made objectively and must convey the anticipated meaning of defined terms. Motives in conceptualization can generate links to the social context in the definition, or account for computational requirements, which ontology must avoid. Formalization makes this possible: choice of axioms, predicates expressing conditions of necessity, and sufficiency.

- *Consistency*: The inferences made based on ontological definitions and axioms should not create contradictions. This must also be verifiable for concepts, even if informally defined.
- *Extensibility*: An ontology should anticipate usages of a shared vocabulary, as well as tasks. Representations should be defined so that they are extensible. New terms should be defined for specific usage based on existing vocabulary without needing to revise initial definitions.
- *Minimal core encoding*: Knowledge representation-oriented conceptualization should not depend on a given symbolic representation system. Minimal core is achieved when representation choices are made in the sole purpose of satisfying representation and implementation needs.

An ontology must not be hastily assimilated to a taxonomic hierarchy of classes. Let us recall that a taxonomy presents a certain fixed image, rendered by fixed terminological definitions, of a particular domain of knowledge. On the contrary, ontology will seek to broaden its expressiveness, in particular, by a certain descriptive richness of the relations between concepts and notions in order to bring additional knowledge about the world. In an ontology, we will endeavor to specify the properties and constraints attached to the notions to facilitate later their interpretation.

2.7.1   On Tracing "Notions" in an Ontology

The following approach allows tracing in the knowledge chain according to an ontologically defined structure: relation between domains (Fig. 2.25). There must be two ways of doing this: a descendent or downward path and an ascendant or upward path. A descendent path which is a deductive approach can be described formally by a succession of implications, postulats and interpretations (according to example of Fig. 2.25).

The knowledge structure of Fig. 2.25 allows two paths:

1. *Descending path: as in a deductive approach, it can be formally described by a succession of implications, by posing: desc$_\downarrow$ :: "descendent of"*

Fig. 2.25 Ontological tree path

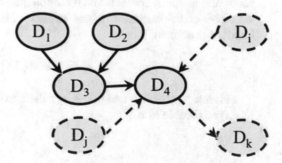

$$\left[\forall x \in \{D_1 \cup D_2\} \Rightarrow (\exists y \in D_3) \wedge desc_\downarrow(y,x)\right] \Rightarrow (\exists z \in D_4) \wedge desc_\downarrow(z,y,x)$$

which we interpret by:

$$desc_\downarrow(z,y,x) :: z \text{ is } \textit{"descendent of"} \text{ both } x \text{ and } y.$$

2. *Ascending path: it is an inductive approach to see whether it is possible for a given z to find ascendants x and y, by posing ascendant:* $asc_\uparrow$ :: *"ascendant"*

$$\left[\forall z \in D_4 \Rightarrow (\exists y \in D_3) \wedge asc_\uparrow(z,y)\right] \Rightarrow (\exists x \in \{D_1 \cup D_2\}) \wedge asc_\uparrow(x,y)$$

which we interpret by:

$$asc_\uparrow(x,y) :: x \text{ is an } \textit{"ascendant"} \text{ of } y$$

Once the existence of ascendants is proven, the second step aims to exploit the relation of ascendance. Depending on choices made at domain creation, several steps can be taken:

- Extending, if necessary, the possible field of ascendants by describing in domains a rule which uses an existential quantifier in its consequent

$$\forall x \in D_i \Rightarrow \left[(\exists\{y_i\} \in D_{i-1}) \wedge asc_\uparrow(\{y_i\},x)\right]$$

- Examining whether specific constraints or properties are attached to ascendant entities that inform about the *"raison d'être"* of this knowledge and consequently extracting additional knowledge to inform about possible lineage between different notions

$$(\forall x \in D_i) \wedge asc_\uparrow(\{y_i\},x) \Rightarrow \left[R(\{y_i\},\{c_i\}) \vee R(\{y_i\},\{p_i\})\right]$$

with R :: *"is attached to,"* c_i :: description of a constraint, p_i :: descriptive of a property, for example to explain its ontological context.
- Triggering general rules associated to the domain containing the selected ascendants, and in which contents and their conditions of existence can be specified. If we consider the condition where explicit rules exist that apply to the entire knowledge domain, then:

$$(\forall x \in D_i) \Rightarrow R[(\exists r_i),x]$$

with R :: *"applies to"* and $[(\exists r_i),x]$ :: there exists at least one (if not a subset of) rule(s) applicable to x.

It is then possible to induce elicitations that can be propagated from one domain D_i to another subordinate domain D_{i-1}, thus clarifying the notions that "frame" the nature as well as the origin of such knowledge:

$$(\exists x_i \in D_i) \wedge \{\exists y_i \in D_{i-1} | asc_\uparrow(y_i, x_i)\} \Rightarrow [(\exists r_i), x_i]$$

All this is only possible insofar as one accepts the validity or legitimacy of several strong hypotheses. In particular, the universe is composed of objects, and it is possible to render them through representation systems based on structured objects. After all, this attitude is not very different from that of solid-state physicists who consider the real world as ontologically decomposed into elementary particles. It should be noted that physicists indeed use approximate models, though this choice is justified by the complexity of accessing real entities and measurement algorithms.

To say that the world consists of objects has the undeniable advantage of offering a solid basis for relying on logics, whatever the purpose, and to have formal reasoning that is more or less efficient. Thus, if we make an inference within the framework of the logic of predicates by applying the *modus ponens*: $\forall x P(x) \Rightarrow Q(x)$ and $x \rightarrow a$, $P(a)$ leads to the conclusion $Q(a)$, which presupposes, of course, the existence of an object a in the world. To those who have doubts about the validity of a hypothesis where a world can be completely described by objects, then it can be argued that things not directly observable become in existence, as soon as they can be named or expressed in words.

2.7.2 Domains of Objects

In a given field, all the objects which carry its knowledge and, which can be represented in a declarative way, constitutes what is generally called the universe of discourse. These objects, as well as the relations that can be described between them, differ little from a representative vocabulary from which a knowledge-based program represents *knowledge*, which implies that in the context of computer science, an ontology can be described as a program by defining a set of representative terms associating:

- Names of entities within the universe of discourse: objects, classes, relations, functions with texts defining their meaning
- Formal axioms that restrict, or at best specify, their use

In doing so, we do nothing but state the basis of a logical theory. The interest in ontology by computer scientists comes from the fact that it sheds light on certain problems posed by the sharing of knowledge and interoperability of domains by facilitating exchanges between different applications; the formalization of knowledge or notional representations; and the elicitation of concepts (making the concepts explicit).

2.7.3   *Implementing the Ontological Dimension*

Based on the above-mentioned ontological structures, it should be possible, practically speaking, to link deductions demonstrating, provided the existence of ascendants for a given object is proven, whether lineage exists, thus explicating the object's nature and relational neighborhood. In predicate calculus, a deduction consists in affirming the truth of a predicate based on another predicate that is considered to be true.

All the rules of inference defined for the calculation of predicates can be used here for:

- Eliminating of the conjunction: reducing the ascendants field
- Introducing disjunction: expanding lineage
- Transitivity: transitioning from a domain to another according to certain formally expressible conditions
- Equivalence

Rules of rewriting can also be useful, notably for transitioning from the universal to the specific and vice versa. For the sake of convenience, we will represent inference with the Gentzen's notation [68].

$$\frac{\rightarrow A}{\rightarrow B}$$

with the following conventions:$\rightarrow A$ above the horizontal line (premises) means that A has the true truth value; $\rightarrow B$ below the horizontal line (conclusion) means that B has also the true truth value. We will say, by simplifying, that the expression which is below the line is deduced (conclusion) from the expression of the top (premises). To illustrate, we will represent the "modus ponens"[22] and the "modus tollens"[23]:

$$\text{(modus ponens)} \qquad \text{(modus tollens)}$$
$$\frac{\rightarrow A \rightarrow (A \Rightarrow B)}{\rightarrow B} \qquad \frac{\rightarrow A \Rightarrow (B \rightarrow \neg A)}{\rightarrow \neg B}$$

Introducing a disjunction in an ascendance:

$$\frac{\rightarrow asc(x,y) \rightarrow asc(x,z)}{\rightarrow asc(x,y) \vee asc(x,z)}$$

can result in:

---

[22] The Latin expression, *modus ponems*, means *affirmative mode*.
[23] The Latin expression, *modus tollens*, means *negative mode*.

$$\frac{[asc(x,y) \vee asc(x,z)],}{[asc(x,y) \vee asc(x,z)]} \Rightarrow \begin{bmatrix} R_1((\exists r_i),y) \wedge R_2((\exists r_i),z)], \\ [R_1((\exists r_i),y) \wedge R_2((\exists r_i),z)] \end{bmatrix}$$

The expression in denominator indicates that it is possible to explain the knowledge of $x \in D_i$ from $y \in D_{i-1}$ and that of $z \in D_{i-2}$.

Eliminating a conjunction:

(*reduction by suppression of B*)	(*reduction by suppression of A*)	Example of transitivity:
$\dfrac{\rightarrow A \wedge B}{\rightarrow A}$	$\dfrac{A \rightarrow B}{\rightarrow B}$	$\dfrac{\rightarrow A \Rightarrow B, \rightarrow B \Rightarrow C}{\rightarrow A \Rightarrow C}$

The following remarks can also be made:

- When an ascendance is true for all the elements of a domain, it is necessarily true, for a given element.

$$\frac{\rightarrow \forall x \in D_i \ asc(a,y) | \exists y \in D_{i-1}}{\rightarrow asc(a,y)}$$

- Inversely, if it is true that an element possesses an ascendance, then it is thus possible to say that there exists at least one element for which it is true:

$$\frac{\rightarrow asc(a,y) | \exists y \in D_{i-1}}{\rightarrow x \ asc(x,y)}$$

In certain cases, it can be important to transition from the universal to the existential and vice versa, which is easily achieved by applying the following rules of rewriting:

$$\frac{\rightarrow \neg \forall x \ asc(x,y)}{\rightarrow \exists \neg asc(x,y)} \quad \text{(universal} \leftrightarrow \text{existential)} \quad \frac{\neg \exists x \ asc(x,y)}{\forall x \ asc(x,y)}$$

2.8   A Need for Formalization

In their everyday use, textual supports often resemble imperfect forms of writing: ambiguity, lack of clarity, etc., which cause misunderstandings. It is not uncommon to experience this kind of difficulty when reading system specifications or product requirements. In the "*Petit Robert*" French dictionary, the lexicographic definition of formalization is the reduction of a knowledge system to its formal structures, and the generalized application of interpretation processes irrespective of content. In the Larousse, formalization consists in explicitly expressing, in deductive theory, rules of formation of expressions or formulas, as well as rules of inference by which we

Table 2.4 Six levels of needs for formalization

Need for	Description
1. Generalization	Designating entities or objects by a collective name or symbol according to their common properties within the same group, class, or category
2. Expressiveness	Formalization of a statement using symbols facilitate its comprehension
3. Reasoning support	Formalization is very useful when forming a basis for reasoning
4. Legibility	Formalization is useful when reducing or eliminating ambiguity from knowledge representation
5. Formal language	In decision-making, formalization is necessary when the human operator delegates elements of decision support to an automaton
6. Coherence	Formalizing a text consists in expressing the main meaningful concepts or elements as non-ambiguous propositions and assigning a symbol to each proposition

reason. The two definitions have in common the notion of organized representation, i.e., a structure and rules, the former highlighting the necessary dissociation between formal representation and content and the latter highlighting inference-based reasoning as the purpose of formal representation. The second definition is reductionist in that reasoning can be other than deductive.

Let us note that these definitions have in common the notion of an organized representation: structure and rules. One brings out the detachment that must exist between the formal representation and the content of the matter to which it refers, whereas the other puts in exerts the reasoning by inference, which becomes the finality of the formal representation. This last one remains, nevertheless, reductive, insofar as the reasoning can be other than deductive. A knowledge chain should meet the double requirement of integrating only the most concise knowledge and excluding where possible knowledge that is likely to generate ambiguity. Formalization is a first step toward this goal. It should ideally address several levels of needs as listed in Table 2.4 and discussed in Barès [1].

1. Generalization

    When entities or objects are combined into a set (class or category) on the basis of common properties or qualities, it is convenient to use a collective name or symbol to represent them. This is a first step in formalization, which facilitates designation and manipulation of combined elements. This approach is similar to using a function to describe a sequence of possible values for a variable within a bounded or unbounded domain.

2. Expressiveness

    The symbolic formalization of a statement facilitates comprehension. Consider the scenario: "*Either she is not home, or she is not answering the phone, therefore she has been kidnapped. If she is not answering the phone, then she is in danger. Therefore, she was either kidnapped, or is in danger.*" The main components of meaning in statements of the text (propositions) can be represented by symbols: "*Either she is not at home (M) or she is not answering the phone (T). If she is not home, then she was kidnapped (K). If she is not answering the phone,*

then she is in danger (D). Therefore, she was either kidnapped or is in danger."
Note that four symbols suffice to express the above text: *M, T, K, D*. Using
classical logic connectors and the two logical operators, "$\neg$" of negation and
"$\vdash$"conclusion (this symbol is called a *turnstile* or an *assertion* sign), the text can
be reduced to argument structure, revealing the relation between the antecedents
and their conclusion as $\neg M \vee \neg T, \neg M \Rightarrow K, \neg T \Rightarrow D \vdash K \vee D$. That can be
alternately written as: $[(\neg M \vee \neg T) \wedge (\neg M \Rightarrow K) \wedge (\neg T \Rightarrow D)] \Rightarrow K \vee D$, and
argument validity can be verified against a truth table.

3. Reasoning Support

Formalization is very useful when forming a basis for reasoning, as shown in
the following example. A race is taking place in 2 weeks and a contestant has
sustained an injury while training. The trainer says: *"If you don't stop immediately, you won't be able to race."* The contestant replies: *"Prove it to me
(formally)."* We will reason hypothetically that the contestant did not stop
running as the trainer advised. Let us formalize the reasoning with the following
symbols:

E :: swollen ankle
C :: continue (running)
G :: heal, D :: race

The premises are:

$$E, E \wedge C \Rightarrow \neg G, \neg G \Rightarrow \neg D$$

and the argument is formulated as:

$$E, E \wedge C \Rightarrow \neg G, \neg G \Rightarrow \neg D \vdash C \Rightarrow \neg D$$

If the premises are asserted, then the conditional conclusion $C \Rightarrow \neg D$ is valid.
Validity can be established from a truth table by rewriting the argument as
follows:

$$E \wedge (E \wedge C \Rightarrow \neg G) \wedge (\neg G \Rightarrow \neg D) \Rightarrow (C \Rightarrow \neg D)$$

Validity is established by what logicians refer to as proof or derivation as in the
example of Table 2.5.

4. Legibility

Formalization is useful when reducing or eliminating ambiguity from knowledge representation, for example, when distinguishing between linguistic forms
that appear the same:

(a) "Bob can be convinced."
(b) "Bob can be convincing."

Here, (a) indicates the result of an action, whereas (b) indicates the role of an
agent. Similarly, a term referencing a category can be distinguished from a term
representing one of its elements:

Table 2.5 Example of reasoning support for *scheme*

Derivation steps	Premises	Assertions	Remarks
#1	E	Asserted facts	
#2	$E \wedge C \Rightarrow \neg G$		
#3	$\neg G \Rightarrow \neg D$		
#4	C	H (hypothesis)	We hypothesize that C :: is true
#5	$E \wedge C$	by #1 & #4	and intersection
#6	$\neg G$	by #2 & #5	and modus ponens
#7	$\neg D$	by #3 & #6	and modus ponens
#8	$C \Rightarrow \neg D$	by #4 & #7	and conditional proof we hypothesize that C :: is true

Adapted from Barès [1]

(a) "The cat that I just petted."

(b) "The cat is a domestic animal."

Here, (a) represents an entity of the feline family, whereas (b) refers to the category of animals represented by the term cat.

5. Formal Language

In decision-making, formalization is necessary when the human operator delegates elements of decision support to an automaton. The automaton must be able to reason "artificially," which obviously presupposes a well-defined formal language based on well-established syntactic and semantic conventions, in order to reduce the literal statements to interpretable entities, e.g., propositions or predicates, which can then be represented by symbols, for simplification. Propositional calculus [69] and predicate calculus [70] are examples of formal languages.

6. Coherence (Consistency)

Formalizing a text consists in expressing the main meaningful concepts or elements as non-ambiguous propositions and assigning a symbol to each proposition. In deductive logic, symbols can be incorporated and structured in argument form. An argument can be validated by proving the coherence of the logical relation between antecedent (premises) and final (conclusion) propositions. Truth Maintenance Systems (TMS) [71] are particularly keen to consistency issues.

2.8.1 Logical Propositions

From a strictly logical point of view, functions[24] or "propositions" are used to incorporate the variables commonly arising in descriptions of reality. Uncertainty

---

[24]The notion of function is much broader in logic than in mathematics: in logic, sin(x) and the equation sin(x) = 1 are functions.

appears in formalization when variables do not have precise values, e.g., assignment of constants, or domain of variation not determined by quantification. The notion of proposition has greatly evolved since antiquity. For Aristotle, a proposition was a judgment assigning an attribute to a subject, expressed by the copula "is, are," while for the Megarian school and the Stoics, a proposition was an entity that included a premise based on logical inference: *"If the student meets the average then he has succeeded."*

A proposition can be defined as a literal statement that can be asserted by:

- Either an affirmation (yes or true)
- Or a negation (no or false)

A proposition can also be used to relate two concepts. Propositions are used to express a given reality of the world, or to make judgments using declarative sentences that can be affirmed or denied, excluding expressions such as orders, exclamations, and interjections. Coherence relations between statements can be established by using classical logic connectors to link propositions within the argument structure. Premises or antecedents on the left support the conclusion on the right.

$$[premise_1, premise_2, \cdots, premise_n] \Rightarrow conclusion$$

It should be noted that a proposition has two basic features:

- *Qualitative:* affirmative or negative
- *Quantitative*: universal, particular, or singular

Universal: holds for all elements in a set. Universality is expressed by the quantifier "for all," $\forall$, or none in the case of a negative proposition.
Particular: holds for some entities only. It is expressed by the quantifier "some," $\exists$.
Singular: an entity is named.

2.8.2   Propositional Transformations

According to Couturat [72], "A proposition is defined as a logical function which, for any value assigned to (a) variable(s), becomes a proposition. A propositional function, unlike a proposition, is undetermined because it can be neither true nor false. It becomes a proposition when a value is substituted for each variable." While propositional functions can be used to express statements of knowledge about a class, they do not yield useful formalizations because they cannot be asserted (are undetermined). A propositional function must therefore be reduced to a proposition.

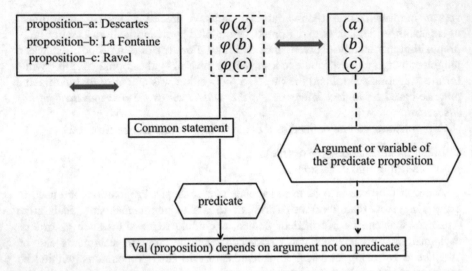

Fig. 2.26 Notion of predicate

2.8.3   Predicate Logic Formalization

Predicates are very useful for formalizing and generalizing a common statement about several propositions (Fig. 2.26). Consider the three propositions: (a) Descartes is a philosopher, (b) La Fontaine is a philosopher, (c) Ravel is a philosopher; a, b, and c have in common the statement "is a philosopher." φ".

It is interesting to note that a predicate[25] can be used in several modes of representation. Defining a set by comprehension amounts to describing it by a common feature or property of its elements, represented by a predicate:

"*is a literature student*" $L :: \{x| \lambda(x)\}$
"*is a philosopher*" $P :: \{x| \varphi(x)\}$

P represents the set of entities x such that $\{\forall x| \text{val}(\varphi(x)) :: \text{true}\}$. P is therefore the extension of the predicate "*is a philosopher*": $\varphi(x)$.

Based on a domain of reference (domain of discourse) defined by a predicate, it is possible to create subsets of interpretation using subordinate predicates of the reference predicate. These predicates make sense only if the variable assigned to them in the argument takes its value in the referent domain (in the above example, the set of university students), also called domain of interpretation. Regular set operations can be applied to sets defined by predicates, as in the inclusion:

$$E\,(x) \supset L\,(x) \supset P\,(x).$$

---

[25] Predicate: state, affirm, or assert (something) about the subject of a sentence or an argument of a proposition.

2.8.4 Formalization by Graphical Representations

A graphical representation is given by a Venn diagram with overlapping circles in which each representative set of a class of entities is represented by a predicate. The graphical representation is made by means of several overlapping circles, in which each representative group of a class of individuals is designated by a predicate (Fig. 2.27). Since these are classes and therefore sets, it is possible to obtain logical relations on these sets with the usual quantifiers: any (all), none (no), some (some). Unlike what is done in propositional calculus, letters or symbols no longer denote sentences but classes, categories, sets. We thus pass from relations between propositions (or sentences) to relations between classes. Using all (All), none (No), some (Some) quantifiers on classes leads to the definition of four relational forms.

- Form: all A are B. This is the assertion that A is a subset of B, all members of A are also members of B. If "all A are B":: true, then "some A are B ":: true, and $A \cap B = A$.
- Form: no A is B. This is the assertion that classes A and B are disjoint, they have no members in common.
- Form: some A are B. This is the assertion that class A shares at least one member with class B. In the usual discourse, the use of the word "some" presupposes that A and B share more than one member. In fact, to say that "some A are B" presupposes that not all A are B. This presupposition being absent in the logical notion of "some" may lead to an imprecise formalization since the quantifier "some" in logic may designate Unity as well as Totality.
- Form: some A are not B. This form can result from several situations indicated by the Fig. 2.27 below.

It may also be interesting to use Boolean symbolism. It is indeed easier at times to formalize by playing with both the complementary sets and the operations on sets, for example, the intersection ∩ as shown in Table 2.6.

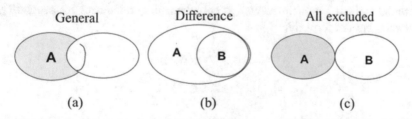

Fig. 2.27 The quantifier "some": (**a**) general, (**b**) difference, and (**c**) all excluded

Table 2.6 Boolean interpretation		
All A are B	$A \cap \sim B = 0$
No A is B	$A \cap B = 0$
Some A are B	$A \cap B \neq 0$
Some A are not B	$A \cap \sim B \neq 0$

2.9   Knowledge Reference Dimension: The Use of Quantification

Before acting, we often need to bear on reference knowledge. Research modalities differ depending on whether you need a simple reference, a subset of referents, or an access to an entire domain of knowledge. The question may also arise whether there is a domain of reference corresponding to a particular need. For the sake of convenience, we will designate the reference domains by predicates.

Example 2.18 Suppose we want to know if there are steel stems references that are round and of 1 meter length. That is set out using the following predicates ($\overset{\text{def}}{=}$ or:: means *defined by*, and the symbol $\therefore$ means *therefore*):

- Being round $\overset{\text{def}}{=} R(x)$ corresponds to set: $\{x | R(x)\}$
- Being long $\overset{\text{def}}{=} L(x)$ corresponds to set: $\{x | L(x)\}$

For this example, $R(x) \wedge L(x)$ defines the set of all references of round pieces of 1 meter. The use of predicates has the great advantage to slot designations of successive references, as in mechanics, when referring to the dependent parts of the same device. In Fig. 2.28, two sets are defined from which it will be convenient to distinguish (sub) sets of references by designation predicates: $R(x) \supset L(x) \leftrightarrow F \supset A$

If you are in a situation where $A \not\subset F$, which means there is no forgeable steel parts, then the predicates are written as:

$$\sim \forall x [(x \in A \Rightarrow x \in F)] \Leftrightarrow \exists x \sim [(x \in A \Rightarrow x \in F)]$$

$$\exists x \sim [\sim (x \in A) \vee (x \in F)] \Leftrightarrow \exists x \sim [(x \in A \Rightarrow x \in F)]$$

$$\exists x \sim [\sim (x \in A) \vee (x \in F)] \Leftrightarrow \exists x [(x \in A) \wedge \sim (x \in F)]$$

The following predicates mean there exist steel parts references of good size but not round and not forgeable, as well as complementary references corresponding to steel parts but not forgeable:

$$\exists x [L(x) \wedge \sim R(x)]$$

$$\text{Compl}_{A/F} \overset{\text{def}}{=} \{x | x \notin A \wedge x \in F\}, \therefore \sim L(x) \wedge R(x)$$

Of course, it is possible to apply basic set operations to referents. Starting from domains of referents E and F of example below defined by the predicate $P(x) \rightarrow E$ and $Q(x) \rightarrow F$ can be led to consider intersected or referents to be diferentiated. Several situations are possible:

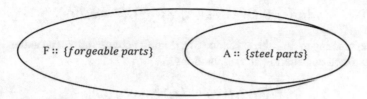

Fig. 2.28 Sets of referents

- *Situation 1*: intersection of sets of referents: $E \cap F \overset{\text{def}}{=} \{x | x \in E \wedge x \in F\} \Leftrightarrow \{x | P(x) \wedge Q(x)\}$.
- *Situation 2*: differentiation of referents: $E - F \overset{\text{def}}{=} \{x | x \in E \wedge x \notin F\} \Leftrightarrow P(x) \wedge \sim Q(x)$.

Two domains of referents having a common intersection may yield the options below.

- *Option 1* is to exclude the common area and access one of the referents to the exclusion of the other, which is equivalent to make a difference, $E - F$ or $F - E$.
- *Option 2* is to keep the common part obtained by symmetrical differentiation:

$$E \vartriangle F \overset{\text{def}}{=} \{x | (x \in E \wedge x \notin F) \vee (x \notin E \wedge x \in F)\}, \{x | [P(x) \wedge \sim Q(x)] \vee [\sim P(x) \wedge Q(x)]\}$$

The use of quantifiers will allow different designations of referents. The universal referent R_u can be obtained either by listing all references or using the universal quantifier $\forall$, noticing then that the predicate P should be applied to all elements without exception of the domain of reference:

$$R_u(P) \overset{\text{def}}{=} \{P(r_1), P(r_2), P(r_3), \ldots\} \Leftrightarrow \forall r P(r)$$

equivalent to the expression,

$$\forall r P(r) \Leftrightarrow P(r_1) \wedge P(r_2) \wedge P(r_3) \wedge \ldots \wedge P(r_n).$$

A particular referent is organized according to the use of the word some, which means that only a number of elements are concerned with the predicate; it corresponds to a subset of references that can contain only one as in the case of a singleton. If we set $R_s(Q) \subset R_u(P)$, where Q is a predicate which is justified within the predicate P. Defining a particular referent then means to write:

$$R_s(Q) \overset{\text{def}}{=} \{Q(r_1), Q(r_2), Q(r_3), \ldots\} \Leftrightarrow \exists r Q(r)$$

equivalent to the expression,

$$\exists r Q(r) \Leftrightarrow Q(r_1) \vee Q(r_2) \vee Q(r_3) \vee \ldots \vee Q(r_n).$$

It is interesting to note the existence of a duality between universal and existential referents when considering all conjunctions:

$$[f(x_1) \wedge f(x_2) \wedge f(x_3) \wedge \ldots \wedge f(x_n)]$$

and its negation,

$$\sim [f(x_1) \wedge f(x_2) \wedge f(x_3) \wedge \ldots \wedge f(x_n)] \Leftrightarrow \sim [\forall x f(x)].$$

Applying Morgan's laws:

$$[\sim f(x_1) \vee \sim f(x_2) \vee \sim f(x_3) \vee \ldots \vee \sim f(x_n)] \Leftrightarrow \sim [\forall x f(x)] \Leftrightarrow \exists x \sim f(x)$$

where the dual expressions:

$$\sim [\forall x f(x)] \Leftrightarrow \exists x \sim f(x)$$
$$\sim \forall x \sim f(x) \Leftrightarrow \exists x \sim \sim f(x) \Leftrightarrow \exists x f(x)$$
$$\sim \exists x \sim f(x) \Leftrightarrow \forall x f(x)$$

2.9.1   Order and Scope of the Quantification of a Referent

If the use of quantifiers undeniably brings a profit in the handling of referents, it must nevertheless be made with caution. How to order them can affect the semantics of the phrase. Consider the case of two referents bound by the following relationship:

$$R \stackrel{\text{def}}{=} \text{``be a teacher of''}$$

where $R(x, y)$ means that x is "be a teacher of" and y can take the following values: *History, English, Chemistry, . . .*, etc.

The two expressions, $\forall x \, \exists \, y \, R(x, y)$ and $\exists x \, \forall \, y \, R(x, y)$, may look similar but they are very different, in the sense that they do not necessarily mean the same thing. Nothing says in effect that a variable used with two different quantifiers denotes the same element. The expression $\forall x \, \exists \, y \, R(x, y)$ means that all teachers teach several (or at least one) disciplines, whereas expression $\exists x \, \forall \, y \, R(x, y)$ indicates that there are several professors (or at least one) who teach all disciplines. In the case of two referents, it is possible to account for their different situations with the following combinations:

$$(\forall x, \forall y); (\forall x, \exists y); (\exists x, \forall y); (\exists x, \exists y)$$

To avoid the drawbacks which have just been mentioned, one should specify on what portion of referents is the object of the quantification. This is called quantification range. In $\forall r\, P(r)$, the variable r is linked to predicate P by the presence of the universal quantifier $\forall$. Similarly, in $\forall r\, Q(r)$, r is a variable linked to Q by $\forall$. Concretely, can we say that the variable r, argument of P, means the same thing than r argument of Q? We cannot say. They are two different variables related to two different predicates. To circumvent that difficulty, we must uniquely connect both propositional functions $P(r)$ and $Q(r)$, writing for example:

$$\forall r\, [P(r) \vee Q(r)].$$

In this case, there is no ambiguity, because the quantifier binds simultaneously the variables to both predicates. The instantiation can be done in a unique way. We shall define the quantifier scope using all occurrences of a variable linked by the same quantifier as the two occurrences $P(r)$, $Q(r)$ of the previous example that depend on $\forall$.

2.9.2   Application of Quantification to Referents

The quantification will help to determine the referents terms of use as illustrated by the following examples.

- All Q references must also be references to

$$P : \forall r\, [Q(r) \Rightarrow P(r)].$$

- All Q references cannot also be references to

$$P : \forall r\, [Q(r) \Rightarrow \sim P(r)].$$

- Some references of Q include those of

$$P : \exists r\, [Q(r) \wedge P(r)].$$

- A command (C) is underway and some Q references are missing (M):

$$C \wedge \exists r\, [Q(r) \wedge M(r)].$$

- Some references of steel parts (A) and forged parts (F) are not missing anymore:

$$\exists r\, [A(r) \wedge F(r)] \wedge \sim M(r).$$

- If steel parts are no longer referenced, then there are no more forged parts in catalog:

$$\forall r | \sim A(r) \Rightarrow \sim \exists r \, [F(r) \wedge A(r)].$$

2.9.3   Relational Predicate Applied to Referents

It often required to use multiple referents concurrently. In the sense of set theory, that means to explore ways to relate elements of different sets. Let us start by the most common case: the relations of two referents. For convenience, we will consider a relation as an extension of a predicate since, after all, a predicative relation will be treated as a predicate with two arguments (arity 2) or more arguments (arity n). However, in the case of a simple predicate $P(x)$, P may designate as well very different statements such as *"be listed," "being a student," "be made of,"*..., etc. When the variable x takes a constant value, then $P(x)$, a propositional function, becomes a proposition: an expression that can be asserted by affirmation or negation of its truth. Thus, the relation $R(a, b)$ where a, b are being constants with precise values is asserted as true. It means that $R(a, b)$ must be considered valid.

If E :: "be made of", then, it is possible to define a domain of interpretation for *crossed references*. *Crossed references* means that all couples make the relation E valid (Fig. 2.29) or in other words:

$$\{(x \in R_1, y \in R_2) | \, E(x, y) :: \text{true}\}$$

with $E(x, y)$ :: true, which corresponds to the domain of interpretation of couples that refer to domains R_1, R_2 with x taking its values in R_1 and y in R_2. There are as many domains of interpretation as there are possible values for x and y. Thus, $E(x, y)$, defined as above, appears to be as a generalization of the notion of *"predicate."* However, the two following remarks are pertinent to that concept:

- Defining a referent by a predicate means to appoint a subset of items that verify the domain of interpretation.

Fig. 2.29 Binary relation between two referents

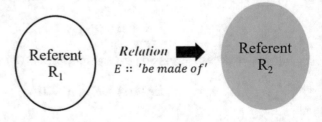

- Defining a predicative relation leads to identify a subset of referent couples that verify the domain of interpretation.

The joined use of *predicates* and *predicative relations* to designate the domains of references creates an interesting potential to formalize sentences from our current natural language related to combinations in which there are indications of various references as it will be seen in the following examples. Let us define the predicates as below:

- The predicate $V(x)$ :: "be a car brand"
- The predicate $C(x)$ :: "be a given type of chassis"
- The relation R :: "be an element of"

The following references can then be formalized with care to the order used with quantifiers ($\wedge$, $\vee$) since their inversion affects the meaning of the statement.

- A car brand V has a chassis type C: $V(a) \wedge C(b) \wedge R(a,b)$.
- All car brands have a chassis type: $\forall x \, \exists \, y \, [V(x) \wedge C(y) \wedge R(x,y)]$
- there are chassis types that all car brands have: $\exists x \, \forall \, y[V(x) \wedge C(y) \wedge R(x,y)]$

We speak of symmetric closure of a predicative relation of R defined on E when we add to all the pairs (x,x) of the diagonal $\Delta E \subset R$ which are not yet in the relation: $R \cup \Delta E$, where ΔE :: $\{(x,x)|x \in E\}$. The notion of predicative relation applied to reference domains is of great interest insofar as it makes it possible to understand the nature of the interrelations associated with them and to see also different modes of manipulation of the reference dimension from different angles such as one can see in the subsections that follow.

2.9.4   Converse Property of a Relational Predicate

If we start from the relation $R(x, y)$ previously defined between the reference domains R_1 and R_2, it may sometimes be more convenient to use the relationship in the "*opposite sense*" of its initial enunciation, for instance: $y \in R_2$ then $x \in R_1$ is called the converse relation of R and it is denoted usually as R_C. Two referential statements are converse if one results from the other by exchanging the terms of the subject and the predicate: "*some E are F*" is the converse of "*some F are E*," and "*all E are F*," is the converse of "*all F are E*." Any inference to get a converse from a statement is called conversion. For example, let us consider direct relations:

$$R(x, y) :: \text{"to be made of"} ; \text{"to be in love with"}$$

and, the converse relations,

$$R_C(y,x) :: \text{``to be a component of''}; \text{``to be loved by''}.$$

Note the importance of the prepositions "in" and "of" in the utterances; their absence would lead to a relation of arity 1, hence to a simple predicate. For any binary relation R, there exists a converse R_C satisfying:

$$\forall x \forall y \left[(R(x,y) \Rightarrow R_C(y,x)) \text{ and } ((R_C(x,y) \Rightarrow (R(y,x)))\right]$$

For a converse relation of R, the image set of R_C is identified with the definition set of R, and inversely the image set of R corresponds to that of definition of R_C.

2.9.5   Reflexivity Property of a Relational Predicate

A relation R defined on E is considered to be *reflexive* if:

$$\forall x \in E | x \in R(x,x).$$

$R ::$ "to be pleased with" constitutes a reflexive relation insofar as it can only be applied to itself. Other examples such as $R ::$ "to be a unique component of" or $R :: \text{``} \| \text{''}$ are also reflexive. On the contrary, R becomes non-reflexive in the case where:

$$\exists x \in E | x \notin R(x,x).$$

For example, $R :: \text{``} \perp \text{''}$ is a non-reflexive relation since no line of a plane can be perpendicular to itself. The relation, $R ::$ "to be the son of", is non-reflexive since no one can be its own son. With the quantifiers $\forall$, $\exists$, the reflexivity can also be expressed by:

$$\sim \forall x R(x,x) \Leftrightarrow \exists \sim R(x,x).$$

2.9.6   Symmetry and Asymmetry Properties of a Relational Predicate

A predicative relation is said to be symmetric if:

$$\forall x \forall y ((x,y) \in R \Rightarrow (y,x) \in R)$$

$$\forall x \forall y ((y,x) \in R \Rightarrow (x,y) \in R)$$

$$\forall x \forall y ((x,y) \in R \Leftrightarrow (y,x) \in R)$$

In the case of $R \subset E \times E$, this means that all the points x, y have a symmetry with respect to the diagonal of the representative matrix of R (in addition to those belonging to the diagonal necessarily symmetrical with themselves). $R ::$ " to be married with" and $R ::$ " to be the cousin of" are examples of symmetrical relations.

Moreover, a predicative relation R defined on E becomes asymmetric if:

$$\exists(x, y) \in E | ((x, y) \in E) \wedge ((y, x) \notin E).$$

In fact, this amounts to denying the symmetry of R:

$$\sim [\forall x \forall y\, R(x, y) \Rightarrow R(y, x)],$$

or with the duality of the quantifiers:

$$\sim [\forall x \forall y\, R(x, y)] \Rightarrow \exists x \exists y \sim [R(x, y) \Rightarrow R(y, x)],$$

the second member can be written:

$$\exists x \exists y \sim [R(x, y) \Rightarrow R(y, x)] \Leftrightarrow \exists x \exists y \sim [\sim R(x, y) \vee R(y, x)],$$

which gives:

$$\exists x \exists y \sim [\sim R(x, y) \vee R(y, x)] \Leftrightarrow \exists x \exists y\, [R(x, y) \wedge \sim R(y, x)].$$

The relation $R ::$ " $\geq$ " will be called non-symmetric: $5 \geq 2 \Rightarrow 2 \ngeq 5$. A relation R is said to be non-symmetric if R is not symmetric and neither asymmetric. In the case of

$$\forall x \forall y [(R(x, y) \wedge R(y, x)) \Rightarrow x = y],$$

R must be considered as antisymmetric. Examples of antisymmetric relations are:

- $R ::$ " be divisor of", if $,n \in \{integers\}$, $(R(p, n) \wedge R(n, p)) \Rightarrow (n = p)$
- $R ::$ " $\subset$ ", defined on E with $P, N \in \wp (E)$, $(P \subset N \wedge N \subset P) \Rightarrow P \leftrightarrow N$.

2.9.7   Negation Property of a Relational Predicate

In some cases, it may be necessary to disregard certain references or to hide others. In this case, we see the predicative relations from a negative angle. Any pair $(x, y) \notin R$ invalidates the relation R and then makes the proposition $R (x, y)$ false:

$$[(x, y) \notin R] \Rightarrow [\text{Val} [R(x, y) :: \text{"False"}]] \Rightarrow [\text{Val} [\sim R(x, y) :: \text{"True"}]]$$

The negation of R will be denoted by R_N:

Table 2.7 An example to illustrate the negation

R :: "to be made of"		E :: {metals}		
		Steel "S"	Bronze "B"	Copper "C"
F :: {parts}	P_1	X		
	P_2		X	
	P_3	X	X	X
	P_4			

$$R_N :: \left\{(x,y) | \sim R(x,y) :: \text{"True"}\right\}$$

Let us examine Table 2.7 as an example.
The graph of R_N for the whole Table 2.7 is:

$$G_N :: \{(P_1, B), (P_1, C), (P_2, S), (P_1, B), (P_2, C), (P_4, S), (P_4, B), (P_4, C)\}$$

We note that G_N is indeed the complement of G, the graph of the relation R, in $E \times F$. As R is a proposition, it will always be allowed to apply the negation:

$$\forall x \forall y [(R(x,y) \Rightarrow \sim R_N(x,y) \wedge R_N(x,y)) \Rightarrow (R(y,x))].$$

If R_{CN} denotes the converse negation of a relation R and R_{NC} the negation of the converse of this same relation, there is identity between the two. It is therefore the same to take the negative of a converse relation as to take the converse of its negation.

2.9.8   Transitivity Property of a Relational Predicate

It can be interesting from the same relation between different domains to see if it is possible to chain them together; if this is the case, then $R :: $ "*to be made of*" is said to be *transitive*:

$$\forall x \forall y \forall z [R(x,y) \wedge R(y,z)] \Rightarrow R(x,z).$$

Noting that $\sim \forall\, xP(x) \Leftrightarrow \exists\, x{\sim}P(x)$, P being a proposition, the absence of *transitivity* can be expressed by:

$$\forall x \forall y \forall z \sim [R(x,y) \wedge R(y,z)] \Rightarrow R(x,z)$$

$$\forall x \forall y \forall z \sim [R(x,y) \wedge R(y,z)] \Rightarrow R(x,z)] \Leftrightarrow \sim [\sim (R(x,y) \wedge R(y,z)) \vee R(x,z)]$$

$$\forall x \forall y \forall z [\sim\sim (R(x, y) \wedge R(y, z)) \wedge \sim R(x, z)] \dots (\text{De Morgan's laws})$$

$$\forall x \forall y \forall z [(R(x, y) \wedge R(y, z)) \wedge \sim R(x, z)]$$

2.10   Conclusion

This chapter has discussed the properties and dimensions of knowledge. Knowledge has been treated in the context of epistemic structures. Notions from semiotics and linguistics have been used. Knowledge representation and acquisition have been discussed with emphasis on the multidimensionality of knowledge. Ontological, semantic, temporal, and reference dimensions of knowledge have been studied. Formalization and quantification of knowledge have been introduced at the end.

References

1. M. Barès, *Maîtrise du savoir et efficience de l'action*: Editions L'Harmattan, 2007.
2. G. Vico, A. Battistini, and A. Pons, *La méthode des études de notre temps*: les Belles lettres, 2010.
3. M. Burgin, *Theory of information: fundamentality, diversity and unification* vol. 1: World Scientific, 2010.
4. J. Piaget, *Structuralism (psychology revivals)*: Psychology Press, 2015.
5. M. Burgin, *Theory of Knowledge: Structures and Processes* vol. 5: World scientific, 2016.
6. C. S. Peirce, *Peirce on signs: Writings on semiotic*: UNC Press Books, 1991.
7. C. K. Ogden, I. A. Richards, B. Malinowski, and F. G. Crookshank, *The meaning of meaning*: Harcourt, Brace & World New York, 1946.
8. F. De Saussure, *Cours de linguistique générale* vol. 1: Otto Harrassowitz Verlag, 1989.
9. F. De Saussure, *Course in general linguistics*: Columbia University Press, 2011.
10. R. Harris, *Language, Saussure and Wittgenstein: How to play games with words*: Psychology Press, 1990.
11. T. L. Short, *Peirce's theory of signs*: Cambridge University Press, 2007.
12. U. Eco, *Semiotics and the Philosophy of Language* vol. 398: Indiana University Press, 1986.
13. C. E. Shannon, *The mathematical theory of communications*. Urbana: The University of Illinois Press, 1949.
14. É. Bossé and B. Solaiman, *Fusion of Information and Analytics for Big Data and IoT*: Artech House, Inc., 2016.
15. P. beim Graben, "Pragmatic information in dynamic semantics," *Mind and Matter,* vol. 4, pp. 169–193, 2006.
16. C. W. Morris, *Foundations of the Theory of Signs* vol. 1: University of Chicago Press, 1938.
17. M. Barès, *Pratique du calcul relationnel*. Paris: Edilivre, 2016.
18. E. Tatievskaia, "Russell on the Structure of Propositions," in *The Paideia Archive: Twentieth World Congress of Philosophy*, 1998, pp. 131–137.
19. M. Burgin and V. Kuznetsov, "Fuzzy sets as named sets," *Fuzzy sets and systems,* vol. 46, pp. 189–192, 1992.
20. M. Burgin, "Named Set Theory Axiomatization: T Theory," *Mathematics Preprint Archive,* vol. 2004, pp. 333–344, 2004.

21. W. D. Blizard, "The development of multiset theory," *Modern logic,* vol. 1, pp. 319–352, 1991.
22. M. Burgin, "Unified foundations for mathematics," *arXiv preprint math/0403186,* 2004.
23. S. Awodey, *Category theory*: Oxford university press, 2010.
24. G. J. Klir and B. Yuan, "Fuzzy sets and fuzzy logic: theory and applications," *Upper Saddle River,* p. 563, 1995.
25. H.-J. Zimmermann, *Fuzzy set theory—and its applications*: Springer Science & Business Media, 2011.
26. J. Le Potier, *Lectures on vector bundles*: Cambridge University Press, 1997.
27. G. G. Chowdhury, *Introduction to modern information retrieval*: Facet publishing, 2010.
28. K. Rein, "Fact, Conjecture, Hearsay and Lies: Issues of Uncertainty in Natural Language Communications," in *Information Quality in Information Fusion and Decision Making*, ed: Springer, 2019, pp. 155–179.
29. D. A. Scheufele and N. M. Krause, "Science audiences, misinformation, and fake news," *Proceedings of the National Academy of Sciences,* vol. 116, pp. 7662–7669, 2019.
30. L. A. Zadeh, "The concept of a linguistic variable and its application to approximate reasoning—I," *Information sciences,* vol. 8, pp. 199–249, 1975.
31. N. Chomsky, *Knowledge of language: Its nature, origin, and use*: Greenwood Publishing Group, 1986.
32. N. Chomsky, *Syntactic structures*: Walter de Gruyter, 2002.
33. W. H. Sulis, "Archetypal Dynamics: An Approach to the Study of Emergence," in *Formal Descriptions of Developing Systems, NATO Science Series Volume 121*, 2003, pp. 185–228.
34. P. VAILLANT, "Glossaire de sémiotique, au sein de l'ouvrage Sémiotique des langages d'icônes, 1999," *Paris: Éditions Honoré Champion, Genève: Éditions Slatkine.*
35. A. Tarski, "The semantic conception of truth: and the foundations of semantics," *Philosophy and phenomenological research,* vol. 4, pp. 341–376, 1944.
36. F. De Saussure, W. Baskin, and P. Meisel, *Course in general linguistics*: Columbia University Press, 2011.
37. C. Goodwin, M. H. Goodwin, and D. Olsher, "Producing sense with nonsense syllables," *The Language of Turn and Sequence, Oxford (CUP),* pp. 56–80, 2002.
38. M. M. Smith, "Producing sense, consuming sense, making sense: perils and prospects for sensory history," *Journal of Social History,* vol. 40, pp. 841–858, 2007.
39. R. C. Hill and M. Levenhagen, "Metaphors and mental models: Sensemaking and sensegiving in innovative and entrepreneurial activities," *Journal of Management,* vol. 21, pp. 1057–1074, 1995.
40. D. A. Gioia and K. Chittipeddi, "Sensemaking and sensegiving in strategic change initiation," *Strategic management journal,* vol. 12, pp. 433–448, 1991.
41. G. Klein, J. K. Phillips, E. L. Rall, and D. A. Peluso, "A data-frame theory of sensemaking," in *Expertise out of context: Proceedings of the sixth international conference on naturalistic decision making*, 2007, pp. 15–17.
42. G. Klein, B. Moon, and R. R. Hoffman, "Making sense of sensemaking 2: A macrocognitive model," *Intelligent Systems, IEEE,* vol. 21, pp. 88–92, 2006.
43. A. J. Greimas, *Sémantique structurale: recherche de méthode*: Presses universitaires de France, 2015.
44. N. Deo, *Graph theory with applications to engineering and computer science*: Courier Dover Publications, 2016.
45. J. Bondy and U. Murty, "Graph theory (graduate texts in mathematics)," ed: Springer New York, 2008.
46. J. F. Sowa, "Semantic networks," *Encyclopedia of Cognitive Science,* 2006.
47. C. S. Peirce, "Existential Graphs: MS 514 by Charles Sanders Peirce with commentary by John Sowa, 1908, 2000," ed, 2010.
48. J. F. Sowa, "Conceptual graphs," *Foundations of Artificial Intelligence,* vol. 3, pp. 213–237, 2008.

49. M. Minsky, "A framework for representing knowledge," *The psychology of computer vision,* vol. 73, pp. 211–277, 1975.
50. M. Fernandez-Lopez and O. Corcho, *Ontological Engineering: with examples from the areas of Knowledge Management, e-Commerce and the Semantic Web*: Springer Publishing Company, Incorporated, 2010.
51. O. Lassila and D. McGuinness, "The role of frame-based representation on the semantic web," *Linköping Electronic Articles in Computer and Information Science,* vol. 6, p. 2001, 2001.
52. E. F. Kendall and M. E. Dutra, "Method and apparatus for frame-based knowledge representation in the unified modeling language (UML)," ed: Google Patents, 2008.
53. R. B. Atman and N. F. Abernethy, "Frame-based knowledge representation system and methods," ed: Google Patents, 2002.
54. J. Bao, D. Caragea, and V. G. Honavar, "Modular Ontologies - A Formal Investigation of Semantics and Expressivity ", S. B. Heidelberg, Ed., ed Berlin / Heidelberg: Springer Berlin / Heidelberg, 2006, pp. 616–631.
55. O. Corcho, M. Fernández-López, and A. Gómez-Pérez, "Ontological engineering: principles, methods, tools and languages," in *Ontologies for software engineering and software technology*, ed: Springer, 2006, pp. 1–48.
56. D. Fensel, "Ontologies," in *Ontologies*, ed: Springer, 2001, pp. 11–18.
57. G. Bārzdiņš, N. Grūzītis, G. Nešpore, B. Saulīte, I. Auziņa, and K. Levāne-Petrova, "Multidimensional ontologies: integration of Frame semantics and ontological semantics," in *Proceedings of the 13th EURALEX International Congress*, 2008, pp. 277–283.
58. É. Bossé, J. Roy, and S. Wark, *Concepts, Models and Tools for Information Fusion*: Artech House, 2007.
59. E. Blasch, É. Bossé, and D. A. Lambert, Eds., *High-Level Information Fusion Management and Systems Design.* Artech House, 2012.
60. A. Dey, "Understanding and Using Context," *Personal and Ubiquitous Computing,* vol. 5, pp. 4–7, 2001.
61. A. Zimmermann, A. Lorenz, and R. Oppermann, "An operational definition of context," in *Modeling and using context*, ed: Springer, 2007, pp. 558–571.
62. M. R. Endsley, "Situation awareness," in *Handbook of human factors and ergonomics*, 434–455, Wiley, 2021.
63. (2015-04-22). *context awareness* Available: http://en.wikipedia.org/wiki/Context_awareness
64. J. Ma and B. Knight, "Reified temporal logics: An overview," *Artificial Intelligence Review,* vol. 15, pp. 189–217, 2001.
65. D. McDermott, "A temporal logic for reasoning about processes and plans," *Cognitive science,* vol. 6, pp. 101–155, 1982.
66. J. F. Allen and P. J. Hayes, "A Common-Sense Theory of Time," in *IJCAI,* 1985, pp. 528–531.
67. T. Gruber, "*A Translation Approach to Portable Ontology Specifications*," *KNOWLEDGE ACQUISITION,* vol. 5, pp. 199–220, 1993.
68. F. Poggiolesi, *Gentzen calculi for modal propositional logic* vol. 32: Springer Science & Business Media, 2010.
69. M. Ardeshir and W. Ruitenburg, "Basic propositional calculus I," *Mathematical Logic Quarterly,* vol. 44, pp. 317–343, 1998.
70. E. W. Dijkstra and C. S. Scholten, *Predicate calculus and program semantics*: Springer Science & Business Media, 2012.
71. J. Doyle, "A truth maintenance system," *Artificial intelligence,* vol. 12, pp. 231–272, 1979.
72. L. Couturat, *Les principes des mathématiques*: Librairie Armand Colin, 1905.

Chapter 3
The Infocentric Knowledge Processing Chain

Abstract This chapter focusses on the information-processing aspect of a knowledge chain. We discuss the relationship between data, information, and knowledge with their associated imperfections (i.e., imperfect world). Distinctions in terms of semantic status, role, and processing are important since it has a significant impact on their conceptual and implementation approaches. The notion of "quality" is of prime importance to any information processing, and particularly for a knowledge processing chain, since the objective is to improve, through processing, data, information, and knowledge quality. Various aspects of Quality of Information (QoI) are then discussed with emphasis on uncertainty-based information.

3.1 Introduction

For any human, acting on or in the environment implies having knowledge from many information elements acquired either timely based on need (factual information) or beforehand from learning and cultural rooting (permanent knowledge). Knowledge is a prerequisite to taking any reasoned action or course of action according to rational rules [1, 2]. It can be described as a superset of domains of knowledge aggregated according to a specific application. Its complex structure is formed through several levels of aggregation:

- Aggregation of signs from the perceived world that are considered relevant for future decision-making contexts.
- Aggregation of data serving as a basis for enactment.
- Aggregation of knowledge more or less expanded and structured according to the cultural level and expertise of the acting entity.

Knowledge is the culmination of a chain of different types of entities whose essence is determined by the perception of signs and signals in the perceived world participating in action control according to the level of signified conveyed (i.e., semantic status). The knowledge chain represents the necessary bridging that must occur at the informational level, between the observable and the cognitive world as globally pictured in Fig. 3.1.

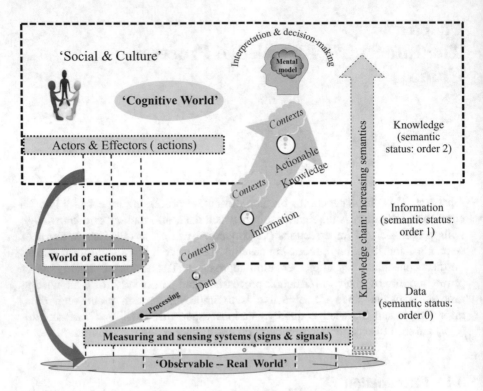

Fig. 3.1 A global contextual illustration of a knowledge chain

The previous chapter presented knowledge, its multidimensionality and structures. This chapter focusses on the information-processing aspect of a knowledge chain. We discuss the relationship between data, information, and knowledge with their associated imperfections (i.e., imperfect world). Distinctions in terms of semantic status, role, and processing are important since it has a significant impact on their conceptual and implementation approaches. The notion of "quality" is of prime importance to any information processing, and particularly for a knowledge processing chain, since the objective is to improve, through processing, data, information, and knowledge quality. Various aspects of Quality of Information (QoI) are then discussed with emphasis on uncertainty-based information.

The three words information, uncertainty, and entropy have distinct meanings for many people in different domains. The sources and characterization of uncertainties in the knowledge chain are quite important. The sources of uncertainty are generally categorized as either aleatory or epistemic. Uncertainties are characterized as epistemic, if one sees a possibility to reduce them by gathering more data or by refining models. Uncertainties are categorized as aleatory if one does not foresee the possibility of reducing them. Identifying epistemic uncertainties that have the potential of being reduced has a pragmatic value for the processing of an *infocentric* knowledge chain.

3.2   Distinctions Between Data, Information, and Knowledge

As already mentioned, the terms data, information, and knowledge are not always clearly distinguished in the literature. However, each refers to ideas and concepts that, after examination, will appear differentiable. A distinction is clearly apparent in current language when talking about data from a physical problem, information brochure on a commodity, while the term of knowledge remains vested in the mastery of a technique or holding an important function. The terms are distinct since they do not reflect the same reality, and they convey different levels of signified [1, 2]. The advances in the development of techniques emphasize that distinction by dedicating to them separated roles and processing. Their conceptual and implementation approaches are different in the architecture of information systems.

3.2.1   Semantic Status

The subject of knowledge, at the heart of human experience, raises many questions: What mechanisms are at play in knowing, how is knowledge memorized and transmitted, and what is the useful ratio between implicit and explicit knowledge for taking reasonable action? It is beyond the scope of our work to address these questions here. Modalities of apprehension and representation of knowledge, specifically in symbolic form, should first be addressed before defining an adapted formalism for establishing the basis for artificial reasoning.

Any automaton providing decision-making support should have reasoning or interpretation capabilities according to the application and context. These capabilities are achieved through an inference mechanism based on existing knowledge that has been appropriately formalized and has gained meaning within the highly structured framework of classical or other logics.

For human beings, knowing is a primordial mental activity because it is intricately connected to action. Especially given that 20% of the human life cycle, including successive stages of education and professional training until the age of 20/25, is devoted to acquiring knowledge. All future action is partially conditioned by permanent knowledge primitively initialized and maintained over time, as well as timely knowledge continuously acquired according to need. Understanding what is learned is necessary to action. This does not imply, however, that action is not necessary to understanding. This refers to a capability to reason, but also to impute meaning to forms and to their sometimes tenuous interconnections; to draw the appropriate links between data, information, and knowledge; and to harness action to enrich knowledge.

As discussed previously, the terms data, information, and knowledge refer to notions and concepts, which upon examination appear to be differentiable. They are differentiated in common parlance, e.g., data relevant to a physics problem, a market product information leaflet, and knowledge of a technique or function. Indeed, differentiation is mandatory because data, information, and knowledge not only refer to different realities but also convey quite different levels of signified (referred to as *semantic status*, Fig. 3.1). The distinction is emphasized by technical developments in specialized functions and processes as well as conceptual approaches with different application procedures within information systems architectures. We thus use the term data fusion distinct from information fusion, knowledge base distinct from data base, data management distinct from information management, etc.

Example 3.1 The set of signs (1 20 075288) 10 borrowed from the decimal alphabet represents data that would appear to have no relevance as it is not situated in a potential application context: Is it a phone number, a (library) call number, or a (clinic) patient ID?

Data does not make sense if is not associated with an application or process or if it is not operational within a context. Its morphology may be indicative of applicable processes and enable the data to be situated within a given operational context. Based on this same data, additional information can be generated (i.e., by the operator) for future exploitation. The informational aspect of the initial data varies. Therefore, as the set of signs begins to make sense within a new context, the term information is substituted for data. The operator enacts the information based on the morphology of the number, resulting here in a semantic elevation of the data. Like data, information reflects a given reality through its different perceivable elements (objects, facts, events), but it imparts a meaning to this reality.

This trivial example highlights the key function of the object-subject pair, in which knowledge or the cognitive level of the knowing subject is fundamental to the apprehension of a given reality because it modulates the subject's modes of appropriation of the knowledge object. In philosophy, there is no knowledge beyond the relationship between knowing subject and knowledge object.

Meaning emerges when information content is matched with or triggers other cognitive elements contained in a permanent memory. That is, information content is enriched in the transition from information to knowledge with the associated increase (elevation and expansion) in semantics. Incidentally, for a decision-maker, maximizing action efficiency is synonymous with amplifying the emergence of meaning.

The basis of knowledge is information gain through human action or semantic gain through intelligent processing. The latter will of course have a larger impact on decision-making. The world of action "exists" through knowledge in that it is formed outside of the more or less conscious worlds of the human mind. To reveal the impact of information content on action and the decision-making context, we introduce the notion of semantic status and justify an order relation: data, information, and knowledge.

Conventionally, the status:

- *Of data is of order 0* if information content is invariable because it can be neither increased nor decreased.

- *Of information is of order 1* if information content is likely to vary because of processing.
- *Of knowledge is of order 2* if information content is likely to be significantly enhanced by human appropriation.

In other words, the semantics of these different entities are enhanced when passing from data to knowledge using refinement of information. Knowledge is formed by information enhancement or refinement through human or machine actions. Potential worlds "exist" through knowledge in which new forms of action are possible. Take the example of an image analyst studying a photo negative on a screen. For a non-specialist, this negative is a set of more or less dark blots, a sequence of points, indeed, of data, which, in the absence of a context of interpretation, have no specific meaning. At first, the analyst starts the processing using tools available in the interface such as brightness adjustment and zoom contours. That gives analyst informational details to assist in his identification task. The information technology tools help give meaning to initial data and provide information to the image analyst. Based upon his expertise, these elements of information allow him to interpret features in the scene.

Along the knowledge chain from data to knowledge, subjectivity prevails due to the analyst cognition in starting his work of interpretation. If, thereafter, there are difficulties in understanding the scene that would require searching for details, so back to revisiting data, then objectivity would become preeminent. In quest of a gain in semantics leads to a preponderance of subjectivity, its reduction implies however a predominant objectivity.

3.2.2   Relationships Between Data, Information, and Knowledge (DIK)

The data-information-knowledge (DIK)[1] pyramid is a popular approach to relate data, knowledge, and information. People still do not achieve unambiguous understanding of relations between these three concepts. Many authors[2] present different views on such relations. For instance, 130 definitions of data, information, and knowledge formulated by 45 scholars are collected in Zins [3]. However, the DIK pyramid has many deficiencies and limitations. Some authors argue that in contrast to the conventional estimation that knowledge is more valuable than information, while information is superior to data, these relations must be reversed. They argue that since only data can effectively be processed by computers, data is the most

---

[1] https://en.wikipedia.org/wiki/DIKW_pyramid

[2] Some researchers also related information to structure of an object. For instance, information is characterized as a property of how entities are organized and arranged, but not the property of entities themselves.

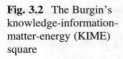

Fig. 3.2 The Burgin's knowledge-information-matter-energy (KIME) square

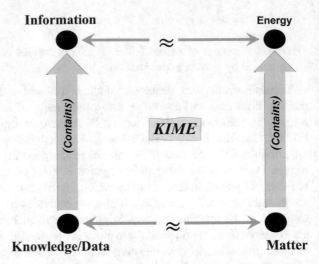

valuable of the three components of the DIK pyramid, which consequently should be turned upside down.

To eliminate these shortcomings, Burgin [4] proposes a knowledge-information-matter-energy (KIME) square (Fig. 3.2). In the current book, we have adopted the KIME square to interpret the relationship between data, information, and knowledge. Burgin states that information is not of the same kind as knowledge and data, which are structures. Figure 3.2 means that information has essentially different nature than knowledge and data, which are of the same kind. It is possible to explain similarities and distinctions between data and knowledge with the two metaphors of Example 3.2. Knowledge and data are structures, while information is only represented and can be carried by structures. The distinction between information and knowledge can be done along three axes:

1. *Multiplicity: Information is piecemeal, fragmented, and specific, while knowledge is structured, coherent, and universal. Knowledge is a structure or a system of structures.*
2. *Time: Information is timely, transitory, even ephemeral, while knowledge is enduring and temporally expansive.*
3. *Space: Information is a flow across spaces, while knowledge is a stock, specifically located, yet spatially expansive.*

Example 3.2

Metaphor A. Data and knowledge are like molecules, but data are like molecules of water, which has two atoms, while knowledge is like molecules of DNA, which unites billions of atoms

Metaphor B. Data and knowledge are like living beings, but data are like bacteria, while knowledge is like a human being.

Fig. 3.3 Burgin's
information triad

Looking at relations between information and knowledge as discussed above suggests that the proposed Burgin's triad of Fig. 3.3 called the data-knowledge triad provides a more appropriate relational model than the DIK pyramid. For the triadic representation of Fig. 3.3, Burgin brings the following interpretation: "Data, under the influence (action) of additional information, become Knowledge. That is, information is the active essence that transforms data into knowledge. It is similar to the situation in the physical world, where energy is used to perform work, which changes material things, their positions and dynamics."

3.2.3   The DIK Semiotic Dimensions

Consider the three semiotic dimensions of Morris [5] discussed previously:

- **Syntactics:** The rules that govern how signs relate to one another in formal structures.
- **Semantics:** The relations of signs to the objects to which the signs are applicable.
- **Pragmatics:** The relation of signs to interpreters.

Lenski [6] makes a very interesting link between data, information, and knowledge and the three semiotic dimensions: syntactics, semantics, and pragmatics. Figure 3.4 sketches his view where data denote the syntactical dimension of a sign, knowledge denotes the semantic dimension of a sign, and information denotes the pragmatic dimension of a sign. Data are a system that is organized by structural and grammatical rules of sign. Knowledge is that which is known, and it exists in the mind of the knower. It can be made tangible by the means of symbolic representations. From this point of view, *represented knowledge* is becoming a particular kind of information. Knowledge is, by definition, subjective. It ends up always in personal "knowing." When socially considered, it still means a sort of aggregation of individual "knowings." Information is tied to a cognitive system which processes signs that constitute data for a possible action exemplified by information operations in Fig. 3.4 and detailed in Table 3.1. Information emphasizes reaction and performance (pragmatics dimension), while knowledge is abstracted from any reference to the actual performance. This is under the DIK pragmatics dimension that we talk about *actionable knowledge*. To perform an action, knowledge has to be represented, and consequently it becomes a particular kind of information. We call that an *infocentric* knowledge processing chain.

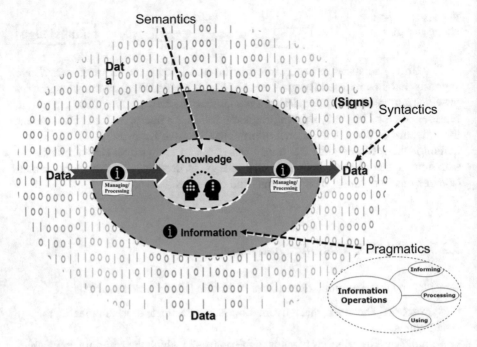

Fig. 3.4 Burgin's DIK triad enriched by Lenski's semiotic dimensions

3.2.4 Infocentricity and the Knowledge Processing Chain

The relationships between data, information, and knowledge can be briefly expressed this way:

- Knowledge is a compressed, goal-oriented structural depository of *refined information*.
- Data is a raw symbolic depository of *information*.
- Conversion of data into knowledge is a refinement of *information*.

Infocentricity means here that the relationship with data, information, and knowledge is centered on information. The transactions to transform data into knowledge are based on some types of information processing as illustrated in Fig. 3.5 and suggested by the KIME model. Just mentioned in the previous section above, *represented* knowledge is in fact *information*. We then call that an' *infocentric knowledge chain*. However, Fig. 3.5 also indicates that the knowledge chain can be reversed since it is possible to create data from knowledge. Again, this reversed chain is also *infocentric* since it utilizes a definite type of information processes.

As already mentioned, conversion of data into knowledge can be thought as a refinement of information to provide knowledge about a specific event or situation. Data, under the influence (action) of additional information, are transformed into knowledge (top of Fig. 3.5). It suggests that information creates knowledge

Table 3.1 The pragmatics dimension of DIK: *information*

Informing	An operation when a portion of information acts on a system *A*	*Information reception* is an operation when *A* gets information that comes from some source
		Information acquisition is an operation when *A* performs work to get existing information
		Information production is an operation when *A* produces new information
Processing	An operation when *A* acts on a portion of Information	*Information transformation* is an operation when changes of Information or its representation take place
		Information transition, or movement, is an operation when information is not changing but the place where the information carrier is changes
		Information storage is an operation when information is not Changing but time for the information carrier changes
Using	An operation when the system *A* acts on Another system *Q* by means of information	*Information reconstruction* is an operation when the initial Information and its initial representation change
		Information construction is an operation when the initial information and its initial representation are not changing but new information and representation are created
		Information biconstruction is an operation when the initial information and its initial representation change and new information and representation are created

Created from Burgin [7], p.28

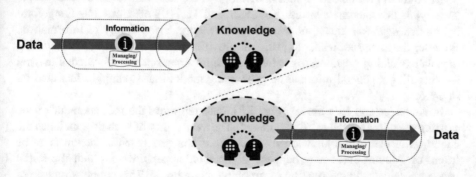

Fig. 3.5 The *infocentric* knowledge chain

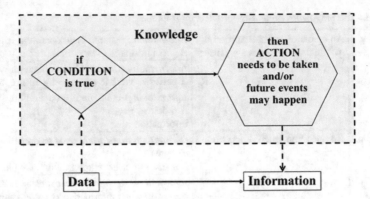

Fig. 3.6 Relationship between data, knowledge, and information according to Kettinger and Li [10]

(*infocentricity*), while at the bottom of Fig. 3.5 indicates that it is possible to create data from knowledge using special types of information processes (*infocentricity*). For instance, when an e-mail is sent, the text, which is usually a knowledge representation, is converted to data packages, using a definite type of information processes (refinement) prior to be transmitted (action) to the recipient. Another example is the management of sensors using a priori knowledge. Obviously, the information processes to convert data into knowledge in Fig. 3.5 are different than those to convert knowledge to data.

This *infocentricity* just expressed above to depict the relationship between the DIK core concepts is in line with the view of Langefors' *infological eq.* [8, 9] suggesting that information is the joint function of data and knowledge. Langefors' theory describes data as the measurement of facts or states. Information is generated from the interaction between the states measured in data and their relationship with future states predicted in knowledge. So, information represents a status of conditional readiness for an action, and knowledge is the relationship between concepts underlying the measured states. Kettinger and Li [10] extended the Langefors' *infological equation* into a model called knowledge-based theory of information. We refer the interested reader to [10] that provides a discussion on various existing models explaining relationships between data, information, and knowledge as well as presenting a general information processing model that is simply illustrated on Fig. 3.6.

In Kettinger and Li's model (Fig. 3.6), data represent the measurement of the existing conditions, such as "It is raining." The availability of such data matches the condition clause in the knowledge structure, so that the potential action is to be taken, or a future event may happen. When this happens, we say that the initial messages deliver the information of grabbing an umbrella. This simple structure of DIK concepts interactivity is a general artificial intelligence (AI) structure where the initial states are converted into goals through procedures. Citing [10] to explain Fig. 3.6, "Data refer to the description of the states, information is the readiness for the goal (e.g., to order or not to order), and knowledge is the framework or process

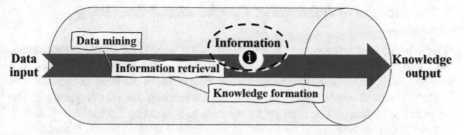

Fig. 3.7 Creation of actionable knowledge

(e.g., the business rule) through which the readiness is triggered by the pre-existing state."

Consider further the process of data transformation into knowledge. The amount of data being collected in databases today grows at a phenomenal rate and far exceeds our ability to reduce and analyze data without the use of automated analysis techniques. Databases are increasing in size in two ways: (1) the number of objects in the database grows and (2) the number of attributes to an object increases. Various methodologies of knowledge discovery (e.g., KDD, knowledge discovery in databases [11], information retrieval, data mining [12, 13]) have emerged to provide people with efficient tools to extract knowledge from these volumes of data. Most of these methodologies are empirical, and some present deficiency of sound theoretical foundations so preventing to achieve sufficient efficiency in processing knowledge by computers. A high-level model of knowledge discovery consists of three levels, data mining, information retrieval, and knowledge formation, as illustrated in Fig. 3.7.

Knowledge is obtained through *information retrieval*, which is based on data collection, mining, and analysis. The three strati, data mining, information retrieval, and knowledge formation, mentioned in Fig. 3.7, represent the generic information processing that both people and computers perform to create knowledge. This stratification of knowledge creation is based on distinctions between data, information, and knowledge. Consequently, there is a difference in processing data, processing information, and acquiring or producing knowledge, but it is all based on information processes.

An important stratus in the creation of actionable knowledge (Fig. 3.7) is called *data mining*, which is the application of specific algorithms for extracting patterns from data. The *information retrieval* stratus takes the raw results from data mining and transforms them into useful and understandable information. This information is retrieved using artificial intelligence (AI) techniques [14]. The retrieved information is transformed into knowledge through the *knowledge formation* process, which places information into a definite knowledge system of a user. *Data mining* does not provide knowledge per se but provides large amounts of relevant/irrelevant data to a user who must convert these data into knowledge. Let us recall that knowledge is formed only inside some knowledge system. It may be the mind of a user or an automated knowledge system on a computer.

3.2.5   Actions of Information Computational Processing

Infocentricity is also present in the actions of information computational processing. According to Vlăduțescu and Smarandache [15], any information processing is a computational processing. Computation can be natural or artificial as explained in [15]: "... natural computation means mental processing, mental computation. The thought is computing. The most important mental processing (mental computation) is information processing. Information represents the form in which thought organizes the cognitive reality. When we think we run computational processes. When we think we realize information computational processing."

Any computation involves operations, actions, mechanisms, and strategies. What do we mean by "actions"? Again here, consider Vlăduțescu and Smarandache [15] who define "the action" the following way: "The composition of mechanisms and strategies supposes a series of operations which are developed together, are stabilized as procedure, are unitarily structured in producing some coherent effects. These performed unitary operations to achieve a unique effect are called: the action."

The actions consist of operations of different genres. They may be perceptual or mental actions (judgment, reasoning). They may involve circulations and transformations of objects, images, or concepts with defined informational content, on the signs and formal-abstract structures. Any action or behavior, automatic or controlled, benefits by an anticipatory informational model, which is part of the knowledge base. The automatic actions or behaviors are determined by knowledge of working memory. The controlled actions and behaviors are the results of knowledge processing and of structuring of goals from the most active part of working memory. The choice and the performing of actions depend on the form of presentation of the communication situation (*objectual, imagistic,* or *symbolic-abstract*).

Vlăduțescu and Smarandache [15] describe five basic informational computational actions: *exploration, grouping, anticipation, schematization,* and *inferential structure of meanings*. The informational actions are described in Table 3.2. The content of Table 3.2 has been mainly excerpted from Vlăduțescu and Smarandache [15] and Piaget [16]. These informational actions have influenced the identification of our Analytics and Information Fusion (AIF) core technological processes that we believe can support the processing of a knowledge chain to create actionable knowledge.

3.3   Quality of Information (QoI) and the Knowledge Chain

The availability of reliable, accurate, and timeliness information is crucial for a knowledge chain to deliver actionable knowledge for any decision-making task. As already mentioned, information is the mean by which data is transformed into actionable knowledge. The concept of information quality is relatively difficult to capture. However, methods developed to assess information quality are progressing,

Table 3.2 Five basic informational computational actions

Action name	Characteristics	Rules and Principles
Exploring	Exploring can be spontaneous-random or selective-directional Involves operations centering (exploration, search, analysis, comparison operations) Space-temporal transpositions of remote elements, ensuring a balancing game between centering effects Presence notification, volume adjusting of processed meanings and connection to a controlled purpose structure of the cognitive system of the informational subject	The rule of the **economy** consists of processing centering on the nearest meaning from semantic point of view, which is in relation to the one previously integrated in the structure of information establishment The rule of **informative areas**, representing the orientation toward centers containing the greatest amount of meanings The rule of **dissymmetry** up-down that supposes exploring from high to low, from general meanings to a particular meaning
Grouping	The action of grouping consists of coordinating some operations of delimitation and association according to criteria that lead to the structuring rules discovery of informational object The law of generalization: Perceiving of an informational form attracts for this grant of a meaning The law of constancy: Good forms tend to conserve characteristics in spite of the presentation mode; an information is structured from fragments "Grouping performs for the first time the balance between things, the assimilation in subject action and the subjective schemata accommodation to changes of things" (Piaget)	The principle of **proximity**: Close elements are perceived as forming a unity, a configuration The principle of **similarity**: Analogous elements are perceived as constituting a form The principle of **continuity**: Oriented elements in the same direction tend to be structured perceptively in the same form The principle of **symmetry**: Symmetric figures beside one or two axes constitute "good forms" and are easier to process The principle of good **continuation**: At the intersection of two contours, their continuation will be perceived after the continuation of the simpler one The principle of **closure**: Processing tends toward well-delimited forms, closed and stable
Anticipation (*feed-forward*)	The concept of anticipation was introduced by J. Piaget A kind of pre-inference, an aspect of a perceptive schema, entraining other through an immediate involvement which modifies the perception Informational act, of any kind, oriented (executed in a task achieving), latitude (executed by pleasure or ludic), analytical, audio-perceptive, video-perceptive, etc. does not represent the absolute beginning of an informational act, but it follows	Repeating generally the same type of act mobilizes an old act that, extracted from memory, becomes anticipation of that which is gradually formed Action of anticipation requires at least two operations: Memory accesses times long or short and projected into the new act of the principal act model data from storage Any cognitive activity requires some anticipation Information without anticipation is nonsense

(continued)

Table 3.2 (continued)

Action name	Characteristics	Rules and Principles
	another act whose achievement was already engraved	
Schematization	From J. Piaget: "…some schemata (…) that must be accommodated constantly by explorations and corrections, situations, even at the same time when they assimilate" Action of schematization is composed of exploratory operations, corrections, and modeling It is the consequence of a previous conceptualization and represents a current act direction The scheme absorbs meanings, within certain limits, and changes and processes them, so as to produce a greater volume of information An informational structure that is a system of interdependent relations In processing, cognitive subject always has a grid that centers its approach Center makes the area of attention focusing to be stronger than the periphery of the field and not confusing	Schematization is based on a repetition Generalization as a common structure or schema of a certain activity as a result of its repeating There are two types of cognitive schemata: Empirical and geometric with two types of processing: **Geometricizing** (those approaching the object of something known) and **empirical** (those locating the object in the proximity of the informational objects familiar to the cognitive subject) Both schemata have therefore deforming effects, leading to the appearance of errors and compensatory effects, **corrections** Examples of specific operations of schematization: Disambiguation of discourses, summarizing of materials of hundreds of pages. Segmentation operations of audio or video fields. Repeated centering on different areas will make semantic space clearer, more accessible, understandable, information more easily structured
Inferential structuring	The basic of the computation is the connection, the inference, and computation is a fundamental human mental process Organizing partial perceptions in an integral one occurs through operations of exploration, comparison, correction, modeling, and control. All of this constitutes the action of inferential structure	From Piaget -- *structuring "successive centering", involving operations of "correction and adjustment" leads also to the internal organization of thinking schemata through pre-inferences* Inference is the assembly of operational steps and elementary transformations applied to information in its internal processing Pre-inferences that structure informational fields are similar to basic axioms, basic rules, the golden rules: Pathways orientation determines the reference systems, the main access roads, and fundamental ways for phenomena understanding Pre-inferences can be of four types:

(continued)

Table 3.2 (continued)

Action name	Characteristics	Rules and Principles
		Inductive, deductive, abductive, and analogical

Created from [15] and J. Piaget [16]

evidenced by practitioners and researchers who express a steadily growing interest. The ever-increasing interconnectivity among information sources (due to the Internet) drives that interest. Several books have appeared in the last decades covering diverse aspects of that complicated issue of information quality.

According to Wikipedia: "Information quality is a measure of the value which the information provides to the user of that information... The fitness for use of the information provided." The two words Information and Quality are themselves not completely defined and standardized. For instance, most information system practitioners use the term information synonymously with data, and several books reflect that by making no distinction between data quality and information quality. The previous section discussed notions such as: what is information? What is the relation between data, information, and knowledge? It makes sense to pursue the discussion on what is quality associated with these notions.

3.3.1   Basic Questions Related to Quality of Information

The subject of information and data quality has been receiving significant attention in recent years in many areas including communications, business processes, personal computing, healthcare, and databases. At the same time, the problem of information quality in the fusion-based human-machine systems for decision-making has attracted less attention. The main body of the literature on information fusion concerns building an adequate uncertainty model without paying much attention to the problem of representing and incorporating other quality characteristics into fusion processes. Many relevant research questions, related to the information quality problem in the designing fusion-based systems, are raised including [17]:

- What is ontology of quality characteristics?
- How to assess information quality of incoming heterogeneous data as well as the results of processes and information produced by users?
- How to combine quality characteristics into a single quality measure?
- How to evaluate the quality of the "quality assessment procedures" and results?
- How to compensate for various information deficiencies?
- How do quality and its characteristics depend on the context?
- How does subjectivity, i.e., user biases, affect information quality?

Nevertheless, there is no unique definition of Information quality. In fact, several definitions of information quality are available in the literature:

- "Quality is the totality of characteristics of an entity that bear on its ability to satisfy stated and implied needs." [18]
- "Quality is the degree to which information has content, form, and time characteristics, which give it value to specific end users."[19]
- "Quality is the degree to which information is meeting user needs according to external, subjective user perceptions." [20]
- "Quality is fitness for use." [21].

While having different emphases, all these definitions point to the fact that information quality is a "user-centric" notion and needs to be measured in terms of the potential or actual value for the users. In the human system context, "users" can be either humans or automated agents and models. Quality of Information (QoI) is "information about information," or meta-information, and the best way of representing and measuring the value of this meta-information is through its attributes since "without clearly defined attributes and their relationships, we are not just unable to assess QoI; we may be unaware of the problem" [22].

These attributes must be considered in relation to specific user objectives, goals, and functions in a specific context (internal/external contexts). Since all users, whether human, semi-automated, or automatic processes, have different data and information requirements, the set of attributes considered and the level of quality considered satisfactory vary with the user's perspective, the type of the models, algorithms, and processes comprising the system. Therefore, the general ontology designed to identify possible attributes and relations between them for a human-machine integrated system will require instantiation in every particular case. Considerable research targeting to study and to classify various aspects of Quality of Information has been conducted in both civilian and military domains. The following sections are devoted to detail this specific issue.

3.3.2 Evaluation of Quality of Information

Quality is defined as an essential and distinguishing attribute of something or someone, as a degree of excellence or worth, and as a characteristic property that molds the apparent individual nature of something or someone. Information quality studies are oriented at the first of these definitions, i.e., information quality is perceived as an essential and distinguishing attribute of information. There is also a pragmatic approach to information quality. In it, information quality is treated as fitness for a purpose. This makes information quality very subjective and relative to specific goals and intentions. In pragmatic terms, information quality means that information items geared toward one set of consumers may be perceived as poor quality when located by a different set of consumers, known to the user or not. Information quality is mostly understood as satisfaction of the user needs. Thus, the quality of information depends on the different perceptions and needs of the information users and is therefore relative. High-quality information would therefore

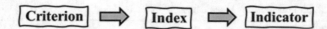

Fig. 3.8 Preparation of information quality evaluation

meet the specific requirements of its intended use. However, in spite of all distinctions between users and their goals and requirements, information quality researchers have tried to find common and comprehensive quality to this theory; the process of measurement/evaluation has three main stages: **preparation, realization**, and **analysis**.

The first stage in evaluation preparation demands to determine a specific criterion for evaluation. Such a criterion describes the goal of evaluation. Criteria of information quality include such properties as reliability, adequacy, exactness, completeness, convenience, user-friendliness, etc. However, such properties are directly immeasurable, and to estimate them, it is necessary to use corresponding indicators or indices. With respect to information quality, such indicators are called general information quality measures or metrics. However, an indicator can be too general for direct estimation. This causes necessity to introduce more specific properties of the evaluated object. To get these properties, quantifiable tractable questions are formulated. Such properties play the role of indices for this criterion. Thus, the second stage of evaluation consists of index selection that reflects criteria. Sometimes, an index can coincide with the corresponding criterion, or a criterion can be one of its indices. However, in many cases, it is impossible to obtain exact values for the chosen indices. For instance, we cannot do measurement with absolute precision. What is possible to do is only to get some estimates of indices. Consequently, the third stage includes obtaining estimates or indicators for selected indices.

Thus, to achieve correct and sufficiently precise evaluation, preparation demands the following operations:

- Choosing evaluation criteria.
- Corresponding characteristics (indices) to each of the chosen criteria.
- Representing characteristics by indicators (estimates).

This shows that a complete process of evaluation preparation has the following structure (Fig. 3.8):

Creation of information quality measures must include the following three stages:

- Setting goals specific to needs in terms of purpose, perspective, and environment.
- Refinement of goals into quantifiable tractable questions.
- Construction of an information quality measure and data to be collected (as well as the means for their collection) to answer the questions.

Information quality measures are useful only when there are corresponding procedures/algorithms of measurements. This process is illustrated in Table 3.3 where Web indicators (measures) have correspondence with the information quality criteria.

Table 3.3 Criteria-indicators correspondence

Information quality criteria	Web indicators
Accuracy	User ratings
Clearness/clarity	User ratings
Applicability	The number of orphaned, i.e., not visited or linked, pages or user ratings
Correctness	User ratings
Consistency	The number of pages with style guide deviations
Currency	Last mutation
Conciseness	The number of deep (highly hierarchic) pages
Timeliness	The number of heavy (oversized) pages/files with long loading time
Convenience	Difficult navigation paths, i.e., the number of lost/interrupted navigation trails
Traceability	The number of pages without the author or source
Interactivity	The number of forms The number of personalized pages
Accessibility	The number of broken links The number of broken anchors
Security	The number of weak log-ins
Maintainability	The number of pages with missing meta-information
Speed	Server and network response time
Comprehensiveness	User ratings

Source: Burgin [7], p.450

3.3.3   Frameworks for Evaluation of Quality of Information (QoI)

Quality is an overly complex property, and to evaluate information quality, it is necessary to have an efficient framework that provides a systematic and concise set of criteria to which information can be evaluated. Information quality is often considered as a measure of the reliability and effectiveness of data that carry this information, especially, in the context of decision-making. Hundreds of tools have been produced for evaluating quality in various application contexts. Various parameters have been included in information quality, such as relevancy, accessibility, usefulness, timeliness, clarity, comprehensibility, completeness, consistency, reliability, importance, and truthfulness. The most popular framework [23] to assess data information quality is illustrated in Fig. 3.9.

These studies are intrinsically related to information quality studies as data are primary sources of information and the most frequent representation of information processed by people. However, data are important not by themselves but exclusively by the virtue of the information they carry. From this perspective, data are a kind of information representation, and thus, the quality of information depends on the quality of its representation.

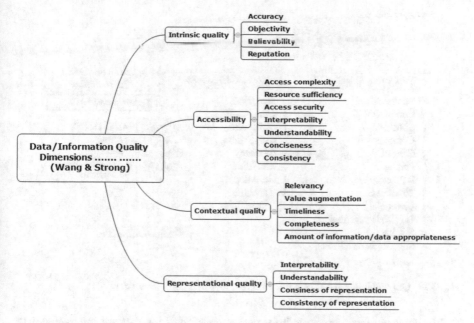

Fig. 3.9 The data-information quality framework of Wang and Strong [20]

In the civilian domain, one of the most cited frameworks for Quality of Information (QoI) is, from [20], reproduced in Fig. 3.9. QoI is defined in terms of certain characteristics of the data itself called data quality dimensions. The authors have classified QoI into four major categories illustrated in Fig. 3.9: intrinsic, contextual, representational, and accessibility. QoI criteria can be identified through contextual system analysis.

- The intrinsic level characterizes an information element in terms of its nature, interpretation, and modeling while considering it as "independent" entity, out of any global fusion context.
- Conversely, the contextual level characterizes an information element in terms of its impact, completeness, relevance, conflict, and redundancy within a global fusion context.
- The representational level addresses the formal representation of the problem (and its context). It encompasses both the inner element representation (e.g., discrete vs continuous output set) and the semantic alignment that allows an information element to be considered as either input or output for the fusion process (i.e., to be involved in a processing elements flow).
- The accessibility level characterizes information on that accessibility. Accessibility might be physical (e.g., with regard to a sensor's availability) as well as organizational (e.g., holding the right to access and process specific information elements).

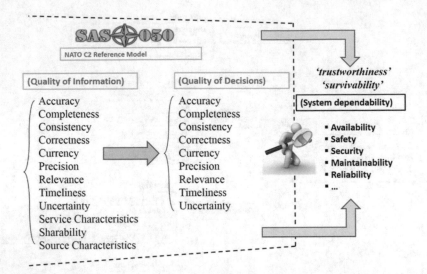

Fig. 3.10   QoI within the NATO C2 reference model (SAS-050). (Source: [32])

The intrinsic facet concerns information characteristics that are independent of the application context in which this information is exploited. On the other side, the contextual facet gathers characteristics allowing the understanding of the information usefulness and its informative contribution (its impact, completeness, relevance, conflict, redundancy) when integrated into a larger information processing context. Representational and accessibility emphasize the importance of the role of systems.

In military domain, a very good representative of QoI body of research is the efforts conducted under the NATO Research and Technology Office-Research Group SAS-050 to define a conceptual reference model of Command and Control (C2) [24–26]. The SAS-050 conceptual reference C2 model consists of a set of key variables and relationships to understanding C2. Among those sets of variables is a set of attributes related to information quality, pictured in Fig. 3.10, which is believed to impact directly the quality of decisions. It consists of twelve attributes or variables including information accuracy, information completeness, information consistency, information correctness, information currency, information precision, information relevance, information timeliness, information uncertainty, information service characteristics, information shareability, and information source characteristics.

The key to decisions is information. Within the NATO C2 reference model, Quality of Information is directly linked with the quality of decisions as shown in Fig. 3.10. Quality of information and decisions impacts directly "trustworthiness," "survivability," and "dependability" in systems, three concepts that are roughly equivalent in goals and addressing similar threats [27, 28]. Dependability [27] in systems and especially in cyber-physical systems is a huge and very challenging domain of application. The detailed description and system implementation workings of these concepts are beyond the scope of this book. There are numerous other books available that explain these concepts in greater details down to the mathematical description by means of equations, statistical formulae, and methods [29 31].

3.3.4 Books on Data and Information Quality: A Guided Tour

Several books have appeared in the last decades covering diverse aspects of that complicated issue of information quality. This section presents a quick survey of existing books on information quality. The list is not exhaustive, and we apologize in advance for having missed an important contribution.

In Piattini et al. book [33] on information quality, several authors address the issues of data quality to improve organizational processes. At that time, organizations were discovering how critical information is to the success of their businesses, and they did not have effective ways of managing the quality of this information. The book focusses on databases, data warehousing systems, and data presentation quality. There is very little distinction between data and information. The book identifies and discusses the most important data issues facing the typical organization at that time and presents the different existing proposals of a conceptual model of quality by looking at the strengths and weaknesses of each one.

One of the first books to address information quality as opposed to data quality is the one by Perry et al.'s [34] (2004) where a methodology is proposed for measuring the Quality of Information and its impact on shared awareness. The context is military. The book presents a methodology—including metrics, formulas for generating metrics, and transfer functions for generating dependencies between metrics—for measuring the Quality of Information and its influence on the degree of shared situational awareness. The results presented are from a working group of scientists who looked at key concepts and related metrics that are necessary to explore information quality and its value for awareness, shared awareness, collaboration, and synchronization, to force effectiveness and mission outcome.

The book of Batini and Scannapieco (2006) [35] still focuses on data quality (DQ) issues. The book provides a systematic and comparative description of the vast number of research issues related to quality of data, and thus to illustrate the state of the art in the area of data quality in 2006. They translated the issues and dimensions of data quality to information quality in a recent book published in 2016 [36]. Additional recent books focusing on data quality issues do exist: (1) Sebastian-Coleman (2013) [37] on measuring DQ, Samitsch [38] on DQ and its impact on decision-making (2014), and Fùber [39] on DQ management with semantic technologies.

The book of Eppler (2006) [40] examines the concept of information quality in the context of knowledge-intensive processes such as product development, market research, strategy development, business consulting, or online publishing. Examples of problems with QoI in many knowledge-intensive processes might be the lack of clarity in a project report, the confusing design of a website, or the lengthy and wordy style of a research book. Various frameworks exist that treat these problems mostly from a static database perspective. They sometimes lack the process dimension, and their view of information as isolated entities does not involve sense-making processes. Eppler proposes to build a versatile framework that takes the specificities of information in knowledge-intensive processes into account, specificities such as ambiguity, the crucial role of context, or the cognitive and behavioral dimension of information: information has to be interpreted, and it should be acted upon to be of value.

In the Al-Hakim [41] book (2007), the relationship between data and information is clarified as well as the distinction between data quality (DQ) and Quality of Information (QoI). The following definitions, originating from [42], have been adopted through the book for data and information:

- *Data:* Items about things, events, activities, and transactions are recorded, classified, and stored but are not organized to convey any specific meaning. Data items can be numeric, alphanumeric, figures, sounds, or images.
- *Information:* Data that has been organized in a manner that gives meaning for the recipient. They confirm something the recipient knows or may have "surprise" value by revealing something not known.

The Al-Hakim's book also states that QoI has no universal definition and that individuals have different ways of considering the Quality of Information as they have different wants and needs, hence, different quality standards. As stated in [41]: *Information users can view QoI from various perspectives,* as *"fitness for intended use," "conformance to specifications,"* or "meeting or exceeding customer expectations." While these perspectives capture the essence of QoI, they are very broad definitions and are difficult to use in the measurement of quality. There is a need to identify the dimensions that can be used to measure QoI. Those dimensions are summarized at the preface in Table 3.4 of [41]. The following topics listed in a bulk manner are being covered: issues associated with QoI processing including QoI metrics for entity resolution, query processing, and attributes of symbolic representation, the challenge of QoI assessment and improvement, and QoI Applications in Manufacturing and Management. Al-Hakim also offered another book in 2006 on managing QoI in service organizations [43].

The book of Baskarada (2009) [44] presents the development of an Information Quality Management Capability Maturity Model, a set of evaluation tools, which are intended to assist in the identification of problems in the collection/storage/use of information and other information management practices. This is intended to provide organizations with a measure of their capability maturity in information quality management, along with recommendations for increasing the level of maturity.

The second edition of Web Wisdom [45] shows how to adapt and apply the five core traditional evaluation criteria (authority, accuracy, objectivity, currency, coverage) to the modern-day Web environment. The book introduces a series of checklists comprised of basic questions to ask when evaluating or creating a particular type of Web page. It also provides important guidance for creators of

Table 3.4 The taxonomy of uncertainties and decisions from [77]

Uncertainty	Objective	*Epistemical* ...caused by gaps in knowledge that can be closed by research	*Decision: Knowledge guided* ...decision-makers must both rely on existing knowledge and reflect on any remaining uncertainties. One strategy in this regard is a comparative risk assessment of similar situations
		Ontological ...caused by the stochastic features of a situation, which will usually involve complex technical, biological, and/or social systems	*Decision: Quasi-rational* ...impossible to make rational decisions...situations are largely unpredictable; ... past experience and probabilistic reasoning provide some guidance on how complex systems will react
	Subjective	*Moral* ...caused by a lack of applicable moral rules, and we call these situations "moral uncertainties"	*Decision: Rule guided* Decision-makers have to fall back on more general moral rules and use them to deduce guidance for the special situation in question
		Rule ...caused by uncertainty in moral rules. This means that we act on the basis of fundamental pre-formed moral convictions in addition to experiential and internalized moral models	*Decision: Intuition guided* ... can make decisions only by relying on our intuition rather than knowledge, or explicit or implicit moral rules

Web-based resources who have information that they want to be recognized as reliable, accurate, and trustworthy. This book benefits to be complemented by the handbook on research web systems quality [46].

Talburt (2011) [47] covers the principles of entity resolution and information quality, both principles that shall be linked to business value. QoI is viewed under the perspective that information is a non-fungible asset of the organization and that its quality is directly related to the value produced by its application. The book also discusses the information product model of QoI and the application of Total Quality Management (TQM) principles to QoI management as well as the close relationship between Entity Resolution (ER) and QoI. Theoretical models of ER and frameworks for the entity-based data integration (EBDI) are proposed for formally describing integration contexts and operators independently of their actual implementation.

In the book (2014) edited by Floridi and Illari [48], various authors present contributions based on this cited motivation from the book itself: "Consideration of information and data quality are made more complex by the general agreement that there are a number of different aspects to information/data quality but no clear agreement as to what these are. Lacking a clear and precise understanding of QoI standards (such as accessibility, accuracy, availability, completeness, currency, integrity, redundancy, reliability, timeliness, trustworthiness, usability, and so forth) causes costly errors, confusion, and impasse and missed opportunities. Part of the difficulty lies in putting together the right conceptual and technical framework necessary to analyze and evaluate QoI."

Kenett and Shmueli (2016) [49] address QoI with a holistic approach from data analysis (using statistical modeling, data mining approaches, or any other data analysis methods) by structuring the main ingredients of what turns numbers into information. Their main thesis is that data analysis and especially the fields of statistics and data science need to adapt to modern challenges and technologies by developing structured methods that provide a broad life cycle view, that is, from numbers to insights. This life cycle view needs to be focused on generating QoI as a key objective. Their book on the potential of data and analytics to generate knowledge offers an extensive treatment of a QoI framework to guide workers in their analytic or statistical work in generating information of high quality.

The book of Batini and Scannapieco (2016) [36] adds significantly new contents to their previous one published in 2006 [35] showing the evolution of research and application issues that led the authors to move from the data quality (DQ) concept to the information quality (QoI) one. The book addresses the questions: What is, in essence, data quality? Which kind of techniques, methodologies, and data quality issues are at a consolidated stage? What are the well-known and reliable approaches? Which problems are open?

The book states that the fundamental technology used in Information Systems (IS) for the last 40 years has been databases. Information systems have been evolving to a network-based structure, where the potential data sources have dramatically increased in size and scope. In networked IS, processes are involved in complex information exchanges and often operate on input obtained from other external sources, frequently unknown a priori. Therefore, the overall data that flow between information systems may rapidly degrade over time if both processes and their inputs are not themselves under strict quality control. On the other hand, the same networked information systems offer new opportunities for data quality management. Problems of information management are much more complex in networked IS since heterogeneities and autonomies usually increase with the number of tiers and nodes. More recently, IS have evolved toward Web IS, adopting a variety of Web technologies, where single data producers are highly autonomous and heterogeneous and have no obligation for the quality of the information produced, and no producer has a global view of the system. According to C. Batini and M. Scannapieco [36], these systems are extremely critical from the point of view of information quality.

This book [36] is highly recommended to those who are interested in research issues on information quality. We cite here the conclusion of Chap. 1 of [36] that is presented as follows: "While information quality is a relatively new research area, other areas, such as statistical data analysis, have addressed in the past some aspects of the problems related to information quality; with statistical data analysis, also knowledge representation, data and information mining, management information systems, and data integration share some of the problems and issues characteristic of information quality and, at the same time, provide paradigms and techniques that can be effectively used in information quality measurement and improvement activities."

Finishing this quick tour is a quite recent book on information quality (Bossé & Rogova) [50] which addresses the problem of information quality in the fusion-based human-machine systems for decision-making. The following excerpt from the preface of [50] explains the problem of information quality for information fusion and decision-making:

> Information fusion is dealing with gathering, processing, and combining a large amount of diverse information from physical sensors (infrared imagers, radars, chemical, etc.), human intelligence reports, and information obtained from open sources (traditional such as newspapers, radio, TV as well as social media such as Twitter, Facebook, Instagram).
>
> That data and information obtained from observations and reports as well as information produced by both human and automatic processes are of variable quality and may be unreliable, of low fidelity, insufficient resolution, contradictory, and/or redundant. Furthermore, there is often no guarantee that evidence obtained from the sources is based on direct, independent observations. Sources may provide unverified reports obtained from other sources (e.g., replicating information in social networks), resulting in correlations and bias. Some sources may have malicious intent and propagate false information through social networks or even coordinate to provide the same false information in order to reinforce their opinion in the system.

The objective of Bossé and Rogova's book [50] is to provide an understanding of the specific problem of information quality in the fusion-based processing and address the challenges of representing and incorporating information quality into the whole processing chain from data to information to actionable knowledge for decisions and actions to support decision-makers in complex dynamic situations.

3.3.5   Toward an Information Quality Ontology

There have been multiple views on information quality ontologies, identifying quality attributes and classifying them into broad categories and relations. In [20] (Wang & Strong), data quality was classified into four categories: intrinsic, contextual, representational, and accessibility. In [51], three categories were enumerated: pragmatic, semantic, and syntactic (semiotics dimension), while in [22], four sets were identified: integrity, accessibility, interpretability, and relevance. In [52], they used relevance, reliability, completeness, and uncertainty. In other words, there is no clear understanding of what dimensions define information quality and how those dimensions are interrelated (*other aspects* in Fig. 3.11). The information quality ontology presented in [17, 53] was one of the first attempt to fill this gap.

More recently, significant efforts [54–58] have been pursued to address the evaluation of techniques of uncertainty reasoning and to define an ontology called URREF (Uncertainty Representation and Reasoning Evaluation Framework) under the International Society of Information Fusion (ISIF) Evaluation of Technologies for Uncertainty Representation Working Group (ETURWG). These efforts contribute toward the understanding of QoI and the formalization of some criteria.

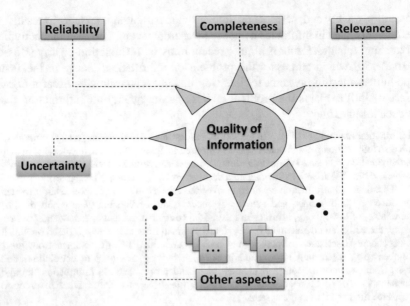

Fig. 3.11 Four aspects of Quality of Information that need formalization

However, research is still ongoing and requires additional efforts to get a QoI ontology. In fact, from all the efforts just mentioned, there are mainly four main aspects of information quality that require formalizations exploitable in computer-based support systems, namely, on uncertainty, reliability, completeness, and relevance as summarized in Fig. 3.11. Research is still needed to make Fig. 3.11 more complete.

3.4 Information, Uncertainty, and Entropy

The presentation of this section mainly follows Chap. 3 of Bossé and Solaiman [32]. First, it appears that the three words information, uncertainty, and entropy have distinct meanings for many people in different domains [59]. For instance, look at the confusion of the ordinary use of the word "information" with the word "information" in "information theory." Information is most of the time associated with the idea of knowledge through its popular use (see previous chapters). Information in information theory, also called "Shannon's information," is narrowly defined and is rather associated with uncertainty and the resolution of uncertainty. Sometimes, authors confuse the two words. The same is true for the word "entropy." Authors will often use the word entropy without saying whether they are talking about thermodynamic entropy or "Shannon's entropy" and vice versa. The definition of the word uncertainty presents the same kind of discrepancies. In the economic literature [60], for instance, there exist different conceptions of uncertainty involving different types and different degrees of uncertainty as well as similar conceptions under different labels. Dequech [60] even suggests not to use the word "uncertainty" alone, but accompanied by qualifiers to specify its type and then reduce misunderstandings.

3.4.1   Information Theory and Entropy

Shannon made use of two different (but related) measures of information: entropy and mutual information. Entropy $H(X)$ is the amount of information of a discrete time discrete alphabet random process $\{X_n\}$. Detailed definition of $H(X)$ can be found in several reference books of information theory such as in [61, 62]. Shannon proved a coding theorem showing that if one wishes to code $\{X_n\}$ into a sequence of binary symbols, a receiver viewing the binary sequence can reconstruct the original process nearly perfectly. Mutual information is a measure of the information contained in one process $\{X_n\}$ about another process $\{Y_n\}$ to make decisions on it: one random process $\{X_n\}$ representing an information source and another $\{Y_n\}$ representing the output of a communication medium wherein the coded source has been corrupted by another random process called noise. Shannon introduced the notion of the average mutual information between the two processes:

$$I(X, Y) = H(X) + H(Y) - H(X, Y)$$

where $I(X, Y)$ is the sum of the two self-entropies minus the pair entropy. Average mutual information can also be defined in terms of *conditional entropies* (see Chap. 3 of [61] for details).

$$I(X, Y) = H(X) - H(X|Y) = H(Y) - H(X|Y)$$

We can read this equation in an alternative way as:

$$H(X|Y) = H(X) - I(X, Y); \text{and } H(X|Y) = H(Y) - I(X, Y).$$

For a random variable X with a probability distribution $p(.)$ defined over a finite set χ, Shannon's entropy is defined as:

$$H(X) = -\sum_{x \in \chi} p(x) \log_2 p(x) \geq 0$$

It quantifies the *unevenness* of the probability distribution $p(x)$. In particular, the minimum $H(X) = 0$ is reached for a constant random variable (i.e., equiprobable random variable). Entropy is also denoted as [63]:

$$S(p) = -\sum_{i=1}^{|\chi|} p(x_i) \log_2 p(x_i)$$

which underlines the fact that entropy is a feature of the probability distribution p. The source generating the inputs $x \in \chi$ is characterized by the probability distribution $p(x)$. Shannon's entropy $S(p)$ appears as the average missing information, that is, the average information required to specify the outcome x when the receiver knows the distribution p. It equivalently measures the amount of uncertainty represented by a probability distribution. In the context of communication theory, it amounts to the minimal number of bits that should be transmitted to specify x. The following excerpt from J. R. Pierce [62] shows the closed link between the amount of information conveyed by a message and the amount of uncertainty: "A message which is one out of ten possible messages conveys a smaller amount of information than a message which is one out of a million possible messages. The more we know about what message the source will produce, the less uncertainty, the less the entropy, and the less the information."

The conditional entropy, $H(X|Y)$, which appears to be the average (over Y) of the entropies of the conditional probability distributions $p(X|Y = y)$:

$$H(X|Y) \equiv H(X, Y) - H(Y) = \sum_{y \in Y} p(y) \left[-\sum_{x \in \chi} p(x|y) \log_2 p(x|y) \right]$$

when the random variables X and Y have the same state space χ, with respective distributions p_X and p_Y, it is possible to consider the relative entropy:

$$S_{rel}(p_X|p_Y) = -\sum_x p_X(x) \log_2 [p_X(x)/p_Y(x)]$$

The opposite of the relative entropy defines the Kullback-Leibler divergence [64] for two probability distributions p and q on the same space χ,:

$$D(p \parallel q) = -S_{rel}(p|q) = \sum_x p(x) \log_2 [p(x)/q(x)] \geq 0$$

Kullback-Leibler divergence is an important quantity that is exploited in defining measures of information or uncertainty. Most of the measures of uncertainty so far are being developed based upon Shannon's entropy [65–68]. However, the notion of entropy as already mentioned may not be sufficient to cover all aspects of uncertainty aggravated when more than one meaning of information creeps into a discussion. The ideas which gave rise to the entropy of physics and the entropy of Shannon are quite different although they can be described in terms of uncertainty in similar mathematical terms. This is still a confusing matter to see if one can established significant and useful relations between both entropies [69, 70].

3.4.2 Uncertainty and Risk

Uncertainty is a term used in subtly different ways in a number of fields, including philosophy, physics, statistics, economics, finance, insurance, psychology, sociology, engineering, and information science. Several uncertainty taxonomies [71–75] exist for each domain, and it often happens that uncertainty is confused with the notion of risk.

In the schematic approach (Fig. 3.12) labelled "igloo of uncertainty," the authors distinguish between open and closed forms of both ignorance and knowledge and delimitate a field of uncertainty that concerns both risk and danger. Dangers are defined in terms of the possible outcomes of a given situation. A decisive difference between both is a danger is present regardless of choice and can therefore be avoided or counteracted, whereas a risk is either optionally accepted or imposed. We refer here the reader to [76, 77] for examples and a more detailed discussion on concepts presented in Fig. 3.12 such as open and closed knowledge and ignorance, risks, and dangers. The authors in [76, 77] present a discussion on the distinction between closed and open knowledge with respect to risk and closed and open ignorance with respect to danger as well as prerequisite for turning dangers to risks.

When a certain situation involves dangers and risks, it is important to reduce gaps in knowledge either by research or by learning. A strategy to address gaps in knowledge first requires insight into the specific type of uncertainty. Table 3.4 presents a "taxonomy of uncertainty," a description of its causes, and its related types of decision. Two layers are presented: two fundamental forms (objective,

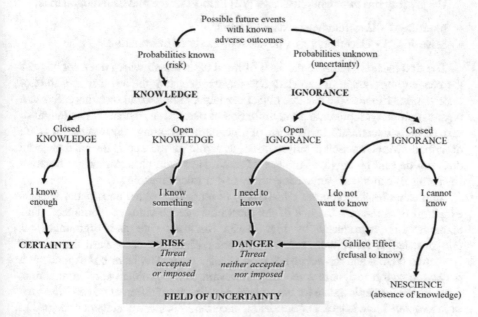

Fig. 3.12 "Uncertainty igloo". (Adapted from [76, 77] (Source: [32]))

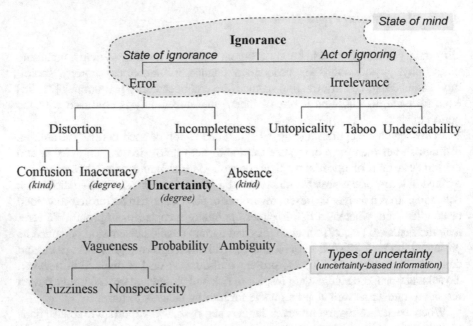

Fig. 3.13 Smithson's taxonomy of ignorance. (Source: [32])

subjective), both of which are divided into two further subforms (epistemical, ontological, moral, rule).

Uncertainty has two main meanings [72] in most of the classical dictionaries:

- **Meaning I -** Uncertainty as a state of mind.
- **Meaning II -** Uncertainty as a physical property of information.

The first meaning refers to the state of mind of an agent, which does not possess the needed information or knowledge to make a decision; the agent is in a state of uncertainty: "I'm not sure that this object is a table." The second meaning refers to a physical property, representing the limitation of perception systems: "The length of this table is uncertain." In theories of uncertain reasoning, uncertainty is often described as imperfection of information, as errors on measures, for example, and does not depend on any kind of state of mind. However, an uncertain information (meaning II) can induce some uncertainty in our mind (meaning I).

Smithson [78, 79] proposes a taxonomy of ignorance where uncertainty appears as a kind of ignorance "... one of the most manageable kinds of ignorance." This taxonomy is reproduced in Fig. 3.13. Smithson interprets ignorance as non-knowledge. He initially separates ignorance into two categories: the state of ignorance (*error*) and the act of ignoring (*irrelevance*). The latter corresponds to a deliberate action to ignore something irrelevant to the problem-solving situation, whereas the former is a state (of ignorance) resulting from different causes (*distorted* or *incomplete* knowledge). For Smithson, uncertainty is incompleteness in degree as compared to *absence* which is incompleteness in kind and is subdivided into three types: *probability*, *vagueness* (being either *nonspecificity* or *fuzziness*), and *ambiguity*.

Uncertainty, as a state of mind (meaning I), corresponds to ignorance in Smithson's hierarchy, while Smithson's concept of uncertainty corresponds to uncertainty-based information (meaning II). Smithson's taxonomy could be an interesting guide for development of computer-based systems that support situation awareness since it accounts for different levels of processing: from physical property of information (*uncertainty*) to mental state (*ignorance*).

3.4.3 Dequech's Typology of Uncertainty in Economics

Dequech [60] proposes a typology of the main concepts of uncertainty used by economists. Dequech insists upon three main distinctions presented in Table 3.5. The first distinction is between substantive and procedural uncertainty. The second distinction is that between weak and strong uncertainty. The third distinction is that between ambiguity and fundamental uncertainty. The definition of ambiguity (from [80]): "ambiguity is uncertainty about probability, created by missing information that is relevant and could be known."

The basic concept of weak uncertainty is that individuals can build unique, additive (sum to 100%), and fully reliable probability distributions. This category can be subdivided into two: Knightian's risk and Savage's uncertainty.

Table 3.5 Dequech's typology of uncertainty in economics [60]

Type of Uncertainty	Weak Uncertainty: Unique, additive, and fully reliable probability distribution.	Strong Uncertainty: Absence of such a distribution
Substantive Uncertainty: Lack of some relevant and good-quality information	**Weak Uncertainty:** uncertainty about which state will obtain.	**Ambiguity:** Uncertainty about probability, caused by missing information that could be known; predetermined list of states
		Fundamental Uncertainty: Possibility of non-predetermined structural change; non-predetermined list of states
Procedural Uncertainty: Complexity in relation to limited computational and cognitive capabilities of the agents.		**Procedural uncertainty**

Table 3.6 Savage's (acts-states-consequences) framework

Acts	States			
	s_1	s_2	...	s_m
a_1	c_{11}	c_{12}	...	c_{1m}
a_2	c_{21}	c_{22}	...	c_{1m}
⋮	⋮	⋮	⋮	⋮
a_n	c_{n1}	c_{n2}	...	c_{nm}

- In *Knightian's risk*, individuals can act on the basis of a probability that is objective (in the sense that any reasonable person would agree on it) and known. This probability may be of one of the following two kinds: a priori or statistical probability. A priori probabilities can be attributed objectively by logical reasoning, without the need to perform any experiments or trials. Statistical probabilities are relative frequencies.
- In *Savage's uncertainty* [81, 82], the notion of subjective probabilities, originally developed by Ramsey and de Finetti [83], is used. Subjective theory of probability treats probability as a degree of belief and makes possible to assign precise numerical probabilities to virtually any proposition or event. Betting rates are the mechanism through which subjective probabilities can be inferred.

Dequech [60] states that Knightian's risk can be seen as a special case of Savage's uncertainty since the latter can handle subjective probabilities, with or without objective probabilities.

Decision-making under weak uncertainty, as viewed by Savage, chooses among acts as in Table 3.6. Each act, a_i, has a consequence, c_{ij}, depending on which state, s_j, occurs. A state of the world is "a description of the world, leaving no relevant aspect undescribed." The set of states is exhaustive and defined independently of the set of acts. States are mutually exclusive, and each state has a probability, p_j. The expected utility of each act, a_i, is the weighted average of the utilities of the several consequences, where the weights are the probabilities of the respective states:

$$u(a_i) = \sum p_j u(c_{ij})$$

In the case of "strong" uncertainty associated with the notion of **procedural uncertainty**, the decision problem is complex and populated by individual or collective agents with limited mental and computational capabilities. In [60], based on literature [84] back in 1997, two problems have been identified on procedural uncertainty: (1) large amounts of information, referred to as *extensiveness*; (2) the other called *intricacy* refers "to the density of structural linkages and interactions between the parts of an interdependent system." Using flavor of the day language, one can easily relate *extensiveness* and *intricacy* to "data deluge" and complex networking, i.e., to the Vs dimensions of Big Data: volume, velocity, variety, veracity, and value.

In a situation involving **ambiguity**, the decision-maker cannot unambiguously assign a definite probability to each and every event because some relevant information that could be known is missing. Dubois [85] explains that Savage's rational decision-making chooses according to expected utility with respect to a subjective probability but that in the presence of incomplete information: (1) decision-makers do not always choose according to a single subjective probability; (2) there are limitations with Bayesian probability for the representation of belief; and (3) a single subjective probability distribution cannot distinguish between uncertainty due to variability and uncertainty due to lack of knowledge. Dubois [85] pursues with eliciting motivations for going beyond pure probability and set representations. He suggests to find uncertainty representations that combine probability and sets such as imprecise probability theory [86, 87] (sets of probabilities); Dempster-Shafer theory [88–90] (random sets); and numerical possibility theory [91, 92] (fuzzy sets). Finally, under the above theories, an event can be represented with a degree of belief (certainty) and a degree of plausibility, instead of a single degree of probability.

Fundamental uncertainty in Table 3.5 is the lack of knowledge that results from the characterization of social reality as subject to non-predetermined structural change [60]: "Future knowledge is not knowable in advance, in the sense that we do not know exactly what we are going to learn over the next years and when we are going to learn it." Innovations, which cause non-predetermined structural change in economic relationships, are very illustrative of fundamental uncertainty brought by human creativity and change in knowledge in the economic reality. To illustrate further, two types of economic reality are being mentioned in [93]: "immutable" and "transmutable." An "immutable" reality is one in which "the future path of the economy and the future conditional consequences of all possible choices are predetermined" also referred to as *ontological uncertainty*. A "transmutable reality" is described as: "the future can be permanently changed in nature and substance by the actions of individuals, groups ... and/or governments, often in ways not completely foreseeable by the creators of change." There is *epistemological uncertainty* when "some limitation on human ability ... prevents agents from using (collecting and analyzing) historical time-series data to obtain short-run reliable knowledge regarding all economic variables."

3.4.4 Dubois and Prade's "Typology of Defects"

A very pertinent contribution on *defining information* in the context of knowledge and information processing is from Dubois and Prade with their *Information: a typology of defects* [94]: "The term information refers to any collection of symbols or signs produced either through the observation of natural or artificial phenomena or by cognitive human activity with a view to help an agent understand the world or the current situation, making decisions, or communicating with other human or artificial agents." This definition suits very well with *actionable knowledge* to support decisions and actions (agents). Table 3.7 has been built from [94] where distinctions between three pairs of information qualifiers are being presented: objective, subjective; quantitative, qualitative; and generic, singular.

These distinctions presented in Fig. 3.14 and Table 3.7 are important for engineering a computer-based support solution to decision-making. The designer can see a possibility to reduce uncertainty by gathering more data, refining models, and improving processing to support the situation awareness processes represented in Fig. 3.14 by the Endsley's model [95]: perception, comprehension, and projection. As described in much more detail in [94], an agent is supposed to be at a certain situation awareness state (epistemic) by having some information about the current world. That is made of three components, i.e., from agent, generic knowledge, singular observations, and beliefs. Beliefs are derived from singular, i.e., related to the current situation and generic knowledge kinds of information. Decision-making requires to add another kind of information: agent preferences (Chaps. 2, 7, and 16 of [96]).

Table 3.7 Distinctions of various types of information according to Dubois et al. [94]

	Informal definitions	Representations/examples
Objective	Information stemming from sensor measurement and the direct perception of events	Quantitative-qualitative e.g., balance sheet data, radar data
Subjective	Information typically uttered by individuals or conceived without resorting to direct observations	Quantitative-qualitative e.g., testimonies, business plan
Quantitative	Information modelled in terms of numbers, typically objective information	Numbers, intervals, functions e.g., sensor measurements, counting processes
Qualitative or symbolic	Subjective information	Logical or graphical e.g., expressed in natural language
Generic	Information refers to a collection or a population of situations or a piece of common sense knowledge	e.g., a physical law, a statistical model built from a representative sample of observations
Singular	Information refers to a particular situation, a response to a question on the current state of affairs	An observation (a patient has fever at a given time point) or a testimony (the crazy driver's car was blue)

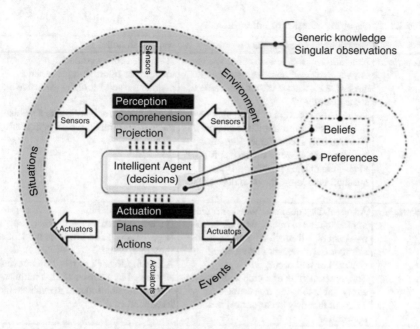

Fig. 3.14 Kinds of information and epistemic state of an intelligent agent. (Source: [32])

In order to represent the epistemic state of an agent, a representation of the states of the world is needed. Let v be the vector of attributes relevant for the agent and Ω the domain of v. Ω is called a *frame*: descriptions of all states of the world. A subset A of Ω, viewed as a disjunction of possible worlds, is called an *event*, to be seen as a proposition that asserts $v \in A$. Dubois and Prade present four kinds of imperfection of information expressible on the frame Ω: incomplete, uncertain, gradual, and granular. Table 3.8 summarizes these four kinds of imperfection without any mathematical details but with simple examples to illustrate their meaning.

The best known and oldest formal framework to reason under uncertainty is the probabilistic one, where uncertainty due to variability is modeled by classical probability distributions. However, a single probability distribution cannot adequately account for incomplete or imprecise information, so alternative theories and frameworks have been proposed. The three main such frameworks are, in decreasing order of generality, imprecise probability theory [86], random disjunctive sets [88, 89, 97], and possibility theory [91, 92]. In addition to [85, 94], a unified overview of various approaches to representations of uncertainty is presented in two parts by Destercke et al. [98, 99].

Table 3.8 Kinds of imperfection of information

	Description and examples	Representations
Incomplete	Not sufficient to allow an agent to answer a relevant question in a specific context, e.g., what is the current value of some quantity v? Imprecision is a form of incompleteness, in the sense that an imprecise response provides only incomplete information, e.g., the term *minor* is imprecise; it provides incomplete information if the question is to know the birth date of the person	A set used for representing a piece of incomplete information is called a disjunctive set: Mutually exclusive elements A conjunctive set represents a precise piece of information: a collection of elements
Uncertain	An agent does not know whether a piece of information is true or false. A primitive item of information being a proposition or the statement that an event occurred or will occur The representation of a proposition as an entity liable to being true or false (or of an event that may occur or not) is a convention	Uncertainty qualifi Ers are, respectively, a number (a probability) and symbolic modalities (possible, certain) Assigning to each proposition or event A, viewed as a subset of Ω, a number $g(A)$ in the unit interval to evaluate the likelihood of A
Gradual	Underlies a ranking, in terms of relevance, of attribute values to which it refers. The propositions are not always Boolean and the transition to the convention of being true or false (or of an event that may occur or not) is gradual The proposition *Pierre is young* could be neither totally true, nor totally false: The meaning of *young* will be altered by linguistic hedges expressing intensity	Fuzzy sets [100] are used when dealing with a piece of information expressed in natural language and referring to a numerical attribute
Granular	Changes in the level of resolution or refinement to which the set Ω of states of affairs can represent the world; that has an impact on the possibility or not to represent relevant information It may happen that Ω does not contain all propositions to completely describe the problem under concern, and if new propositions are added to modify the set Ω, this is called a *granularity change* for the representation	The probabilistic representation of incomplete information by means of the insufficient reason principle does not resist to a change of granularity, but possibilistic representation does (Sect. 3.2.4 of [94])

3.4.5   Typologies of Uncertainty

In 1993, Krause and Clark [101] proposed an alternative typology to Smithson's, centered on the concept of uncertainty. Krause and Clark distinguish two aspects: *unary* (i.e., uncertainty applied to individual propositions) and *set theoretic* (i.e., uncertainty applied to sets of propositions). Both categories lead to either *conflict*

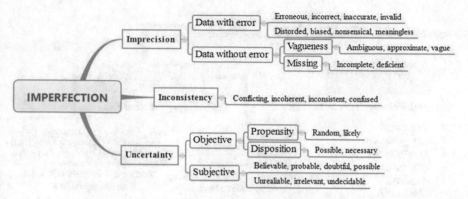

Fig. 3.15 Adapted from Smets' structured thesaurus on imperfection [72]

(conflicting knowledge) or *ignorance* (lack of knowledge). As subcategories, we find *vagueness*, *confidence*, *propensity*, *equivocation*, *ambiguity*, *anomaly*, *inconsistency*, *incompleteness*, and *irrelevance*. Compared to Smithson's taxonomy, Krause and Clark added the unary/set theoretic dichotomy, in order to introduce the concept of *inconsistency* and to remove the concept of *incompleteness* from the unary branch to the set theoretic one. Krause and Clark's classification concerns meaning II of uncertainty from a formal point of view since the distinction is based on propositions. This approach is a straightforward way to process propositional belief, a concept central to situation analysis.

Bouchon-Meunier and Nguyen [102] propose a model for uncertainty. They refer to uncertainty as "imperfection on knowledge" and denote then three main types of imperfection: (1) probabilistic uncertainty, (2) incompleteness in knowledge (belief, general laws, imprecision), and (3) vague and imprecise description. The schema of Fig. 3.15 is a good way to make the distinction between the two main general meanings of uncertainty. Reading the graph from right to left, uncertainty appears as a final state of mind (meaning I) possibly caused by belief, general laws, imprecision, vagueness, or incompleteness.

Instead of a typology of uncertainty, Smets built a typology of *imperfection of information* [90] avoiding the confusion between the two meanings of uncertainty. The model proposed by Smets distinguishes three main categories of imperfect information (Fig. 3.15):

- **Imprecision**: related to the content of the statement—several worlds satisfy the statement.
- **Inconsistency**: no world satisfies the statement.
- **Uncertainty**: induced by a lack of information, by some imprecision.

Smets considers imperfection as a central term, uncertainty being a kind of imperfection. Uncertainty can be either *objective* (property of the information, i.e., meaning II) or *subjective* (property of the observer, i.e., meaning I).

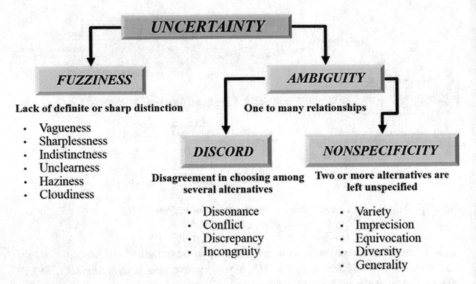

Fig. 3.16 Klir and Yuan's typology. (Created from [103, 104])

The typology proposed by Klir and Yuan [103, 104] is built upon the different existing mathematical theories of uncertainty and directly linked to measures of uncertainty. Klir conceives information in terms of uncertainty reduction and consequently introduces the term of **uncertainty-based information**. For Klir et al. [103, 104], uncertainty can be either *fuzziness* or *ambiguity* (two types of uncertainty). Ambiguity can itself be either *nonspecificity* or *discord*. These concepts can be related to some previously used in the other classifications: fuzziness is close to vagueness, discord is a synonym of conflict, and nonspecificity means principally imprecision or generality. In Fig. 3.16 typology, they integrate the main key terms used by Smithson (fuzziness, nonspecificity, ambiguity) as well as the set theoretical aspect introduced by Krause and Clark (discord). Klir et al. [103, 104] do not mention knowledge and thus stay at a lower level of processing, i.e., at the information level.

Klir et al.'s conception of uncertainty is closely related to quantitative theories of uncertainty as depicted on Fig. 3.17 such as probability, possibility, fuzzy sets, and evidence theories and leads to corresponding measures of uncertainty, i.e., of uncertainty-based information [104]. The typology presented in Fig. 3.17 is an extended version of Klir and Yuan [103]. It highlights four main types of uncertainty pertaining in the different theories: *fuzziness*, *nonspecificity*, *dissonance*, and *confusion*. The last two terms correspond to two aspects of conflict. The distinction imprecision-inconsistency of the model of Smets appears also: imprecision gathers fuzziness and nonspecificity, and conflict corresponds to inconsistency.

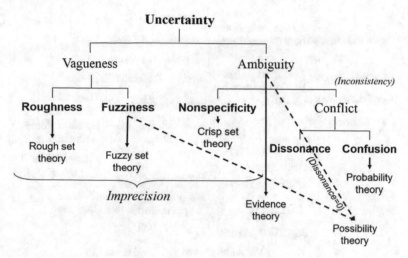

Fig. 3.17 Uncertainty theories related to types of uncertainty. (Source: [32])

3.4.6   *Representations of Uncertainty*

Material in this section has been arranged from mainly two references: [105] and [71]. As discussed above, *uncertainty* and its other face *information* present quite a challenge if we look at the three semiotics dimensions: syntactics, semantics, and pragmatics. The representation of uncertainty is an important issue in many areas of science and engineering, especially in the systems design and control. To model uncertainty, many mathematical tools have been developed, being either qualitative (modal logic, non-monotonic reasoning, etc.) or quantitative approaches (probability theory, fuzzy sets, rough sets, random sets, belief functions, etc.). Figure 3.18 presents a list of theories regrouped into three main categories, quantitative, qualitative, and hybrid/graphical approaches, to process data, information, and knowledge. All these techniques are making use of relations in some sort, and some of them are applications of relational calculus. Each of the approaches in Fig. 3.18 is often compared in terms of their different strengths and weaknesses, on their correspondence to a particular type of uncertainty, on their requirement of prior knowledge, on their computational time, and on the need of independence on data (e.g., Chaps. 7, 8, and 9 of [105]). There is a plethora of books on the methods and techniques listed in Fig. 3.18, but this is beyond the scope of the current book to explain how authors have been using the concepts of *relation* to develop their approach.

For instance, some of these approaches have been discussed and compared. The unified and very instructive overview on uncertainty presented in two parts by Destercke et al. [98, 99] as well as contributions from references such as [96, 104, 106, 107] can certainly help in facing the semiotics challenge. Especially the treatise on decision-support by Bouyssou et al. [96] where a multidisciplinary view (psychologists, economists, sociologists, mathematicians, computer scientists, and decision sciences) is presented. This is fundamental and instrumental for the

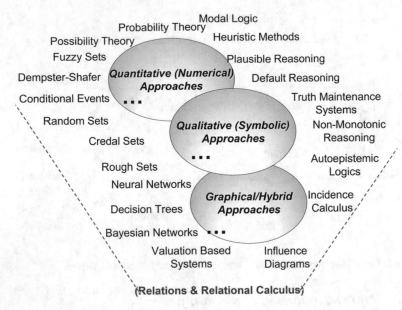

Fig. 3.18 - Mathematical and logical tools to perform "processing" in a knowledge chain. (Adapted from [32])

development of computer-based supporting tools to create and generate actionable knowledge.

3.4.7 General Uncertainty Principles in Data and Information Transformations

Fundamental principles must guide the processing in a knowledge chain. It is inevitable that information is lost or gained in the process. The three uncertainty principles presented by Klir et al. in [104] and reproduced on Fig. 3.19 become essential: maximum uncertainty, minimum uncertainty, and uncertainty invariance.

The *principle of minimum uncertainty* is basically an arbitration principle that is applied mainly for simplification problems. When a system is simplified, it is usually unavoidable to lose some information contained in the system. The amount of information that is lost in this process results in the increase of an equal amount of relevant uncertainty. A sound simplification of a given system should minimize the loss of relevant information while achieving the required reduction of complexity. When properly applied, the principle of minimum uncertainty guarantees that no information is wasted in the process of simplification. There are many simplification strategies, which can perhaps be classified into three main classes:

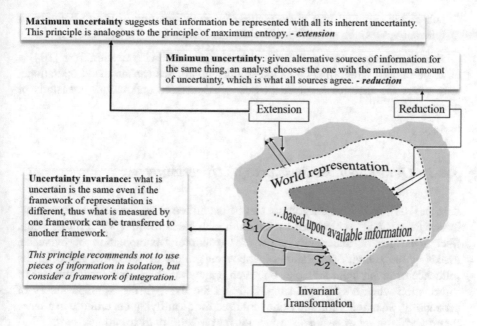

Fig. 3.19 Klir's three uncertainty principles. (Source: [32])

- Simplifications made by eliminating some entities from the system (variables, subsystems, etc.)
- Simplifications made by aggregating some entities of the system (variables, states, etc.)
- Simplifications made by breaking overall systems into appropriate subsystems ([104], p.108).

The *principle of maximum uncertainty* is essential for inductive reasoning and requires that in any inductive inference to use all information available but make sure that no additional information is unwittingly added. That is, that conclusions resulting from any inductive inference maximize the relevant uncertainty within the constraints representing the premises. This principle guarantees that our ignorance be fully recognized and that our conclusions are maximally noncommittal with regard to information not contained in the premises ([104], p.110).

To transform the representation of a problem-solving situation in one theory, $\mathfrak{T}_1$, into an equivalent representation in another theory, $\mathfrak{T}_2$, the principle of *uncertainty invariance* requires that:

- The amount of uncertainty associated with the situation be preserved when we move from $\mathfrak{T}_1$ into $\mathfrak{T}_2$ to guarantee and that no uncertainty is added or eliminated solely by changing the mathematical theory by which a particular phenomenon is formalized.
- The degrees of belief in $\mathfrak{T}_1$ be converted to their counterparts in $\mathfrak{T}_2$ by an appropriate scale, at least ordinal to guarantee that certain properties, which are

considered essential in a given context (such as ordering or proportionality of relevant values), be preserved under the transformation ([104], p.120).

Finally, as an introduction to the next section, the following citation from [104] is quite pertinent: "The principle of uncertainty invariance can be made operational only when the theories involved possess well-justified and unique measures of uncertainty."

3.4.8 Quantification: Measures of Uncertainty

Klir, in [103], poses that uncertainty can be quantified in an equivalent manner that information can be as proposed by Shannon with his known theory of information. Such measures of uncertainty allow, for example, "extrapolation of evidence, assessing the strength of relationship between groups of variables, assessing the influence of given input variables on given output variables, measuring the loss of information when a system is simplified and the like." In the classical theory of probability, Shannon's entropy is the tool used for quantifying uncertainty in a sense that it verifies a set of desirable properties for probability distributions.

In the presence of incomplete and vague information, the probabilistic representation is inadequate so that an imprecise probability theory can be used. The following list of imprecise theories is mainly from Abellán [108]: imprecise probabilities [86], Dempster-Shafer's theory (DST) [88, 89], interval-valued probabilities [109], order-two capacities [110], upper-lower probabilities [111–114], or general convex sets of probability distributions also called credal sets [111, 115–119].

Events in probability theory, as well as focal elements in evidence theory, are crisp subsets of the universe of discourse or the frame of discernment. Evidence theory can thus be generalized to fuzzy evidence theory. Zadeh first proposed such an extension [120], and it is worth mentioning other contributions, for example, Smets [121], Yager [122], Yen [123], and Lucas and Araabi [124]. In the area of generalized information theory (GIT), gathering among other all the theories mentioned above, three main types of uncertainty are usually mentioned: fuzziness, nonspecificity, and discord (or randomness). In an equivalent way that information is quantified by Shannon's entropy in the probability framework, or by the Hartley measure [125] in classical set theory, information or preferably uncertainty-based information [104] can be quantified by different measures commonly called measures of uncertainty. In order to quantify the uncertainty representations, Shannon's entropy has been used as the starting point, and the most common approach to justification is the axiomatic approach, i.e., by assuming a set of necessary basic properties that a measure must verify [104].

Although there are significant contributions [65, 66, 108, 115, 116, 126] on measures of uncertainty recently, great challenges are still unresolved:

- More investigations are required to attach meaning to those measures. The *pragmatics* has not been established yet. What those measures are really measuring?
- Meeting a set of mathematical properties may not mean a meaningful translation to real-world systems. Empirical studies need to be conducted on numerous scenarios to attach a sense to those properties.

3.5 Conclusion

Knowledge is the culmination of a chain of different types of entities whose essence is determined by the perception of signs and signals in the perceived world participating in action control according to the level of signified conveyed (i.e., semantic status). The knowledge chain represents the necessary bridging that must occur at the informational level, between the observable and the cognitive world.

This chapter emphasized the information-processing aspect of a knowledge chain. We discuss the relationship between data, information, and knowledge with their associated imperfections. Information processing plays a key role. The notion of "quality" is of prime importance to any information processing, and particularly for a knowledge processing chain, since the objective is to improve, through processing, data, information, and knowledge quality. Various aspects of Quality of Information (QoI) have then been discussed with emphasis on uncertainty-based information.

The three words information, uncertainty, and entropy have distinct meanings for many people in different domains. The sources and characterization of uncertainties in the knowledge chain are quite important. The sources of uncertainty are generally categorized as either aleatory or epistemic. Uncertainties are characterized as epistemic, if one sees a possibility to reduce them by gathering more data or by refining models. Uncertainties are categorized as aleatory if one does not foresee the possibility of reducing them. Identifying epistemic uncertainties that have the potential of being reduced has a pragmatic value for the processing of an *infocentric* knowledge chain.

References

1. M. Barès, *Maîtrise du savoir et efficience de l'action*: Editions L'Harmattan, 2007.
2. M. Barès, *Pratique du calcul relationnel*. Paris: Edilivre, 2016.
3. C. Zins, "Conceptual approaches for defining data, information, and knowledge," *Journal of the American society for information science and technology,* vol. 58, pp. 479-493, 2007.
4. M. Burgin, "Data, Information and Knowledge," *INFORMATION-YAMAGUCHI-,* vol. 7, pp. 47-58, 2004.
5. C. W. Morris, *Foundations of the Theory of Signs* vol. 1: University of Chicago Press, 1938.

6. W. Lenski, "Remarks on a Publication Based Concept of," in *New Developments in Electronic Publishing of Mathematics: Joint Proceedings of the ECM 4 Satellite Conference on Electronic Publishing at KTH, Stockholm, June 25-27, 2004, and the Special Session on Electronic Publications at the Joint AMS SMM International Meeting, Houston, May 13, 2004*, 2005, p. 123.

7. M. Burgin, *Theory of information: fundamentality, diversity and unification* vol. 1: World Scientific, 2010.

8. B. Langefors, "Infological models and information user views," *Information Systems*, vol. 5, pp. 17-32, 1980.

9. B. Langefors, "Information systems theory," *Information Systems*, vol. 2, pp. 207-219, 1977.

10. W. J. Kettinger and Y. Li, "The infological equation extended: towards conceptual clarity in the relationship between data, information and knowledge," *European Journal of Information Systems*, vol. 19, pp. 409-421, 2010.

11. G. Piatetsky-Shapiro, "Knowledge discovery in databases: 10 years after," *Acm Sigkdd Explorations Newsletter*, vol. 1, pp. 59-61, 2000.

12. A. Adhikari and J. Adhikari, *Advances in knowledge discovery in databases*: Springer, 2015.

13. A. Azevedo, "Data mining and knowledge discovery in databases," in *Advanced Methodologies and Technologies in Network Architecture, Mobile Computing, and Data Analytics*, ed: IGI Global, 2019, pp. 502-514.

14. G. G. Chowdhury, *Introduction to modern information retrieval*: Facet publishing, 2010.

15. Ş. Vlăduţescu and F. Smarandache, "Five Computational Actions in Information Processing," *Social Sciences and Education Research Review*, vol. 2, pp. 29-44, 2014.

16. J. Piaget, "Sagesse et illusions de la philosophie [1965]," *Paris: PUF*, vol. 3, 1992.

17. G. Rogova and É. Bossé, "Information quality in information fusion," in *Proceedings of the 13th International Conference on Information Fusion (FUSION2010)* Edinburg, UK, 2010.

18. I. Standard, "8402," *Terminology. Good mark*, vol. 30, 1994.

19. J. A. O'brien and G. M. Marakas, *Introduction to information systems* vol. 13: McGraw-Hill/ Irwin, 2005.

20. R. Y. Wang and D. M. Strong, "Beyond accuracy: what data quality means to data consumers," *Journal of Management Information Systems*, vol. 12, pp. 5-34, 1996.

21. J. M. Juran and J. A. De Feo, *Juran's quality handbook: the complete guide to performance excellence*: McGraw Hill, 2010.

22. M. Bovee, R. P. Srivastava, and B. Mak, "A conceptual framework and belief-function approach to assessing overall information quality," *International journal of intelligent systems*, vol. 18, pp. 51-74, 2003.

23. R. Y. Wang and D. M. Strong, "Beyond accuracy: What data quality means to data consumers," *Journal of management information systems*, vol. 12, pp. 5-33, 1996.

24. P. Eggenhofer-Rehart, J. Barath, P. S. Farrell, R. K. Huber, J. Moffat, P. W. Phister Jr, *et al.*, "C2 CONCEPTUAL REFERENCE MODEL VERSION 2.0."

25. D. S. Alberts, "Agility, focus, and convergence: The future of command and control," DTIC Document2007.

26. D. S. Alberts, "The agility imperative: Precis," *Unpublished white paper] CCRP. Retrieved from http://www.dodccrp.org*, 2010.

27. A. Avizienis, J.-C. Laprie, and B. Randell, *Fundamental concepts of dependability*: University of Newcastle upon Tyne, Computing Science, 2001.

28. A. Avizienis, J.-C. Laprie, B. Randell, and C. Landwehr, "Basic concepts and taxonomy of dependable and secure computing," *Dependable and Secure Computing, IEEE Transactions on*, vol. 1, pp. 11-33, 2004.

29. L. Petre, K. Sere, and E. Troubitsyna, *Dependability and Computer Engineering: Concepts for Software Intensive Systems*: IGI Global, 2012.

30. E. A. Lee and S. A. Seshia, *Introduction to embedded systems: A cyber-physical systems approach*: Lee & Seshia, 2011.

31. P. Marwedel, *Embedded system design: Embedded systems foundations of cyber-physical systems*: Springer Science & Business Media, 2010.
32. É. Bossé and B. Solaiman, *Information fusion and analytics for big data and IoT*: Artech House, 2016.
33. M. G. Piattini, C. Calero, and M. F. Genero, *Information and database quality* vol. 25: Springer Science & Business Media, 2001.
34. W. L. Perry, D. Signori, and E. John Jr, *Exploring information superiority: a methodology for measuring the quality of information and its impact on shared awareness*: Rand Corporation, 2004.
35. C. Batini and M. Scannapieco, *Data Quality: Concepts, Methodologies and Techniques. Data-Centric Systems and Applications*: Springer, 2006.
36. C. Batini and M. Scannapieco, "Data and information quality," *Cham, Switzerland: Springer International Publishing. Google Scholar,* 2016.
37. L. Sebastian-Coleman, *Measuring data quality for ongoing improvement: a data quality assessment framework*: Newnes, 2012.
38. C. Samitsch, *Data Quality and Its Impacts on Decision-making: How Managers Can Benefit from Good Data*: Springer, 2014.
39. C. Fürber, *Data quality management with semantic technologies*: Springer, 2015.
40. M. J. Eppler, *Managing information quality: Increasing the value of information in knowledge-intensive products and processes*: Springer Science & Business Media, 2006.
41. L. Al-Hakim, *Information quality management: theory and applications*: IGI Global, 2007.
42. E. Turban, J. Aronson, and T. Liang, "Decision Support System, 7th Editio," ed: Pearson/ Prentice Hall, 2005.
43. L. Al-Hakim, *Challenges of managing information quality in service organizations*: IGI Global, 2006.
44. S. Baškarada, *Information quality management capability maturity model*: Springer, 2009.
45. M. A. Tate, *Web wisdom: How to evaluate and create information quality on the Web*: CRC Press, 2009.
46. C. Calero, *Handbook of research on Web Information Systems quality*: IGI Global, 2008.
47. J. R. Talburt, *Entity resolution and information quality*: Morgan Kaufmann, 2011.
48. L. Floridi and P. Illari, *The philosophy of information quality* vol. 358: Springer, 2014.
49. R. S. Kenett and G. Shmueli, *Information quality: The potential of data and analytics to generate knowledge*: John Wiley & Sons, 2016.
50. É. Bossé and G. L. Rogova, *Information Quality in Information Fusion and Decision Making*: Springer, 2019.
51. M. Helfert, "Managing and measuring data quality in data warehousing," in *Proceedings of the World Multiconference on Systemics, Cybernetics and Informatics, Florida, Orlando*, 2001, pp. 55-65.
52. E. Lefebvre, M. Hadzagic, and É. Bossé, "On Quality of Information in Multi-Source Fusion Environments," *Advances and Challenges in Multisensor Data and Information Processing*, p. 69, 2007.
53. G. Rogova and E. Bossé, "Information quality effects on information fusion," DRDC Valcartier, Tech Rept2008.
54. E. Blasch, P. Valin, and É. Bossé, "Measures of Effectiveness for High-Level Information Fusion," in *High-Level Information Fusion Management and Systems Design*, E. Blasch, É. Bossé, and D. A. Lambert, Eds., ed Boston: Artech House, 2012, pp. 331-348.
55. E. Blasch, P. C. Costa, K. B. Laskey, H. Ling, and G. Chen, "The URREF ontology for semantic wide area motion imagery exploitation," in *Aerospace and Electronics Conference (NAECON), 2012 IEEE National*, 2012, pp. 228-235.
56. E. Blasch, A. Josang, J. Dezert, P. C. Costa, and A.-L. Jousselme, "URREF self-confidence in information fusion trust," in *Information Fusion (FUSION), 2014 17th International Conference on*, 2014, pp. 1-8.

57. E. Blasch, K. B. Laskey, A.-L. Jousselme, V. Dragos, P. C. Costa, and J. Dezert, "URREF reliability versus credibility in information fusion (STANAG 2511)," in *Information Fusion (FUSION), 2013 16th International Conference on*, 2013, pp. 1600-1607.

58. P. C. Costa, K. B. Laskey, E. Blasch, and A.-L. Jousselme, "Towards unbiased evaluation of uncertainty reasoning: the URREF ontology," in *Information Fusion (FUSION), 2012 15th International Conference on*, 2012, pp. 2301-2308.

59. S. Uminsky. (2015-03-10). *Information Theory Demystified*. Available: http://www.ideacenter.org/contentmgr/showdetails.php/id/1236

60. D. Dequech, "Uncertainty: a typology and refinements of existing concepts," *Journal of economic issues,* vol. 45, pp. 621-640, 2011.

61. R. M. Gray, *Entropy and information theory*: Springer Science & Business Media, 2011.

62. J. R. Pierce, *An introduction to information theory: symbols, signals and noise*: Courier Corporation, 2012.

63. A. Lesne, "Shannon entropy: a rigorous notion at the crossroads between probability, information theory, dynamical systems and statistical physics," *Mathematical Structures in Computer Science,* vol. 24, p. e240311, 2014.

64. S. Kullback and R. A. Leibler, "On information and sufficiency," *The annals of mathematical statistics,* pp. 79-86, 1951.

65. C. Liu, D. Grenier, A.-L. Jousselme, and É. Bossé, "Reducing algorithm complexity for computing an aggregate uncertainty measure," *Systems, Man and Cybernetics, Part A: Systems and Humans, IEEE Transactions on,* vol. 37, pp. 669-679, 2007.

66. C. Liu, A.-L. Jousselme, É. Bossé, and D. Grenier, "Measures of uncertainty for fuzzy evidence theory " Technical Report DRDC-Valcartier TR2010-223, 2011 2010.

67. C. Liu, " A general measure of uncertainty-based information," Ph.D, Electrical and Computer Engineering, Université Laval, Québec, 2004.

68. A. Garrido, "Classifying entropy measures," *Symmetry,* vol. 3, pp. 487-502, 2011.

69. A. J. Short and S. Wehner, "Entropy in general physical theories," *New J. Phys.,* vol. 12, p. 033023, // 2010.

70. A. Ben-Naim, *Entropy demystified: the second law reduced to plain common sense*: World Scientific, 2008.

71. É. Bossé, A. Jousselme, and P. Maupin, "Knowledge, uncertainty and belief in information fusion and situation analysis," *NATO Science Series III Computer and Systems Sciences,* vol. 198, pp. 61-81, 2005.

72. A.-L. Jousselme, P. Maupin, and É. Bossé, "Uncertainty in a situation analysis perspective," in *Proceedings of the Sixth International Conference of Information Fusion*, 2003.

73. P. K. Han, W. M. Klein, and N. K. Arora, "Varieties of Uncertainty in Health Care A Conceptual Taxonomy," *Medical Decision Making,* vol. 31, pp. 828-838, 2011.

74. A. J. Ramirez, A. C. Jensen, and B. H. Cheng, "A taxonomy of uncertainty for dynamically adaptive systems," in *Software Engineering for Adaptive and Self-Managing Systems (SEAMS), 2012 ICSE Workshop on*, 2012, pp. 99-108.

75. H. M. Regan, M. Colyvan, and M. A. Burgman, "A taxonomy and treatment of uncertainty for ecology and conservation biology," *Ecological applications,* vol. 12, pp. 618-628, 2002.

76. M. Faber, R. Manstetten, and J. L. Proops, "Humankind and the environment: an anatomy of surprise and ignorance," *Environmental values,* vol. 1, pp. 217-241, 1992.

77. C. Tannert, H. D. Elvers, and B. Jandrig, "The ethics of uncertainty," *EMBO reports,* vol. 8, pp. 892-896, 2007.

78. M. Smithson, *Ignorance and uncertainty: Emerging paradigms*: Springer-Verlag Publishing, 1989.

79. G. Bammer and M. Smithson, *Uncertainty and risk: multidisciplinary perspectives*: Routledge, 2012.

80. C. Camerer and M. Weber, "Recent developments in modeling preferences: Uncertainty and ambiguity," *Journal of risk and uncertainty,* vol. 5, pp. 325-370, 1992.

81. L. J. Savage, *The foundations of statistics*: Courier Corporation, 1972.

82. M. Friedman and L. J. Savage, "The expected-utility hypothesis and the measurability of utility," *The Journal of Political Economy,* pp. 463-474, 1952.
83. M. C. Galavotti, "The notion of subjective probability in the work of Ramsey and de Finetti," *Theoria,* vol. 57, pp. 239-259, 1991.
84. G. M. Hodgson, "The ubiquity of habits and rules," *Cambridge journal of economics,* vol. 21, pp. 663-684, 1997.
85. D. Dubois, "Uncertainty Theories: A Unified View," in *SIPTA school 08 - UEE 08,* Montpellier, France, 2008.
86. P. Walley, "Statistical Reasoning With Imprecise Probabilities," ed: Chapman and Hall, 1991.
87. P. Walley, "Towards a unified theory of imprecise probability," *International Journal of Approximate Reasoning,* vol. 24, pp. 125-148, 2000.
88. G. Shafer, *A Mathematical Theory of Evidence*: Princeton University Press, 1976.
89. A. P. Dempster, "Upper and lower probabilities induced by a multivalued mapping," *The annals of mathematical statistics,* pp. 325-339, 1967.
90. P. Smets, "Imperfect information: Imprecision and uncertainty," in *Uncertainty Management in Information Systems,* ed: Springer, 1997, pp. 225-254.
91. D. Dubois and H. Prade, *Possibility Theory: An Approach to Computerized Processing of Uncertainty*: Plenum Press, 1988.
92. L. Zadeh, "Fuzzy Sets as the Basis for a Theory of Possibility," *Fuzzy Sets and Systems,* vol. 1, pp. 3-28, 1978.
93. D. Dequech, "Uncertainty: individuals, institutions and technology," *cambridge Journal of Economics,* vol. 28, pp. 365-378, 2004.
94. D. Dubois and H. Prade, "Formal representations of uncertainty," *Decision-Making Process: Concepts and Methods,* pp. 85-156, 2009.
95. M. R. Endsley and D. J. Garland, *Situation awareness analysis and measurement*: CRC Press, 2000.
96. D. Bouyssou, D. Dubois, H. Prade, and M. Pirlot, *Decision Making Process: Concepts and Methods*: John Wiley & Sons, 2013.
97. I. Molchanov, *Theory of random sets*: Springer Science & Business Media, 2006.
98. S. Destercke, D. Dubois, and E. Chojnacki, "Unifying practical uncertainty representations–I: Generalized p-boxes," *International Journal of Approximate Reasoning,* vol. 49, pp. 649-663, 2008.
99. S. Destercke, D. Dubois, and E. Chojnacki, "Unifying practical uncertainty representations: II. Clouds," *arXiv preprint arXiv:0808.2779,* 2008.
100. L. A. Zadeh, "Fuzzy sets," *Information and control,* vol. 8, pp. 338-353, 1965.
101. P. Krause and D. Clark, *Representing uncertain knowledge: an artificial intelligence approach*: Kluwer Academic Publishers, 1993.
102. B. Bouchon-Meunier and H. T. Nguyen, *Les incertitudes dans les systèmes intelligents*: Presses universitaires de France, 1996.
103. G. Klir and B. Yuan, *Fuzzy sets and fuzzy logic* vol. 4: Prentice Hall New Jersey, 1995.
104. G. J. Klir and M. J. Wierman, *Uncertainty-Based Information: Elements of Generalized Information Theory*: Physica-Verlag HD, 1999.
105. É. Bossé, J. Roy, and S. Wark, *Concepts, models, and tools for information fusion*: Artech House, Inc., 2007.
106. B. M. Ayyub and G. J. Klir, *Uncertainty modeling and analysis in engineering and the sciences*: CRC Press, 2006.
107. G. J. Klir, *Uncertainty and information: foundations of generalized information theory*: John Wiley & Sons, 2005.
108. J. Abellán and A. Masegosa, "Requirements for total uncertainty measures in Dempster–Shafer theory of evidence," *International journal of general systems,* vol. 37, pp. 733-747, 2008.

109. L. M. DE CAMPOS, J. F. HUETE, and S. MORAL, "Probability intervals: a tool for uncertain reasoning," *International Journal of Uncertainty, Fuzziness and Knowledge-Based Systems,* vol. 2, pp. 167-196, 1994.

110. M. Grabisch and C. Labreuche, "A decade of application of the Choquet and Sugeno integrals in multi-criteria decision aid," *Annals of Operations Research,* vol. 175, pp. 247-286, 2010.

111. J. Abellán and S. Moral, "Upper entropy of credal sets. Applications to credal classification," *International Journal of Approximate Reasoning,* vol. 39, pp. 235-255, 2005.

112. I. Couso and L. Sánchez, "Upper and lower probabilities induced by a fuzzy random variable," *Fuzzy Sets and Systems,* vol. 165, pp. 1-23, 2011.

113. J.-Y. Jaffray and F. Philippe, "On the existence of subjective upper and lower probabilities," *Mathematics of Operations Research,* vol. 22, pp. 165-185, 1997.

114. P. Suppes and M. Zanotti, "On using random relations to generate upper and lower probabilities," *Synthese,* vol. 36, pp. 427-440, 1977.

115. J. Abellán, G. Klir, and S. Moral, "Disaggregated total uncertainty measure for credal sets," *International Journal of General Systems,* vol. 35, pp. 29-44, 2006.

116. J. Abellan and S. Moral, "Maximum of entropy for credal sets," *International Journal of Uncertainty, Fuzziness and Knowledge-Based Systems,* vol. 11, pp. 587-597, 2003.

117. J. Abellán and S. Moral‡, "Difference of entropies as a non-specificity function on credal sets †," *International journal of general systems,* vol. 34, pp. 201-214, 2005.

118. A. Karlsson, "Evaluating credal set theory as a belief framework in high-level information fusion for automated decision-making," 2008.

119. Z.-g. Liu, Q. Pan, J. Dezert, and G. Mercier, "Credal classification rule for uncertain data based on belief functions," *Pattern Recognition,* vol. 47, pp. 2532-2541, 2014.

120. L. A. Zadeh, "Toward a theory of fuzzy information granulation and its centrality in human reasoning and fuzzy logic," *Fuzzy sets and systems,* vol. 90, pp. 111-127, 1997.

121. P. Smets, "The degree of belief in a fuzzy event," *Information sciences,* vol. 25, pp. 1-19, 1981.

122. R. R. Yager, "Generalized probabilities of fuzzy events from fuzzy belief structures," *Information sciences,* vol. 28, pp. 45-62, 1982.

123. J. Yen, "Generalizing the Dempster-Schafer theory to fuzzy sets," *Systems, Man and Cybernetics, IEEE Transactions on,* vol. 20, pp. 559-570, 1990.

124. C. Lucas and B. N. Araabi, "Generalization of the Dempster-Shafer theory: a fuzzy-valued measure," *Fuzzy Systems, IEEE Transactions on,* vol. 7, pp. 255-270, 1999.

125. R. V. Hartley, "Transmission of information1," *Bell System technical journal,* vol. 7, pp. 535-563, 1928.

126. J. Abellán and S. Moral, "Measuring total uncertainty in Dempster-Shafer theory of Evidence: properties and behaviors," in *Fuzzy Information Processing Society, 2008. NAFIPS 2008. Annual Meeting of the North American,* 2008, pp. 1-6.

Chapter 4
Preliminaries on Crisp and Fuzzy Relational Calculus

Abstract This chapter presents preliminaries of crisp and fuzzy relational calculus to support the discussion in the subsequent chapters. Under the terms crisp and relational calculi, we group here the basic elements allowing to perform different operations on the calculation of the mathematical relations and also to examine the interest of their properties according to their respective contexts of use.

4.1 Introduction

Everything has its structure, and the concept of structures is of prime importance for understanding the phenomenon of information and knowledge. Knowledge is always about the structure of a phenomenon rather than the essence of it. Relation is an ontological element of that structure (Fig. 4.1). Previously, we defined a *structure* as a representation of a complex entity that consists of components in relation to each other. This definition implies that two structures are identical if and only if (a) their nonrelational components are the same, (b) their relation components are the same, and (c) corresponding components stand in corresponding relations. The concept of a *named set*, also called a fundamental triad or a fundamental structure, has been introduced in the previous chapters where "relation" is being an inherent element of that fundamental mathematical construction (Fig. 4.1).

Whenever named sets are being discussed as the fundamental structure (triad), the relationship between the elements is the crucial thing to address. Relations may exist between objects of the same set or between objects of two or more sets. Fuzzy relations appear as a generalization of crisp relations. While a crisp relation determines the presence or absence of interconnectedness between the elements of two or more sets, fuzzy relations supply additional information for degrees of membership between elements. Zadeh [1] was the first to look at relations as fuzzy sets [2] on the universe $X \times X$. Zadeh [2] introduced the concept of fuzzy relation, defined the notion of equivalence, and gave the concept of fuzzy ordering. Compared with crisp relations, fuzzy relations have greater expressive power and broader utility. They are considered as softer models for expressing the strength of links between elements. They also permit to manipulate values that can be specified in linguistic terms.

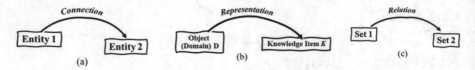

Fig. 4.1 Named sets: (**a**) fundamental structure, (**b**) a specific knowledge structure, (**c**) a set-theoretical named set

The importance of the theory of fuzzy relational equations is best described by Zadeh in the preface of the monograph by Di Nola et al. (1989) [3]:

> Human knowledge may be viewed as a collection of facts and rules, each of which may be represented as the assignment of a fuzzy relation to the unconditional or conditional possibility distribution of a variable. What this implies is that the knowledge may be viewed as a system of fuzzy relational equations. In this perspective, then, inference from a body of knowledge reduces to the solution of a system of fuzzy relational equations.

Since the 1960s, fuzzy relations have been defined, investigated, and applied in many ways, e.g., in fuzzy modeling, fuzzy diagnosis, and fuzzy control. In systems, the relationship between input and output parameters can be modelled by fuzzy relation between input and output spaces [4]. Fuzzy relational calculus is a powerful tool to study the behavior of such systems. Fuzzy relations and fuzzy relational calculus have many applications in pure and applied mathematics, artificial intelligence, psychology, medicine, economics, and sociology. They are implemented in all inference forward or backward chain reasoning schemes. The enumeration of application fields listed in Peeva and Kyosev [5] advocates strongly on the importance of implementing fuzziness in systems:

> The most valuable implementations are in expert systems and in artificial intelligence areas - approximate reasoning, inference systems, knowledge representation, knowledge acquisition and validation, learning, in information processing, in pattern analysis and classification, in fuzzy system science for fuzzy control and modelling, in decision making, in engineering for fault detection and diagnosis, in management, etc.

In this chapter, our aim is to present preliminaries of crisp and fuzzy relational calculus to support the discussion in the subsequent chapters. Under the terms crisp and relational calculi, we group here the basic elements allowing to perform different operations on the calculation of the mathematical relations and also to examine the interest of their properties according to their respective contexts of use. The notion of relation is ubiquitous when we establish links between objects, doing matching, similarities, various analogies, and so on. It is also in the universe of our discourse, with frequent statements eliciting the dependencies, causal links, between ideas, or concepts. The material in this chapter is essentially from the following sources [5–10].

4.2 Relations and their Properties

The origin of the word "relation" is from the Latin word *relatio* that characterizes the relationship between two objects or beings considered one to the other. In an extended sense, the notion of relation will bring out the idea of relation, connection, or correlation. In this respect, let us quote C. Bernard[1]: "on the essence of things being unknown, one can only know the relations of these things." Note those different appellations, which can be considered synonymous, cover the notion of relation, thus particularizing the domains in which it is applied: causality, correspondence, similarity, belonging to, analogy with, equivalence between, relation to, matching with, connection to, etc. It is equally appropriate to consider the role of the notion of relation in the evocation of determinism.[2]

In the perspective of modern mathematics, it is the relations considered in themselves and studied for themselves, as they unite mathematical entities that become predominant. In this new perspective, a theory of relations and its operations has taken shape with a specific symbolism, thereby permitting numerous fields of application as, for instance, in knowledge and information processing. It should be noted that the study of relations implies a careful examination of their specific properties: *reflexivity*, *symmetry*, *transitivity*, *inversion*, etc. The notion of relation thus becomes ubiquitous in mathematics as well as in formal logic according to Borel[3]: "Mathematics appears as the science that studies the relationships between certain abstract objects." According to Bourbaki[4], mathematical entities taken in themselves are of little importance; what matters is their relations. A set is formed of elements that possess certain properties and of having certain relations between them, or with the elements of other sets.

4.2.1   Binary Relations

Consider two arbitrary sets B and E. The set of all ordered pairs (b, e) where $b \in B$ and $e \in E$ is called the product, or Cartesian product, of B and E. A short designation of this product is $B \times E$. When the set (relation, respectively) is taken in conventional sense, we call it crisp set (crisp relation, respectively).

Definition 4.1 Cartesian product

$$B \times E = \{(b, e) | b \in B \wedge e \in E.\}$$

---

[1]Original citation: *L'essence des choses devant nous rester toujours ignorée, nous ne pouvons connaître que les relations de ces choses, et les phénomènes ne sont que des résultats de ces relations.*

   (https://dicocitations.lemonde.fr/citations/citation-98396.php)

[2]The doctrine that all events, including human action, are ultimately determined by causes external to the will

[3]Original citation: *De plus en plus, les mathématiques apparaissent comme la science qui étudie les relations entre certains êtres abstraits définis d'une manière arbitraire, sous la seule condition que ces définitions n'entraînent pas de contradiction.* (http://www.maphilo.net/citations.php?cit=6668)

[4]Nicolas Bourbaki is the collective pseudonym of a group of (mainly French) mathematicians.

Fig. 4.2 Illustration of a
binary relation R as a subset
of the Cartesian product
$B \times E$

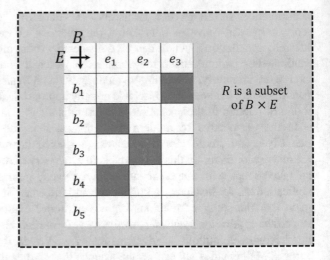

R is a subset
of $B \times E$

Example 4.1 Cartesian product B:: {bolts of 5 different types}, E:: {nuts of 3 given types}; suppose that the ordered pairs according to the threads are (b_1, e_3), (b_2, e_1), (b_3, e_2), *and* (b_4, e_1) which correspond to intersections in Fig. 4.2. In fact, they particularize a subset $R \subset B \times E$, defining a binary relation R (the binary term is often omitted).

Definition 4.2 If R is an arbitrary binary relation on $X \times Y$, the set $\{ y \in Y \mid xRy$ for some $x \in X\}$ is called the *image*, or the *range*, of R. Dually, the set $\{\in X \mid xRy$ for some $y \in Y\}$ is called the *domain* of R.

The *domain* Dom(R) of R is defined to be the set:

$$\text{Dom}(R) = \{x \mid x \in X \wedge (\exists y \in Y)(xRy)\}$$

The *image* (or *range*) of R is defined to be the set:

$$\text{Rng}(R) = \text{Im}(R) = \{y \mid y \in Y \wedge (\exists x \in X)(xRy)\}$$

The *complement* relation coR of R is the relation from Y to X defined as:

$$coR = \{(y, x) \mid (y, x) \in (Y \times X) \wedge (x, y) \notin R\}$$

Definition 4.3 Let R be any relation from a set A to a set B. The inverse of R, denoted by R^{-1}, is the relation from B to A which consists of those pairs which, when reversed, belong to R; that is,

$R^{-1} = (b, a) \mid (a, b) \in R$, or:

$$R^{-1} = \{(b, a) \mid (b, a) \in (B \times A) \wedge (aRb)\}$$

Example 4.2 Let $A = 1, 2, 3$ and $B = x, y, z$. Then the inverse of $R = (1, y), (1, z)$, $(3, y)$ is $R^{-1}(y, 1), (z, 1), (y, 3)$.

Whenever a pair (a, b) exists, the existence of $(a, b) \in R$ is denoted by $R(a, b)$, or by aRb. If a is not in relation to b or the pair (a, b) does not satisfy the relation R, then $(a, b) \notin R$. Since $\varnothing$ and $E \times E$ are subsets of $E \times E$, a relation can be empty: $R \rightarrow \varnothing$ or universal: $R \rightarrow E \times E$. There are several ways to represent the pairings corresponding to the need expressed by the relation R, in other words, to show the existence of ordered pairs in the relation.

The set of ordered pairs satisfying the relation R constitutes the graph G of the relation R. Defining a relation then results in taking two sets A, B and its graph $G \subseteq A \times B$. Any pair $(a, b) \in A \times B$ verifies:

$$R \subseteq A \times B \text{ ssi } (a, b) \in G.$$

Note also that two relations $R :: (A, B, G)$ and $R' :: (A', B', G')$ are equivalent ssi $A \leftrightarrow A', B \leftrightarrow B', G \leftrightarrow G'$. The way of considering the use of a relation can have an influence on its most adapted representation (the matrix form for the computation). A Boolean representation consists of putting 1 s at intersections when the pair (b_i, e_j) exists and 0 in the other cases, which gives Fig. 4.3 for the preceding example:

This representation facilitates the readability for the computation to be carried out within the context of the compositions of several relations between them, each relation being able to be represented by a Boolean matrix. For a Cartesian product $CP :: E \times E$, the presence of the 1 s on the matrix indicates the existence of the arcs on a graphical representation, as indicated in Fig. 4.4.

We can thus visually verify certain properties such as:

- Reflexivity: presence of 1 s on the diagonal of the matrix.
- Transitivity: for example, the existence of pairs (e_1, e_2), (e_2, e_3), and (e_1, e_3), indicated by "1" at the corresponding intersections, indicates the existence on the graph of an oriented path, as shown in Fig. 4.4.

Fig. 4.3 A Boolean representation of a relation R

B E $\downarrow$	e1	e2	e3	e4	e5
b1	0	0	1	0	0
b2	1	0	0	0	0
b3	0	1	0	0	0
b4	1	0	0	0	0
b5	0	0	0	0	0

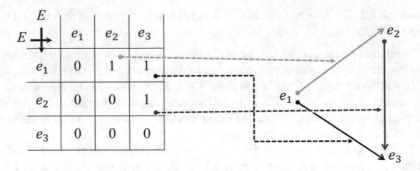

Fig. 4.4 An equivalent graphical representation

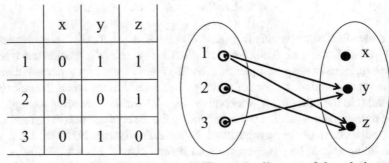

	x	y	z
1	0	1	1
2	0	0	1
3	0	1	0

a) The matrix of the relation   **b) The arrow diagram of the relation**

Fig. 4.5 Two ways of picturing a relation

There are two ways of picturing a relation R from A to B:

- Form a rectangular array (matrix) whose rows are labeled by the elements of A and whose columns are labeled by the elements of B. Put a 1 or 0 in each position of the array according as $a \in A$ is or is not related to $b \in B$. This array is called the *matrix of the relation* (Fig. 4.5a).
- Write down the elements of A and the elements of B in two disjoint disks, and then draw an arrow from $a \in A$ to $b \in B$ whenever a is related to b. This picture will be called *the arrow diagram of the relation* (Fig. 4.5b).

Finally, there is an important way of picturing a relation R on a finite set called the *directed graph* of the relation (Fig. 4.6). First, we write down the elements of the set, and then we draw an arrow from each element x to each element y whenever x is related to y.

Example 4.3 Consider a relation R on the set $A = \{1, 2, 3, 4\}$: $R = \{(1, 2), (2, 2), (2, 4), (3, 2), (3, 4), (4, 1), (4, 3)\}$. Fig. 4.6 is illustrating its corresponding directed graph.

Fig. 4.6 A relation
R represented by a directed
graph

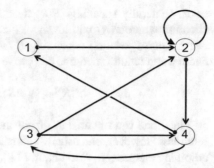

Table 4.1 Properties of a relation

Property: a relation R is ...	Equations	Definitions: If ...	Examples using $=$, $<$, and $\leq$ on integers
Reflexive	xRx	Every element of it is related with itself	$=$ is reflexive $(2 = 2)$ $R = (1, 1), (2, 2), (3, 3)$
Symmetric	$xRy \Rightarrow yRx$	For instance, pairs $(1, 2)$ and $(2, 1)$ are both present	$=$ is symmetric $((x = 2) \Rightarrow (2 = x))$
Transitive	$xRy \wedge yRz \Rightarrow xRz$	For instance, (a, b) and (b, c) are present, then (a, c) must be present	$<$ is transitive $((2 < 3) \wedge (3 < 5) \Rightarrow (2 < 5))$
Irreflexive	$xRy \Rightarrow x \neq y$	Every element of it is not related with itself	$<$ is irreflexive $((2 < 3) \Rightarrow (2 \neq 3))$
Antisymmetric	$xRy \wedge yRx \Rightarrow x = y$	For instance, a pair $(1,2)$ is present, and a pair $(2,1)$ is never present	$\leq$ is antisymmetric $(x \leq y \wedge y \leq x \Rightarrow x = y)$

4.2.2 Basic Properties of a Relation: Reflexive, Symmetric, and Transitive

Let R be a relation on E, and let $x, y, z \in E$. Table 4.1 below lists the main properties of a relation.

4.2.3 Closure Properties

The material in this section is essentially from [9]. Consider a given set A and the collection of all relations on A. Let P be a property of such relations, such as being symmetric or being transitive. A relation with property P will be called a P-relation. The P-closure of an arbitrary relation R on A, written $P(R)$, is a P-relation such that $R \subseteq P(R) \subseteq S$ for every P-relation S containing R. We will write reflexive (R), symmetric (R), and transitive (R) for the reflexive, symmetric, and transitive closures of R. Generally speaking, $P(R)$ needs not to exist. However, there is a general situation where $P(R)$ will always exist. Suppose P is a property such that there is at least one P-relation containing R and that the intersection of any P-relations is again a P-relation. Then one can prove that:

$$P(R) = \cap(S|S \text{ is a } P(R) \text{ and } R \subseteq S).$$

Thus, one can obtain $P(R)$ from the "top-down," that is, as the intersection of relations. However, one usually wants to find $P(R)$ from the "bottom-up," that is, by adjoining elements to R to obtain $P(R)$. This we do below.

4.2.3.1   Reflexive and Symmetric Closures

The next theorem tells us how to obtain easily the reflexive and symmetric closures of a relation. Here, $A = (a, a) \mid a \in A$ is the diagonal or equality relation on A.

Definition 4.4 Let R be a relation on a set A. Then:

 (i) $R \cup \Delta_A$ is the reflexive closure of R.
(ii) $R \cup R^{-1}$ is the symmetric closure of R.

In other words, reflexive(R) is obtained by simply adding to R those elements (a, a) in the diagonal which do not already belong to R, and symmetric (R) is obtained by adding to R all pairs (b, a) whenever (a, b) belongs to R.

Example 4.4 Consider the relation $R = (1, 1), (1, 3), (2, 4), (3, 1), (3, 3), and (4, 3)$ on the set $A = 1, 2, 3, 4$. Then reflexive $(R) = R \cup (2, 2), (4, 4)$ and symmetric $(R) = R \cup (4, 2), (3, 4)$.

4.2.3.2   Transitive Closure

Consider R be a relation on a set A. Recall that $R^2 = R \circ R$ and $R^n = R^{n-1} \circ R$. We define:

$$R^* = \overset{\infty}{\underset{i=1}{\cup}} R^i$$

The following definitions (theorems) apply:

Definition 4.5 R^* is the transitive closure of R. Suppose A is a finite set with n elements. We can show on graphs that $R = R \cup R^2 \cup \ldots \cup R^n$.

Definition 4.6 Let R be a relation on a set A with n elements. Then transitive $(R) = R \cup R^2 \cup \ldots \cup R^n$.

Example 4.5 Consider the relation $R = (1, 2), (2, 3), (3, 3)$ on $A = 1, 2, 3$. Then:

$$R^2 = R \circ R = (1, 3), (2, 3), (3, 3), \text{ and}$$

$$R^3 = R^2 \circ R = (1, 3), (2, 3), (3, 3).$$

Accordingly:

$$\text{transitive } (R) = (1, 2), (2, 3), (3, 3), (1, 3)$$

4.2.4 Finitary Relations

Definition 4.7 A *finitary* relation over sets $X_1, X_2, \cdots, X_n$ is a subset of the Cartesian product $X_1 \times X_2 \times \cdots \times X_n$; that is, it is a set of n-tuples $(x_1, \ldots, x_n)$ consisting of elements $x_i \in X_i$. With n sets in the Cartesian product, the relation is called *n-dimensional*. Relations that are two-dimensional have special significance; they are usually called *binary relations*.

The integer n (non-negative) is called the *arity*, or *degree* of the relation. A relation with n "places" is variously called a *n*-ary relation, or a relation of degree n. By an *n*-ary relation, we mean a set of ordered n-tuples. Relations with a finite number are called *finitary relations*.[5] When all dimensions of a fuzzy relation are finite sets, any n-dimensional fuzzy relation may conveniently be represented by an n-dimensional array whose entries are real numbers in the unit interval [0, 1]. For binary relations, clearly, the arrays become matrices.

From the standpoint of fuzzy relations, ordinary fuzzy sets may be viewed as one-dimensional fuzzy relations. This implies that all concepts introduced for ordinary fuzzy sets are also applicable to fuzzy relations. Fuzzy relations involve additional operations that emerge from their multidimensionality. These additional operations include projections, extensions, compositions, joins, and inverses of fuzzy relations. Projections and extensions are applicable to any fuzzy relations, whereas compositions, joins, and inverses are applicable only to binary relations.

In the following, we will discuss binary relations only, so the term "relation" will be used instead of "binary relation" and "fuzzy relation" instead of "binary fuzzy relation" with "binary" being most of the time suppressed.

---

[5] It is also possible to generalize the concept to *infinitary relations* with infinite sequences.

4.3   Classes of Relations

Using properties of relations, we can then consider some important classes of
relations.

4.3.1   Equivalence Relations

Consider a non-empty set S. A relation R on S is an equivalence relation if R is
reflexive, symmetric, and transitive. That is, R is an equivalence relation on S if it has
the following three properties:

1. For every $a \in S$, aRa.
2. If aRb, then bRa.
3. If aRb and bRc, then aRc.

Example 4.6 The general idea behind an equivalence relation is that it is a classification of objects which are in some way "alike." In fact, the relation "$=$" of equality
on any set S is an equivalence relation; that is: (1) $a = a$ for every $a \in S$; (2) if $a = b$,
then $b = a$; and (3) if $a = b$, $b = c$, then $a = c$.

Consider the relationship between equivalence relations and partitions on a
non-empty set S. Recall first that a partition P of S is a collection A_i of non-empty
subsets of S with the following two properties:

1. Each $a \in S$ belongs to some A_i.
2. If $A_i \neq A_j$, then $A_i \cap A_j = \emptyset$.

In other words, a partition P of S is a subdivision of S into disjoint non-empty sets.
Suppose R is an equivalence relation on a set S. For each $a \in S$, let $[a]$ denote the set
of elements of S to which a is related under R; that is: $[a] = x \mid (a, x) \in R$. We call $[a]$
the equivalence class of a in S; any $b \in [a]$ is called a representative of the
equivalence class. The collection of all equivalence classes of elements of S under
an equivalence relation R is denoted by S/R, that is, $S/R = \{[a] \mid a \in S\}$. It is called
the *quotient set* of S by R. The fundamental property of a quotient set is contained in
the following theorem [9].

Definition 4.7 Let R be an equivalence relation on a set S. Then S/R is a partition of
S. Specifically:

  (i) For each a in S, we have $a \in [a]$.
 (ii) $[a] = [b]$ if and only if $(a, b) \in R$.
(iii) If $[a] \neq [b]$, then $[a]$ and $[b]$ are disjoint.

Table 4.2 Definitions of diverse terms related to *poset*

Upper bound	An element $a \in P$ is called an *upper bound* of the subset H if $h \leq a$ for all $h \in H$
Least upper bound (lub)	An upper bound $a \in P$ of H is called *least upper bound* (*lub*) of H if whenever $a' \in P$ is an upper bound of if it holds $a \leq a'$, in symbols $a = \sup H$. (sup = supremum) or $a = \sqcup H$
Lower bound	An element $b \in P$ is called a *lower bound* of the subset H if $b \leq h$ for all $h \in H$
Greatest lower bound (glb)	A lower bound $b \in P$ of H is called *greatest lower bound* (*glb*) of H if whenever $b' \in P$ is a lower bound of H, it holds $b' \leq b$. In this case, we write $b = \inf H$. (inf = infimum) --- $b = \sqcap H$
Greatest element of	By *a greatest element of* a *poset P*, we mean an element $c \in P$ such that $x \leq c$ for all $x \in P$; *the least element order* of P is defined dually
Least and greatest elements of	The (unique) *least and greatest elements* of P, when they exist, are called *universal bounds* of P and are denoted by 0 and 1

Conversely, given a partition A_i of the set S, there is an equivalence relation R on S such that the sets A_i are the equivalence classes.

4.3.2   Order and Partial Ordering Relations

Partial order theory plays an important role in many disciplines of computer science and engineering such as in distributed computing, programming language semantics, data mining (concept analysis), and many other disciplines of logics and mathematics. This section introduces only the vocabulary and some basic notions and definitions.

A crisp relation R on a set S is called a *partial ordering or a partial order* of S if R is reflexive, antisymmetric, and transitive. A set S together with a partial ordering R is called a *partially ordered set* or a *poset*. Excerpted and rearranged from [5], we will be using the same kind of symbols so, the symbol $\leq$ for *partial order relation* on a *poset P*. When P is a *poset* with respect to $\leq$, we write also $(P, \leq)$ or simply P. A partial order relation R on P is called *total (or linear) order* if for each $x, y \in P$, either $(x, y) \in R$ or $(y, x) \in R$. Let $(P, \leq)$ be a *poset* and $H \subseteq P$. Table 4.2 lists some related definitions of posets.

Finite posets are often depicted graphically using Hasse diagrams. To define Hasse diagrams, we first define a relation *covers* as follows. For any two elements $a, b \in A$, b *covers* a if $a < b$ and

$$\forall z \in A : a \leq z < b \Rightarrow z = a.$$

If b *covers* a, then there should not be any element z with $a \leq z < b$. We use $a<_c b$ to denote that b *covers* a. A Hasse diagram of a poset is a graph with the property that there is an edge from a to b iff $a<_c b$ (a is drawn lower than b when b *covers* a). This

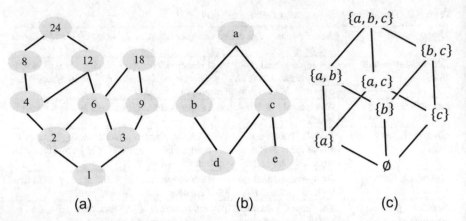

Fig. 4.7 Hasse diagrams of *posets* ordered: (**a**) by "*x divides y*", (**b**) by $d \leq b$, $d \leq a$, $e \leq c$, (c) by *inclusion,* $\subseteq$

allows us to suppress the directional arrows in the edges. To further illustrate, consider the Hasse diagrams of posets of Fig. 4.7 explained in Example 4.7.

Example 4.7 Examples of Hasse diagrams of posets (From [9], Chap. 14):

- Let $A = \{1, 2, 3, 4, 6, 8, 9, 12, 18, 24\}$ be ordered by the relation "*x divides y*." The diagram of A is given in Fig. 4.7a. (Unlike rooted trees, the direction of a line in the diagram of a poset is always upward).
- Let $B = \{a, b, c, d, e\}$. The diagram in Fig. 4.7b defines a partial order on B in the natural way. That is, $d \leq b$, $d \leq a$, $e \leq c$, and so on.
- Any powerset with $\subseteq$ forms an ordered set. The powerset of a, b, c which is $\emptyset$, $c, b, b, c, a, a, c, a, b, a, b, c$ ordered by $\subseteq$ is pictured in Fig. 4.7c.

Sum and product operations can be done on any set of structures to obtain new structures. Below are three examples of definitions of these kinds of operations (excerpted from [11]).

Example 4.8 (Disjoint Sum) Given two posets P and Q, their disjoint sum, denoted by $P + Q$, is defined as follows. Given any two elements x and y in $P + Q$, $x \leq y$ iff (1) both x and y belong to P and $x \leq y$ in P or (2) both x and y belong to Q and $x \leq y$ in Q. The Hasse diagram of the disjoint sum of P and Q can be computed by simply placing the Hasse diagram of P next to Q

Example 4.9 (Cross Product of Posets) Given two posets P and Q, the cross product denoted by $P \times Q$ is defined as $(P \times Q, \leq_{P \times Q})$ where $(p_1, q_1) \leq (p_2, q_2) \stackrel{\text{def}}{=} (p_1 \leq_P p_2) \wedge (q_1 \leq_Q q_2)$. See Fig. 4.8 for an example. The definition can be extended to an arbitrary indexing set

Example 4.10 (Ordinal Sum) Given two posets P and Q, the ordinal sum denoted by $P \oplus Q$ is defined as

$$x \leq_{P \oplus Q} y \text{ iff } (x \leq_P y) \vee (x \leq_Q y) \vee [(x \in P) \wedge (y \in Q)].$$

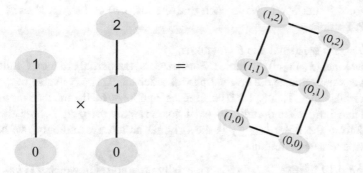

Fig. 4.8 Hasse diagrams of cross product of *posets*

4.4 Lattices

Lattices are algebraic structures which are useful for encoding logical information. In particular, lattices are generalizations of structures like Boolean algebra, sets, hyperpower sets, or *concept lattice.*[6] Concept lattices are frameworks widely used for deriving ontologies through *formal concept analysis.*[7] Note also that the lattice theory provides the most suitable tool for solving problems in fuzzy relational calculus. It is an important mathematical tool in fuzzy sets theory as exemplified by the *L*-fuzzy sets introduced and studied in Goguen [12].

Definition 4.8 Lattice A *lattice* is a *poset* P or $(X, \leq)$ that any two of whose elements x and y have a greatest lower bound (*glb*) denoted by $x \sqcap y$ or by $\sqcap(x, y)$ and a least upper bound (*lub*) or *join* denoted by $x \sqcup y$ or by $\sqcup(x, y)$. In short, P is a *lattice* iff $\forall\, x, y \in X : (x \sqcup y) \wedge (x \sqcap y)$ exists. (see Fig. 4.9) We write:

$$\mathbb{L} = (P, \sqcup, \sqcap, 0, 1)$$

Or simply $\mathbb{L}$ for a bounded chain over the underlying poset, P, with operations *join*, $\sqcup$, and *meet*, $\sqcap$, and with universal bounds 0 and 1. The notations $\varnothing$ or 0 (respectively, Ω or 1) are also used instead of $\perp$ (respectively, $\top$).

---

[6] A *concept lattice* of a formal context (table) is a collection of all formal concepts (conceptual clusters) equipped with a subconcept-superconcept hierarchy. This is the principal notion of *formal concept analysis.*

[7] *Formal concept analysis* (FCA) is a method mainly used for the analysis of data, i.e., for deriving implicit relationships between objects described through a set of attributes on the one hand and these attributes on the other. Formal concept analysis (FCA) is a principled way of deriving a concept hierarchy or formal ontology from a collection of objects and their properties. (Source: Wikipedia)

Definition 4.9 Let P be a *poset* such that the inf(a, b) and sup(a, b) exist for any a, b in P. Letting:

$a \sqcap b = $ inf (a, b) and $a \sqcup b = $ sup (a, b)

we have that $(P, \sqcap, \sqcup)$ is a lattice. Furthermore, the partial order on P induced by the lattice is the same as the original partial order on P. The converse of the above theorem is also true. That is, let L be a lattice, and let $\leq$ be the induced partial order on L. Then inf(a, b) and sup(a, b) exist for any pair (a, b) in L, and the lattice obtained from the poset $(L, \leq)$ is the original lattice. Accordingly, we have the following alternate definition.

Definition 4.10 Lattice A lattice is a partially ordered set in which $a \sqcap b = $ inf (a, b) and $a \sqcup b = $ sup (a, b) exist for any pair (a, b). We note first that any linearly ordered set is a lattice since inf$(a, b) = a$ and sup$(a, b) = b$ whenever $a \leq b$.

All lattices are *posets*, but most *posets* are not lattices. For instance, a *poset* such that $x \leq y \Rightarrow x = y$, is a lattice iff it has at most one element. The set 1, 2, 3, 6 partially ordered by divisibility is a lattice, the set 1, 2, 3 so ordered is not a lattice because the pair $\{2, 3\}$ lacks a join, and it lacks a meet in 2, 3, 6. Below are several examples and illustrations of posets and lattices adapted from [9, 11].

Example 4.11 Examples of posets

- The set of natural numbers under the relation divides forms a lattice. Given any two natural numbers, the greatest common divisor (*gcd*) and the least common multiple (*lcm*) of those two numbers correspond to the sup and inf, respectively.
- The family of all subsets of a set a, b, c, under the relation $\subseteq$ (i.e., the poset $(a, b, c, \leq)$ forms a lattice as in Fig. 4.7c. Given any two subsets Y, Z of X, the sets $Y \cup Z$ and $Y \cap Z$ (corresponding to sup and inf) are always defined.

Among the *posets* illustrated in Fig. 4.9, only (c) is not a lattice since b, c has three upper bounds, d, e, and f, and no one of them precedes the other two, that is, sup(b, c) does not exist. Figure 4.9b shows a *poset* which is a chain of length 5. Note that the infimum is unique whenever it exists. Table 4.3 lists different kinds of lattices, and Table 4.4 summarizes some axioms related to lattice operators.

Definition 4.11 Lattice morphisms Let $\mathbb{L}_1$ and $\mathbb{L}_2$ be complete lattices and X be a non-empty subset of L_1. A map $f : L_1 \rightarrow L_2$ is called a morphism according to operations (join and meet).

Example 4.12 Example of a morphism from ([5], p.6). The real closed interval $[0, 1]$ with operations $\sqcup = max$ and $\sqcap = min$ is a bounded chain:

$$\mathbb{I} = ([0, 1], max, min, 0, 1)$$

written for short as $\mathbb{I}$. If $\mathbb{L} = (L, \sqcup, \sqcap, 0, 1)$ is a bounded chain as described above and $f : L \rightarrow [0, 1]$ is a map defined with $f(a \sqcap b) = min (f(a), f(b))$ and $f(a \sqcup b) = max (f(a), f(b))$, then $f : L \rightarrow [0, 1]$ determines a morphism of lattices. Table 4.5 lists the various kinds of morphisms.

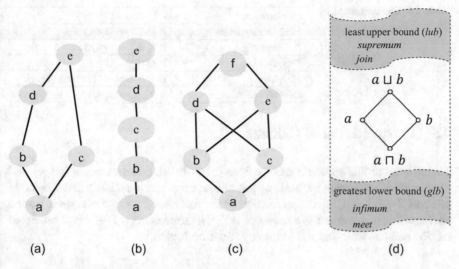

Fig. 4.9 Examples of posets and lattices: (**a**) and (**b**) posets that are lattices, (**c**) a poset that is not a lattice, (**d**) notations used in posets and lattices

Table 4.3 Definitions of diverse kinds of lattices

A chain	A totally ordered *poset* is called a *chain*. A chain with universal bounds 0 and 1 is called *bounded chain*
Complete lattice	A lattice is called *complete* if any of its subsets X has a lub (***join***), denoted by $\sqcup X$ and a g.l.b. (***meet***) denoted by $\sqcap X$
Distributive lattice	A lattice L is called distributive if for all $x, y, z \in L$ the following equality holds: $x \sqcap (y \sqcup z) = (x \sqcap y) \vee (x \sqcup z)$
Complement	A lattice L with universal bounds 0 and 1 is called *complemented* if each element $x \in L$ has a complement. An element $y \in L$ is a complement of $x \in L$, if $(x \sqcap y = 0)$ and $(x \sqcup y = 1)$

there are more pairs of similar operations: *intersection* and *union* for sets, *and* and *or* in logic. *Meet* and *join* are abstractions of these pairs and so have abstract names. In context of algebraic lattices, these are the primitive operations; in equivalent context of order theory, these are names for *inf* and *sup* of two elements. Also note the similarity in symbols for operations $\wedge, \vee, \cap, \cup, \wedge, \vee, \cap, \cup$ and in connection symbols for order $<, >, \subset, \supset, <, >, \subset, \supset$.

Table 4.4 Main axioms defining a *lattice* under meet,$\sqcap$, and join, $\sqcup$, operations

Commutative law	$x \sqcap y = y \sqcap x$	$x \sqcup y = y \sqcup x$
Associative law	$(x \sqcap y) \sqcap z = x \sqcap (y \sqcap z)$	$(x \sqcup y) \sqcup z = x \sqcup (y \sqcup z)$
Absorption law	$x \sqcap (x \sqcup y) = x$	$x \sqcup (x \sqcap y) = x$
Duality law	$x \sqcap (y \sqcup x) = (x \sqcup x)$	$x \sqcup (y \sqcap x) = (x \sqcap x)$
Idempotent law	$x \sqcap x = x$	$x \sqcup x = x$

Table 4.5 Lattice morphisms

⊔- morphism	If $f(\sqcup X) = \sqcup f(X)$
Dual ⊔- morphism	If $f(\sqcup X) = \sqcap f(X)$
⊓- morphism	If $f(\sqcap X) = \sqcap f(X)$
Dual ⊓- morphism	If $f(\sqcap X) = \sqcup f(X)$
Morphism	If *it is both* : ⊔, ⊓ morphisms
Dual morphism	If *it is both dual* ⊔, ⊓ morphisms

4.5   Crisp Relational Calculus

The material of this section is mainly from [5, 6]. In relational calculus, images and compositions are among the basic operations on crisp relations. Let X and Y be crisp sets and $R \subseteq X \times Y$ be a crisp relation as defined above: a crisp relation R from a universe X to a universe Y is a subset of $\times Y$. The statement $(x, y) \in R$ is abbreviated as xRy, and one says that x is in (binary) relation R with y.

4.5.1   Images in Crisp Relational Calculus

Based on this definition, the following four kinds of images of a set under a relation can be enumerated.

The *afterset* $xR \subseteq Y$ for the element $x \in X$ is the set of all elements $y \in Y$ such that $(x, y) \in R$ (see Fig. 4.10a):

$$xR = \{y | y \in Y \land xRy\}$$

The *foreset* $Ry \subseteq X$ for the element $y \in Y$ is the set of all elements $x \in X$ such that $(x, y) \in R$, (Fig. 4.10b):

$$Ry = \{x | x \in X \land xRy\}$$

Let $A \subseteq X$ be a crisp set and $R \subseteq X \times Y$ be a crisp relation. Crisp images are defined as follows: *Direct image* $R(A) \subseteq Y$ of A under R is the set of all elements $y \in Y$, such that for each $y \in R(A)$ there exists element $x \in A$, that is in relation R with y, i.e., $A \cap Ry \neq \varnothing$ (Fig. 4.11a):

$$R(A) = \{y | y \in Y \land (\exists x \in X)(x \in A \land xRy)\} = \{y \in Y | A \cap Ry \neq \varnothing\}$$

Lower or subdirect image $R_{\vartriangleleft}(A) \subseteq Y$ of A under R is the set of all elements $\in Y$, such that $x \in A$ implies xRy, i.e., $A \subseteq Ry$, (Fig. 4.11b):

$$R_{\vartriangleleft}(A) = \{y | y \in Y \land (\forall x \in X)(x \in A \Rightarrow xRy)\} = \{y \in Y | A \subseteq Ry\}$$

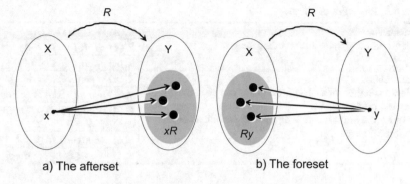

a) The afterset   b) The foreset

Fig. 4.10 Illustrations. (Adapted from [5]): (**a**) an *afterset*, (**b**) a *foreset*

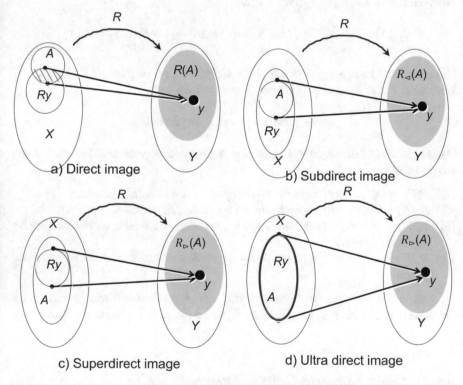

a) Direct image   b) Subdirect image

c) Superdirect image   d) Ultra direct image

Fig. 4.11 Illustration of four types of images. (Adapted from [5]): (**a**) *direct image*, (**b**) *subdirect image*, (**c**) *superdirect image*, (**d**) *ultra-direct image*

Upper or superdirect image $R_\triangleright(A) \subseteq Y$ of A under R is the set of all elements $y \in Y$, such that if xRy then $x \in A$, i.e., $Ry \subseteq A$ (Fig. 4.11c):

$$R_\triangleright(A) = \{y | y \in Y \wedge (\forall x \in X)(xRy \Rightarrow x \in A)\} = \{y \in Y | Ry \subseteq A\}$$

Table 4.6 Properties of images

Containment: Concerns the refining nature of the images	$R_\circ(A) = R_\lhd(A) \cap R_\rhd(A)$
	$R_\circ(A) \subseteq R_\lhd(A) \subseteq R(A)$
	$R_\circ(A) \subseteq R_\rhd(A) \subseteq R(A)$

Relationships:
  How a subdirect or a superdirect image can be converted to a classical direct image
  Express the direct image in terms of the subdirect or the superdirect image
	$R_\lhd(A) = co((coR)(A))$, if $A \neq 0$

	$R_\rhd(A) = co(R(coA)) \cap rng(R)$
	$R_\circ(A) = co((coR)(A)) \cap co(R(-coA))$, if $A \neq 0$
	$R(A) = co((coR)_\lhd(A))$, if $A \neq 0$
	$R(A) = co(R_\rhd(coA)) \cap rng(R)$

Ultra-image $R_\circ(A) \subseteq Y$ of A under R is the set of all elements $y \in Y$, such that if xRy then $x \in A$, i.e., $Ry = A$ (Fig. 4.11d):

$$R_\circ(A) = \{y | y \in Y \wedge (\forall x \in X)(xRy \Rightarrow x \in A)\} = \{y \in Y | Ry = A\}$$

Example 4.13 Image types (from [13]) Consider a set of patients X and a set of symptoms Y. Let R be a relation from X to Y defined by:

$$pRs \Leftrightarrow \text{patient } p \text{ shows the symptom } s.$$

Further, let $F \subseteq X$ be the set of all female patients in the population. Then the images are given by:

  (i) $R(F)$ = the set of symptoms shown by at least one female patient.
 (ii) $R_\lhd(F)$ = the set of symptoms shown by all female patients.
(iii) $R_\rhd(F)$ = the set of symptoms shown by at least one female patient and not by any male patient.
(iv) $R_\circ(F)$ = the set of symptoms shown by all female patient and not by any male patient.

    Table 4.6 lists properties of images that can be used to perform operations on direct and indirect images. Detailed proofs can be found in De Baets and Kerre [14].

4.5.2   Compositions in Crisp Relational Calculus

The composition of relations is the forming of a new binary relation $R \circ S$ from two given binary relations R and S. For the Cartesian product $X \times Y$, we denote by pr_1 and pr_2 its first and second projections, respectively. It means $pr_1(X \times Y) = X$ and $pr_2(X \times Y) = Y$. Relations $R \subseteq X \times Y$ and $S \subseteq Y \times Z$ with $pr_2(X \times Y) = pr_1(Y \times Z) = Y$ are called *composable*.

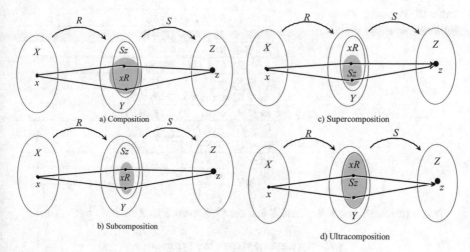

Fig. 4.12 Compositions of relations. (Adapted from [5]): (**a**) Composition $R \circ S$; (**b**) subcomposition $R \lhd S$; (**c**) supercomposition $R \rhd S$; (**d**) ultracomposition $R \bullet S$

Definition 4.12 Compositions of relations Consider a relation R from X to Y and a relation S from Y to Z. We define the following four types of binary compositions of relations $R \subseteq X \times Y$ and $\subseteq Y \times Z$:

(1) $R \circ S$; (2) $R \lhd S$; (3) $R \rhd S$; (4) $R \bullet S$.

1. Composition, $R \circ S$, is illustrated in Fig. 4.12a and given as:

$$R \circ S = \{(x,z)|(x,z) \in X \times Z \wedge (\exists y \in Y)(xRy \wedge ySz)\} = \{(x,z) \in X \times Z \,|\, xR \cap Sz \neq \varnothing\}$$

2. Subcomposition, $R \lhd S$, is illustrated in Fig. 4.12b and given as:

$$R \lhd S = \{(x,z)|(x,z) \in X \times Z \wedge (\forall y \in Y)(xRy \Rightarrow ySz)\} = \{(x,z) \in X \times Z \,|\, xR \subseteq Sz\}$$

3. Supercomposition, $R \rhd S$, is illustrated in Fig. 4.12c and given as:

$$R \rhd S = \{(x,z)|(x,z) \in X \times Z \wedge (\forall y \in Y)(ySz \Rightarrow xRy)\} = \{(x,z) \in X \times Z \,|\, Sz \subseteq xR\}$$

4. Ultracomposition, $R \bullet S$, is illustrated in Fig. 4.12d and given as:

$$R \bullet S = \{(x,z)|(x,z) \in X \times Z \wedge (\forall y \in Y)(ySz \Rightarrow xRy)\} = \{(x,z) \in X \times Z \,|\, Sz = xR\}$$

Example 4.14 Types of compositions of relations (from [13]) Again let $X =$ the set of all patients p and $Y =$ the set of all symptoms s and $Z =$ the set of all illnesses i.

Table 4.7 Properties of compositions: (see proofs in [14])

$R \circ S = (R \lhd S) \cap (R \rhd S)$
$R \circ S \subseteq (R \lhd S) \subseteq (R \circ S)$
$R \circ S \subseteq (R \lhd S) \subseteq (R \circ S)$
$R \circ (S \circ T) = (R \circ S) \circ T$
$R_1 \subseteq R_2 \Rightarrow (R_1 \circ S) \subseteq (R_2 \circ S)$
$R_1 \subseteq R_2 \Rightarrow (R_1 \lhd S) \subseteq (R_2 \lhd S)$
$R_1 \subseteq R_2 \Rightarrow (R_1 \rhd S) \subseteq (R_2 \rhd S)$
$(R \circ S)^{-1} = S^{-1} \circ R^{-1}$
$(R \lhd S)^{-1} = S^{-1} \rhd R^{-1}$
$(R \rhd S)^{-1} = S^{-1} \lhd R^{-1}$
$(R \circ S)^{-1} = S^{-1} \circ R^{-1}$

R is a relation from X to Y and S is a relation from Y to Z defined by:

$$pRs \Leftrightarrow \text{patient } p \text{ shows the symptom } s$$

$$:$$ (and)

$$sSi \Leftrightarrow s \text{ is a symptom of illness } i.$$

The compositions of R and S are given by:

(i) $p(R \circ S)i$ patient p shows at least one symptom of illness i.
(ii) $p(R \lhd S)i$ all symptoms shown by patient p are symptoms of illness i, and patient p shows at least one symptom.
(iii) $p(R \rhd S)i$ patient p shows all symptoms of illness i.
(iv) $p(R \circ S)i$ all symptoms shown by patient p are exactly those of illness i.

Table 4.7 lists properties of images that can be used to perform operations on direct and indirect images. Detailed proofs can be found in De Baets and Kerre [14].

Finally, let us cite [15], p.293, on the above operations: "The above operations, simple yet powerful, are used in a variety of application areas. Their very nature is typical for many mathematical theories: the evaluation of non-empty intersections versus inclusions. We mention, for instance, possibility theory, rough set theory, mathematical morphology, set valued analysis, Dempster-Shafer theory of evidence, topology, etc."

4.5.3   Characteristic Functions of Relations

The material in this section is from Beg [6]. A relation R from X to Y can be identified with its characteristic mapping, $R : X \times Y \to [0, 1]$, defined as:

$$R(x,y) = \begin{cases} 1 \text{ if } (x,y) \in R \\ 0 \text{ otherwise.} \end{cases}$$

1. The characteristic mapping of round composition is:

$$R \circ S(x, z) = \sup_{y \in X}(R(x, y) \wedge_B R(y, x))$$

2. The characteristic mapping of the subcomposition is:

$$R \triangleleft S(x, z) = \left(\inf_{y \in Y} R(x, y) \Rightarrow_B S(y, z) \right) \wedge_B \left(\sup_{y \in Y} R(x, y) \right)$$

3. The characteristic mapping of the supercomposition is:

$$R \triangleright S(x, z) = \left(\inf_{y \in Y} S(y, z) \Rightarrow_B R(x, y) \right) \wedge_B \left(\sup_{y \in Y} S(y, z) \right)$$

4. The characteristic mapping of the ultracomposition is:

$$R \circ S(x, z) = \left(\inf_{y \in Y} R(x, y) \Leftrightarrow_B S(y, z) \right) \wedge_B \left(\sup_{y \in Y} R(x, y) \right)$$
$$= \left(\inf_{y \in Y} R(x, y) \Leftrightarrow_B S(y, z) \right) \wedge_B \left(\sup_{y \in Y} S(y, z) \right)$$

Similarly, the characteristic mappings of lower and upper images are as follows:

$$R_\triangleleft(A)(y) = \inf_{x \in X} A(x) \Rightarrow_B R(x, y)$$
$$R_\triangleright(A)(y) = \inf_{x \in X} R(x, y) \Rightarrow_B A(x)$$

where, $\wedge_B$, $\Rightarrow_B$, and $\Leftrightarrow_B$ stand for Boolean conjunction, implication, and equivalence, respectively. The characteristic functions of relations help in manipulating equations.

4.5.4   Resolution of Forward and Inverse Problems in Crisp Relational Calculus

The resolution of forward and inverse problems plays an important role in automatic and situational control, medical diagnostics, pattern recognition, prediction, image processing, multi-criteria estimation, seismology, machine learning, and other *decision-making* tasks. A *forward problem or direct problem* is understood as a look into

Fig. 4.13 Problems
classification into *forward*
and *inverse*

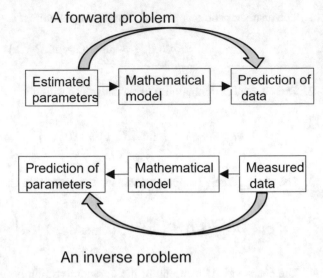

A forward problem

An inverse problem

the future, i.e., a prediction of the effects (outputs) as a result of the observed causes (inputs). An *inverse problem* is understood as a look into the past, i.e., a renewal of the causes (inputs) through the observed effects (outputs).

When a mathematical model is available, we can predict, given some model parameters (known) that describe a physical system, the behavior of the system. The solution of the mathematical model's equation is the state of the physical system (e.g., in optimal control theory, the state equations). In several cases, we are just interested in knowing the effects of the physical state on some objects, i.e., predicting the effects or results for a set of known parameters. Consequently, we need a mapping operation (the forward map) to convert the state of the physical system into the portion/aspect of interest (the set of known parameters). The solution, to this so-called forward problem, consists of two steps:

- Determination of the state of the system from the physical parameters that describe it.
- Application of the observation mapping to the estimated state of the system to predict the behavior of what we want to observe.

When solving an *inverse problem*, we start with the measured data (Fig. 4.13) and then infer what parameters "caused" those measurements. Unlike in a forward problem, the inverse problem has no unique solution, i.e., multiple (in theory infinite) set of parameters might explain the same measurement data. Usually, a first stage is done with some coarse model formed which approximates the input-output interconnection and contains adjustable parameters. At the second stage, parameter values are chosen such that they minimize the distances between the model and experimental outputs. This last stage amounts to the use of various optimization methods. The only difficulty is in finding the global minimum of nonlinear functions with divergence of theory and experimentation and in computing complexity growth with the growth of the adjustable parameters number. The first stage, i.e., the choice of the adjustable model, considerably depends upon the knower, his/her qualifications, expertise, domain understanding, and upon other subjective factors.

For the crisp relation calculus, solutions to forward and inverse problems are described as follows. Classification of the problems in *forward* and *inverse* for crisp relational calculus leads to a scheme that is valid for fuzzy relational calculus as well.

Let the crisp relations $R \subseteq X \times Y$ and $S \subseteq X \times Z$ and the crisp sets $A \subseteq X$ and $B \subseteq Y$ be given.

- Computing $R(A)$, $R_\lhd(A)$, $R_\rhd(A)$:

$$R(A) = \{y \in Y | A \cap Ry \neq \varnothing\}$$

$$R_\lhd(A) = \{y \in Y | A \subseteq Ry\}$$
$$R_\rhd(A) = \{y \in Y | Ry \subseteq A\}$$

is called *forward problem* resolution for direct image, subdirect image, and superdirect image, respectively.

- Equations of the type $R(A) = B$, $R_\lhd(A) = B$, or $R_\rhd(A) = B$, when the unknown set is A or the unknown relation is R are called *image equations*. Solving them is called *inverse problem* resolution for direct image, subdirect image, and superdirect image, respectively.
- Computing composition $R \circ S$, $R \lhd S$, $R \rhd S$, according to:

$$R \circ S = \{(x, z) \in X \times Z \mid xR \cap Sz \neq \varnothing\}$$

$$R \lhd S = \{(x, z) \in X \times Z \mid xR \subseteq Sz\}$$
$$R \rhd S = \{(x, z) \in X \times Z \mid Sz \subseteq xR\}$$

is called *forward problem* resolution for the corresponding composition.

- Equations of the type $R \circ S = T$, $R \lhd S = T$, $R \rhd S = T$, when the unknown relation is S (or R), are called *composite relational equations*. Solving them in the unknown relation is called *inverse problem resolution* for the corresponding relational equation.

4.6 Fundamentals of Fuzzy Sets

The concept of a *set*, and set theory, is a powerful tool in mathematics. Unfortunately, a sine qua non condition underlying set theory, i.e., that an element can either belong to a set or not, is often not applicable in real life where many vague terms as "large," "high," "moderate," "reliable," etc. are extensively used. Unfortunately, such imprecise descriptions cannot be adequately handled by conventional mathematical tools. Atanassov [16] provides a brief chronology of the development of fuzzy sets and their extensions as:

> In 1965, Lotfi Zadeh introduced the concept of fuzzy sets [1]. Since then, fuzzy sets have been firmly established as a fruitful area of research, as well as of a tool for the evaluation of different objects and processes in nature and society. Fuzzy sets became an object of extensions since the sixties [12, 15, 17–19]. Chronologically, the first of these extensions, L-fuzzy sets, were introduced by Goguen in 1969 [12]. The second extension was from Zadeh [20], who introduced the idea of interval-valued fuzzy sets. The third extension is the rough sets, which were defined by Z. Pawlak in 1981 [21]. The fourth was intuitionistic fuzzy sets (IFSs), introduced in 1983 [22]. In recent years, many other extensions of fuzzy sets have been proposed. It is now clear that a significant part of these are trivial modifications of other existing fuzzy set extensions which have been "re-branded" under new names.

4.6.1   Crisp Sets

The concept of a set can be described as a collection of objects possibly linked through some properties.

A classical set has clear boundaries, i.e., $x \in A$ or $x \notin A$ exclude any other possibility.

Definition 4.5 Crisp sets Consider X to be a set and A a subset X $(A \subseteq X)$. Then the function:

$$\chi_A(x) = \begin{cases} 1 \ \text{if} \ x \in A \\ 0 \ \textit{if} \ x \notin A \end{cases}$$

is called the characteristic function of the set A in X.

Classical sets and their operations can be represented by their characteristic functions. Indeed, let us consider the union $A \cup B = \{x \in X | x \in A \text{ or } x \in B\}$. Its characteristic function is:

$$\chi_{A \cup B}(x) = \max\{\chi_A(x), \chi_B(x)\}.$$

For the intersection $A \cap B = \{x \in X | x \in A \text{ and } x \in B\}$, the characteristic function is:

$$\chi_{A \cap B}(x) = \min\{\chi_A(x), \chi_B(x)\}.$$

If we consider the complement of A in X, $A^c = \{x \in X | x \notin A\}$, it has the characteristic function:

$$\chi_{A^c}(x) = 1 - \chi_A(x).$$

Crisp sets satisfy the laws in Table 4.8.

Table 4.8 Laws of the algebra of sets

Idempotent laws	$A \cup A = A$	$A \cap A = A$
Associative laws	$(A \cup B) \cup C = A \cup (B \cup C)$	$(A \cap B) \cap C = A \cap (B \cap C)$
Commutative laws	$(A \cup B) = (B \cup A)$	$(A \cap B) = (B \cap A)$
Distributive laws	$A \cup (B \cap C) = (A \cup B) \cap (A \cup C)$	$A \cap (B \cup C) = (A \cap B) \cup (A \cap C)$
Identity laws	$A \cup \varnothing = A$	$A \cap U = A$
Absorption	$A \cup U = U$	$A \cap \varnothing = \varnothing$
Involution laws	$(A^c)^c = A$	
Complement laws	$A \cup A^c = U$	$A \cap A^c = \varnothing$
	$U^c = \varnothing$	$\varnothing^c = U$
De Morgan's laws	$(A \cup B)^c = A^c \cap B^c$	$(A \cap B)^c = A^c \cup B^c$

[9], p.7

4.6.2   Fuzzy Sets Introduction

Fuzzy set theory, introduced by Zadeh [1], is an extension of classical set theory. Contrary to the classical concept of a set, or crisp set, the boundary of a fuzzy set is not precise. That is, the change from non-membership to membership in a fuzzy set may be gradual rather than abrupt. This gradual change is expressed by a membership function, which completely and uniquely characterizes a particular fuzzy set. The membership function of a fuzzy set A is denoted by $\mu_{A(x)}$ and usually has the form:

$$\mu_A : X \rightarrow [0, 1],$$

where X denotes the universal set under consideration and A is a label of the fuzzy set defined by this function. The universal set is always assumed to be a crisp set. For each $x \in X$, the value $\mu_{A(x)}$ expresses the degree (or grade) of membership of element x of X in fuzzy set A. The symbol A of a fuzzy set is also used to denote the membership function of A. Since each fuzzy set is completely and uniquely defined by one particular membership function, then, $A(x)$ in the second notation has the same meaning as μ_A in the first notation. In this book, we use both notations.

Crisp sets may be viewed from the standpoint of fuzzy set theory as special fuzzy sets, in which $A(x)$ is either 0 or 1 for each $x \in X$ using the same notation for fuzzy sets and crisp sets. Fuzzy sets whose membership functions have various kinds of shapes are illustrated in Fig. 4.14. These functions may be considered as candidates for representing the meaning of a linguistic expression in the context of a given application. The width of each of these functions is strongly dependent on the application context. In general, a membership function that is supposed to capture the intended meaning of a linguistic expression in the context of a particular application must be somehow constructed [7].

The choice of membership functions depends both on contexts and observers. This choice of [0, 1] initially taken to evaluate the degrees of membership is too restrictive in some applications. The unit interval, [0, 1], is totally ordered and does

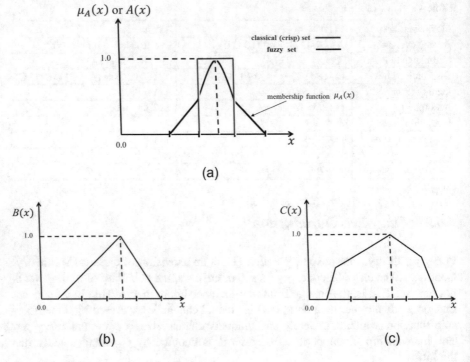

Fig. 4.14 Illustrations of membership functions

not allow incomparable degrees of membership. Goguen [12] has used a complete lattice to evaluate the degrees of membership introducing the L-fuzzy sets concept.

Definition 4.13 Let X be a non-empty, ordinary set, and let L be a complete lattice. A L-fuzzy subset, on X, is a mapping:

$$A : X \to L$$

That is, the family of all L-fuzzy subsets, of X, is just L^X consisting of all mappings from X to L. Here, L^X is called an L-fuzzy space, X is called the carrier domain of each L-fuzzy subset of it, and L is called the value domain, of each L-fuzzy subset of X. In the special case, in which $L = [0, 1] \times [0, 1]$ is equipped with the order $\leq$, where $(x_1, y_1) \leq (x_2, y_2)$, if and only if $x_1 \leq x_2$ and $y_1 \geq y_2$, we then talk about an *intuitionistic fuzzy set* [23].

Definition 4.14 An intuitionistic fuzzy set (IFS) A, in X, is an object of the following form:

$$A = \{ \langle x, \mu_A(x), \nu_A(x) \rangle : x \in X \}$$

where the functions:

$$\mu_A : X \to [0, 1]$$

and:

$$\upsilon_A : X \to [0, 1]$$

define the degree of membership and the degree of non-membership, of an element $x \in X$ and, even more, for each $x \in X$:

$$0 \le \mu_A(x) + \upsilon_A(x) \le 1$$

Now, if $\pi_A(x) = 1 - \mu_A(x) - \upsilon_A(x)$, then $\pi_A(x)$ is called the *degree of non-determinance* of an element $x \in X$ to the set A, where $\pi_A(x) \in [0, 1]$, $\forall\, x \in X$. It can be easily verified that each fuzzy set is a particular case of the intuitionistic fuzzy set, and if A is a fuzzy set, then $\pi_A(x) = 0$, $\forall\, x \in X$.

Given two fuzzy sets A, B defined on the same universal set X, A is said to be a *subset* of B ($A \subseteq B$) if and only if $A(x) \le B(x)$ for all $x \in X$. The set of all fuzzy subsets of X is called the *fuzzy power set* of X and is denoted by $F(X)$. Observe that this set is crisp, even though its members are fuzzy sets. For any fuzzy set A defined on a finite universal set X, its scalar cardinality, $|A|$, is defined by:

$$| A |= \sum_{x \in X} A(x)$$

Scalar cardinality is sometimes referred to in the literature as a *sigma count*.

4.6.3   α-Cuts

Among the most important concepts of standard fuzzy sets are the concepts of an α-cut and a *strong* α-cut as illustrated in Fig. 4.15. Given a fuzzy set A defined on X and a particular number α in the unit interval $[0, 1]$, the α-cut of A, denoted by $^{\alpha}A$, is a crisp set that consists of all elements of X whose membership degrees in A are greater than or equal to α. This can formally be written as:

$$^{\alpha}A = \{x | A(x) \ge \alpha\}.$$

The strong α-cut, $^{\alpha+}A$, has a similar meaning, but the condition "$\ge$" is replaced with the stronger condition ">". Formally:

$$^{\alpha+}A = \{x | A(x) > \alpha\}.$$

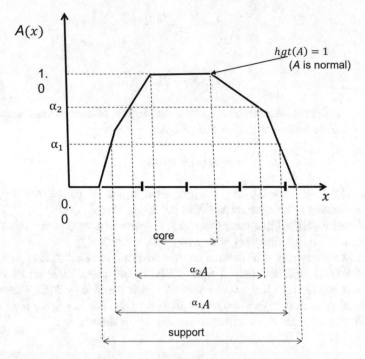

Fig. 4.15 Illustration of the concept of α-cut. (Adapted from [8], Chap. 2)

The set ^{0+}A is called the *support* of A, and the set ^1A is called the *core* of A. When the core A is not empty, A is called *normal*; otherwise, it is called *subnormal*.

Two membership grades are especially important in fuzzy sets, the largest and the smallest. The largest value of A is called the *height* of A, and it is denoted by hgt(A). The smallest is called the *plinth* and is denoted by plth(A). Consider a fuzzy set A in X; then:

Definition 4.15 *Height* The *height* of a fuzzy set A, hgt(A), is the maximum or supremum membership grade achieved by $A(x)$ over the domain X:

$$\mathrm{hgt}(A) = \sup_{x \in X} A(x)$$

Definition 4.16 (*Plinth*) The *plinth* of a fuzzy set A, plth(A), is the minimum or infimum, membership grade achieved by $A(x)$ over the domain X:

$$\mathrm{plth}(A) = \inf_{x \in X} A(x)$$

The *height* and *plinth* operators occur frequently in fuzzy set theory, for instance, in the fuzzy relational calculus and in the generalization of classical inference rules.

Their function is mainly to reduce the size of formulas and to facilitate the study of these formulas by exploiting their properties.

Figure 4.15 illustrates the α-cuts concepts just introduced. We can see that:

$$^{\alpha_1}A \subseteq {}^{\alpha_2}A$$

$$^{\alpha_1+}A \subseteq {}^{\alpha_2+}A$$

when $\alpha_1 \geq \alpha_2$. This implies that the set of all distinct α-cuts (as well as strong α-cuts) is always a *nested family* of crisp sets. When α is increased, the new α-cut (strong α-cut) is always a subset of the previous one. Clearly, $^0A = X$ and $^{1+}A = \emptyset$. It is well established [7] that each fuzzy set is uniquely represented by the associated family of its α-cuts via the formula:

$$A(x) = \sup\{\alpha \cdot {}^{\alpha}A(x) | \alpha \in [0, 1]\}$$

or by the associated family of its strong α-cuts via the formula:

$$A(x) = \sup\{\alpha \cdot {}^{\alpha+}A(x) | \alpha \in [0, 1]\}$$

where sup denotes the supremum of the respective set and $^{\alpha}A$ (or $^{\alpha+}A$) denotes for each $\alpha \in [0, 1]$ the characteristic function of the α-cut (or strong α-cut, respectively). Also presented in ([8], Chap. 2), the significance of the α-cut (or strong α-cut) representation of fuzzy sets is that it connects fuzzy sets with crisp sets where each fuzzy set is a collection of nested crisp sets that are conceived as a whole.

The α-cuts of A are nested in the sense that $\alpha > \beta$ implies that $^{\alpha}A \subseteq {}^{\beta}A$ In particular:

$$^{\alpha}A = \bigcap_{\beta < \alpha} {}^{\beta}A$$

Going from the level cut representation to the membership function and back is easy. The membership function can be recovered from the level cuts as follows:

$$A(u) = \sup\{\alpha : u \in {}^{\alpha}A\} = \sup_{\alpha \in [0, 1]} \min(\alpha, A(\alpha)),$$

with $^{\alpha}A(u) = 1$ if $u \in {}^{\alpha}A$ and 0 otherwise.

Negoita and Ralescu [24] obtained a representation theorem according to which, a given fuzzy set can be represented by a combination of its α-level cuts, and conversely given a family of crisp sets, one can construct a fuzzy set from them under certain conditions.

Definition 4.16 Cutworthy property (from Klir [7]) Any property that is extended from classical set theory into the domain of fuzzy set theory is called a *cutworthy property*. This requires that the classical property be satisfied by all α-cuts of the fuzzy set concerned.

Example 4.15 Cutworthy (convexity) When convexity of fuzzy sets is defined by the requirement that all α-cuts of a fuzzy convex set be convex in the classical sense, this conception of fuzzy convexity is *cutworthy*. Other important examples are the concepts of a fuzzy partition, fuzzy equivalence, fuzzy compatibility, and various kinds of fuzzy orderings that are *cutworthy*.

It is important to realize that many (perhaps most) properties of fuzzy sets, perfectly meaningful and useful, are not *cutworthy*. These properties cannot be derived from classical set theory. Finally, another way to connect classical set theory and fuzzy set theory is to fuzzify the functions as discussed in the following section.

4.6.4 Fuzzified Functions

The material of this section is adapted from [8], Chapter 2. Given a function:

$$f : X \rightarrow Y,$$

where X and Y are crisp sets, we say that the function is *fuzzified* when it is extended to act on fuzzy sets defined on X and Y. Formally, the fuzzified function, F, has the form:

$$F : \mathcal{F}(X) \rightarrow \mathcal{F}(Y),$$

where $\mathcal{F}(X)$ and $\mathcal{F}(Y)$ denote the fuzzy power set of X and Y, respectively. To qualify as a fuzzified version of f, function F must conform to f within the extended domain $\mathcal{F}(X)$ and $\mathcal{F}(Y)$. This is guaranteed when applying the *extension principle*. According to this principle:

$$B = F(A)$$

is determined for any given fuzzy set $A \in \mathcal{F}(X)$ via the formula[8]:

$$B(y) = \max_{x|y=f(x)} A(x)$$

for all $y \in Y$. The inverse function:

$$F^{-1} : \mathcal{F}(Y) \rightarrow \mathcal{F}(X)$$

of F is defined, according to the *extension principle*, for any given $B \in F(Y)$, by the formula:

---

[8] When the maximum does not exist, it is replaced with the supremum.

$$\left[F^{-1}(B) \right](x) = B(y)$$

for all $x \in X$, where $y = f(x)$. Clearly:

$$F^{-1}[F(A)] \supseteq A$$

for all $A \in \mathcal{F}(X)$, where the equality is obtained when f is a one-to-one function.

The use of the extension principle is illustrated in Fig. 4.16, where it is shown how fuzzy set A is mapped to fuzzy set B via function F that is consistent with the given function f. That is, $B = F(A)$. For example, since $b = f(a_1) = f(a_2) = f(a_3)$, then we have $B(b) = \max \{A(a_1), A(a_2), A(a_3)\}$, by evaluating:

$$B(y) = \max_{x|y=f(x)} A(x)$$

Conversely:

$[F^{-1}(B)](a_1) = [F^{-1}(B)](a_2) = [F^{-1}(B)](a_3) = B(b)$ by evaluating, $[F^{-1}(B)]$ $(x) = B(y)$.

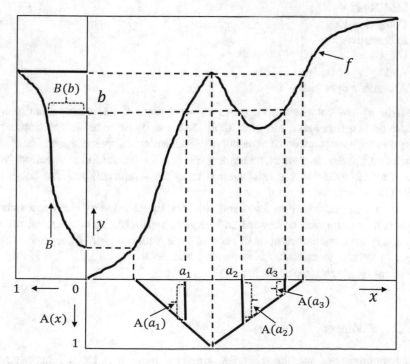

Fig. 4.16 Illustration of the extension principle to obtain a fuzzified function. (Adapted from [8], Chap. 2)

The *extension principle* is basically described by $B(y) = \max\limits_{x|y=f(x)} A(x)$ and $[F^{-1}(B)](x) = B(y)$. These equations are direct generalizations of similar equations describing the *extension principle* of classical set theory. In the latter, symbols A and B denote characteristic functions of crisp sets.

4.6.5   Operations on Fuzzy Sets

The material from this section is partially excerpted from these two references [7, 25]. Let $\mathcal{F}(X)$ denote the collection of fuzzy sets on a given universe of discourse X. The basic connectives in fuzzy logic and fuzzy set theory are inclusion, union, intersection, and complementation. In fuzzy set theory, these operations are performed on the membership functions which represent the fuzzy sets. Zadeh based the union and intersection connectives on the max and min operations.

Operations on fuzzy sets possess a considerably greater variety than those on classical sets. In fact, most operations on fuzzy sets do not have any counterparts in classical set theory. The following six types of operations on fuzzy sets are currently recognized:

(a) Modifiers
(b) Complements
(c) Intersections
(d) Unions
(e) Averaging operations
(f) Arithmetic operations

Modifiers and complements operate on one fuzzy set. Intersections and unions operate on two fuzzy sets, but their application can be extended to any number of fuzzy sets via their property of associativity. The averaging operations, which are not associative, operate, in general, on n fuzzy sets ($n \geq 2$). In addition to these five types of operations, special fuzzy sets referred to as *fuzzy intervals* are also subject to arithmetic operations.

The topic of operations on fuzzy sets has been investigated by many researchers, and it is now quite well developed and very extensive. In this chapter, we present only a very brief characterization of each of the introduced types of operations. The reader is invited to consult the numerous excellent books (e.g., [7, 15, 25]) on operations on fuzzy sets for further study.

4.6.5.1   Modifiers

Modifiers are unary operations whose primary purpose is to modify fuzzy sets to account for linguistic hedges, such as *very*, *fairly*, *extremely*, *moderately*, etc., in representing expressions of natural language. Each modifier, m, is an increasing (and usually continuous) one-to-one function of the form:

$$m : [0, 1] \rightarrow [0, 1],$$

which assigns to each membership grade $A(x)$ of a given fuzzy set A a modified grade $m(A(x))$. The modified grades for all $x \in X$ define a new, modified fuzzy set. Denoting conveniently this modified set by MA, we have:

$$m(A(x)) = MA(x)$$

The function m is totally independent of elements x to which values $A(x)$ are assigned; it depends only on the values themselves. In describing its formal properties, we may thus ignore x and assume that the argument of m is an arbitrary number a in the unit interval $[0, 1]$. In general, a modifier increases or decreases values of the membership functions to which it is applied but preserves the order. That is:

if $a \leq b$ then $m(a) \leq m(b)$ for all $a, b \in [0, 1]$ or,

recognizing the meaning of a and b:

if $A(x) \leq A(y)$ for some $x, y \in X$, then $MA(x) \leq MA(y)$.

Sometimes, it is also required that $m(0) = 0$ and $m(1) = 1$. Modifiers are basically of three types, depending on which values of the membership functions they increase or decrease:

(i). Modifiers that increase all values
(ii). Modifiers that decrease all values
(iii). Modifiers that increase some values and decrease other values

4.6.5.2   Complements

Complements of fuzzy sets may be defined via appropriate *unary* operations on $[0, 1]$. While modifiers preserve the order of membership degrees, complements reverse the order. Each fuzzy complement, c, must satisfy at least the following two requirements:

$(Req. 1)$ $c(0) = 1$ and $c(1) = 0$
$(Req. 2)$ $\forall a, b \in [0, 1]$, if $a \leq b$, then $c(a) \geq c(b)$

Requirement $(Req. 1)$ guarantees that all fuzzy complements collapse to the unique classical complement for crisp sets. Requirement $(Req. 2)$ guarantees that increases in the degree of membership in A do not result in increases in the degree of membership in the complement of A. This is essential since any increase in the degree of membership of an object in a fuzzy set cannot simultaneously increase the degree of non-membership of the same object in the same fuzzy set.

When used as a fuzzy complement, function c is always applied to membership degrees $A(x)$ of some fuzzy set A. It depends only on the values $A(x)$ and not on the objects x to which the values are assigned. For the purpose of characterizing fuzzy complements, we may thus ignore these objects and observe only how function c depends on numbers in $[0, 1]$. This is the reason why no reference is made to specific degrees $A(x)$ in the requirements (*Req.* 1) and (*Req.* 2).

However, when function c defines a complement of a particular fuzzy set A, we must keep track of the relevant objects x to make the connection between $A(x)$ and c $[A(x)]$. Although requirements (*Req.* 1) and (*Req.* 2) are sufficient to characterize the largest class of acceptable fuzzy complements, two additional requirements are imposed on fuzzy complements by most applications of fuzzy set theory:

> (*Req.* 3) c is a continuous function
> (*Req.* 4) $c(c(a)) = a, \ \forall \, a \in [0, 1]$

Requirement (*Req.* 3) guarantees that infinitesimal changes in the argument do not result in discontinuous changes in the function. Requirement (*Req.* 4) guarantees that fuzzy sets are not changed by double complementation. Other classes of fuzzy complements (e.g., Yager's class) are discussed by Klir and Yuan [7].

4.6.5.3   Intersections and Unions

Intersections and unions of fuzzy sets, denoted by i and u, respectively, are generalizations of the classical operations of intersections and unions of crisp sets. They may be defined via appropriate functions that map each pair of real numbers from $[0, 1]$, (representing degrees $A(x)$ and $B(x)$ of given fuzzy sets A and B for some $x \in X$), into a single number in $[0, 1]$, (representing membership degree $(A \cap B)(x)$ of the intersection of A and B or membership degree of the union of A and B for the given x).

Hence:

$$(A \cap B)(x) = i[A(x), B(x)]$$

and:

$$(A \cup B)(x) = u[A(x), B(x)]$$

for all $x \in X$. To discuss properties of functions i and u, which do not depend on x, we may view i and u as functions from $[0, 1] \times [0, 1]$ to $[0, 1]$.

Contrary to their classical counterparts, fuzzy intersections and unions are not unique. This is a natural consequence of the well-established fact that the linguistic expressions "x is a member of A and B" and "x is a member of A or B" have different meanings when applied by human beings to different vague concepts in different contexts. To be able to capture the different meanings, we need to characterize the classes of fuzzy intersections and fuzzy unions as broadly as possible.

Table 4.9 Requirements for *t-norms* and *t-conorms* [8]

	t-norms/fuzzy intersections	*t-conorms*/fuzzy unions
Boundary requirement	$i(a, 1) = a$	$u(a, 0) = a$
Monotonicity	$b \leq d$ implies $i(a, b) \leq i(a, d)$	$b \leq d$ implies $u(a, b) \leq u(a, d)$
Commutativity	$i(a, b) = i(b, a)$	$u(a, b) = u(b, a)$
Associativity	$i(a, i(b, d)) = i(i(a, b), d)$	$u(a, u(b, d)) = u(u(a, b), d)$

It has been established that operations known in the literature as *triangular norms* or *t-norms* and *triangular conorms* or *t-conorms*, which have been extensively studied in mathematics, possess exactly those properties that are requisite, on intuitive grounds, for fuzzy intersections and fuzzy unions, respectively. The class of *t-norms*/fuzzy intersections is characterized by four requirements; the class of *t-conorms*/fuzzy unions is also characterized by four requirements, three of which are identical with the requirements for *t-norms*. In the list of Table 4.9, the requirements for *t-norms*/fuzzy intersections i are paired with their counterparts for *t-conorms*/fuzzy unions u and must be satisfied for all $a, b, d \in [0, 1]$:

It is easy to see that the first three requirements for i ensure that fuzzy intersections collapse to the classical set intersection when applied to crisp sets:

- $i(0, 1) = 0$ and $i(1, 1) = 1$ follow directly from the boundary requirement.
- $i(1, 0) = 0$ and $i(0, 0) = 0$ follow then from commutativity and monotonicity.

Similarly, the first three requirements for u ensure that fuzzy unions collapse to the classical set union when applied to crisp sets.

- Commutativity ensures that fuzzy intersections and unions are symmetric operations, indifferent to the order in which sets to be combined are considered.
- Commutativity and monotonicity guarantee that fuzzy intersections and unions do not decrease when any of their arguments are increased, and do not increase when any arguments are decreased.
- Associativity offers the extension of fuzzy intersections and unions to more than two sets, in perfect analogy with their classical counterparts.

Table 4.10 presents examples of some common fuzzy intersections and fuzzy unions with their usual names (each defined for all $a, b \in [0, 1]$).

It is easy to verify that the inequalities:

$$imin(a, b) \leq i(a, b) \leq min(a, b)$$

$$max(a, b) \leq u(a, b) \leq umax(a, b)$$

are satisfied for all $a, b \in [0, 1]$ by any fuzzy intersection i and any fuzzy union u, respectively. These inequalities specify, in effect, the full ranges of fuzzy intersections and fuzzy unions. It should be mentioned that various other classes of fuzzy intersections and unions have been examined in the literature (e.g., *Yager's classes of intersections and unions* [26]) . Moreover, special procedures are now available by which new classes of fuzzy intersection and unions can be generated.

Table 4.10 Common fuzzy intersections and fuzzy unions [8]

Types of fuzzy intersections and unions	Equations
Standard fuzzy intersection	$i(a,b) = min\,(a,b)$
Algebraic product	$i(a,b) = ab$
Bounded difference	$i(a,b) = max\,(0, a + b - 1)$
Drastic intersection	$imin(a,b) = \begin{cases} a \text{ when } b = 1 \\ b \text{ when } a = 1 \\ 0 \text{ otherwise} \end{cases}$
Standard fuzzy union	$u(a,b) = max\,(a,b)$
Algebraic sum	$u(a,b) = a + b - ab$
Bounded sum	$u(a,b) = min\,(1, a + b)$
Drastic union	$umax(a,b) = \begin{cases} a \text{ when } b = 0 \\ b \text{ when } a = 0 \\ 1 \text{ otherwise} \end{cases}$

Among the great variety of fuzzy intersections and unions, the standard operations possess certain desirable properties namely from the computational point of view.

1. The standard intersection is the weakest one among all fuzzy intersections; the standard union is the strongest one among all fuzzy unions.
2. The standard operations are the only *cutworthy* operations among all fuzzy intersections and unions.
3. The standard operations are the only operations among fuzzy intersections and unions that are idempotent: $i_{standard}(a, a) = u_{standard}(a, a) = a, \quad \forall\, a \in [0, 1]$.
4. The nonstandard fuzzy intersections are only subidempotent: $i_{standard}(a, a) < a, \forall\, a \in [0, 1]$.
5. The nonstandard fuzzy unions are only superidempotent: $u_{standard}(a, a) > a, \forall\, a \in [0, 1]$.
6. When using the standard fuzzy operations, errors of the operands do not combine.
7. Combination of fuzzy counterparts of the three classical set-theoretic operations (complement, intersection, union) inevitably violates some properties of the classical operations (underlying Boolean algebra) due to imprecise boundaries of fuzzy sets.
8. The standard fuzzy operations violate only the law of excluded middle and the law of contradiction. Some other combinations preserve these laws but violate distributivity and idempotence [7].

It is worth noting that logic operations of negation, conjunction, and disjunction are defined in exactly the same way as the operations of complementation, intersection, and union, respectively. In fact, one can write [6]:

Definition 4.17 If $A, B \in F(x)$, then for all $x, y \in X$ the following fuzzy sets were defined by Zadeh:

(i) $A \cup B(x) = max\,(A(x), B(x))$.
(ii) $A \cap B(x) = min\,(A(x), B(x))$.
(iii) $A^c(x) = 1 - A(x)$.

These operations follow De Morgan's laws, idempotency laws, commutativity laws, associativity laws, absorption laws, and distributivity laws, but the law of contradiction and the law of excluded middle no more hold. Anyhow, the overlap between a set and its complement is bounded by the following way:

$$\forall x \in X, A \cap A^c(x) \leq 0.5$$

and the corresponding weakened law of excluded middle is:

$$\forall x \in X, A \cup A^c(x) \geq 0.5$$

Definition 4.18 The triangular norm (*t-norm*) T and triangular conorm (*t-conorm*) δ are increasing, associative, commutative, and $[0,1]^2 \rightarrow [0,1]$ mappings satisfying $T(1, x) = x$ and $\delta(x, 0) = x$ for all $x \in [0,1]$. To every *t-norm* T there corresponds a *t-conorm* T^* called the dual *t-conorm*, defined by:

$$T^*(x, y) = 1 - T\left(1 - x, 1 - y\right)$$

The following is a list of some popular choices for *t-norms*:

(i)   The minimum operator M : $M(x, y) = \ \min(x, y)$
(ii) The Lukasiewicz's t-norm W : $W(x, y) = \ \max(x + y - 1, 0)$
(iii) The product operator P : $P(x, y) = xy$

The corresponding dual t-conorms are:

(i)   The maximum operator M^* : $M^*(x, y) = \ \max(x, y)$
(ii) The bounded sum W^* : $W^*(x, y) = \ \min(x + y, 0)$
(iii) The probabilistic sum P^* : $P^*(x, y) = x + y - xy$

Table 4.11 gives us information about the relationship of these fuzzy conjunctions and disjunctions with laws satisfied by their crisp counterparts. It is important to note that if a system disobeys some law, most of the times, it satisfies some weaker version of the law. An important concept associated with *t-norms* and *t-conorms* is that of compatibility.

Table 4.11 Fuzzy conjunctions and disjunctions with laws satisfied by their crisp counterparts

	(M, M^*)	(W, W^*)	(P, P^*)
De Morgan laws	Y	Y	Y
Commutativity	Y	Y	Y
Associativity	Y	Y	Y
Idempotency	Y	N	N
Absorption law	Y	N	N
Distributivity laws	Y	N	N

Definition 4.19 Compatibility [27] Two crisp sets are compatible if they have at least one element in common. Similarly, in fuzzy context, it is defined as the maximum degree of overlap, i.e., as the height of the intersection of these fuzzy sets. Consider two fuzzy sets A and B in X and a *t-norm T*. The degree of compatibility $\text{Com}_T(A, B)$ of A and B is:

$$\text{Com}_T(A, B) = \sup_{x \in X} T(A(x), B(x))$$

4.6.5.4   Averaging Operations

Fuzzy intersections (t-norms) and fuzzy unions (t-conorms) are special types of operations for aggregating fuzzy sets. Excerpted from Beg [6]: "Given two or more fuzzy sets, they produce a single fuzzy set, an aggregate of the given sets. While they do not cover all aggregating operations, they cover all aggregating operations that are associative. Because of the lack of associativity, the remaining aggregating operations must be defined as functions of n arguments for each n $\geq$ 2. These remaining aggregation operations are called averaging operations. They average in various ways membership functions of two or more fuzzy sets defined on the same universal set. They do not have any counterparts in classical set theory. Indeed, an average of several characteristic functions of classical sets is not, in general, a characteristic function." The treatment of averaging operations to aggregate fuzzy sets for $n \geq 2$ is beyond the scope of this book. We refer the reader to the rather deep treatment of Beliakov et al. in [28, 29].

4.6.5.5   Arithmetic Operations

The material of this section is excerpted and rearranged from Klir et al. ([7] Chap. 4 and [8] Chap. 2). Arithmetic operations are applicable only to special fuzzy sets that are known as *fuzzy intervals*. These are standard and normal fuzzy sets defined on the set of real numbers, $\mathbb{R}$, whose α-cuts for all $\alpha \in (0, 1]$ are closed intervals of real numbers and whose supports are bounded. Any fuzzy interval A for which $A(x) = 1$ for exactly one $x \in \mathbb{R}$ is called a *fuzzy number*.

Every fuzzy interval A may conveniently be expressed for all $x \in \mathbb{R}$ in the canonical form:

$$A(x) = \begin{cases} f_A(x) & \text{when } x \in [a, b) \\ 1 & \text{when } x \in [b, c] \\ g_A(x) & \text{when } x \in (c, d] \\ 0 & \text{otherwise,} \end{cases}$$

where a, b, c, d are specific real numbers such that $a \leq b \leq c \leq d$, f_A is a real-valued function that is increasing and right-continuous, and g_A is a real-valued function that is decreasing and left continuous.

Among the various types of fuzzy sets, of special significance are fuzzy sets that are defined on the set $\mathbb{R}$ of real numbers. These fuzzy sets should capture the intuitive notions of approximate numbers or intervals, such as "numbers that are around a given interval of real numbers." Such concepts are essential for characterizing states of fuzzy variables and, consequently, play an important role in many applications, including fuzzy control, decision-making, approximate reasoning, optimization, and statistics with imprecise probabilities.

To qualify as a *fuzzy number,* a fuzzy set A on $\mathbb{R}$ must possess at least the following three properties:

(i) A must be a normal fuzzy set.
(ii) $^{\alpha}A$ must be a closed interval for every $\alpha \in (0, 1]$.
(iii) The support of A, ^{0+}A, must be bounded.

Special cases of fuzzy numbers include ordinary real numbers and intervals of real numbers, as illustrated in Fig. 4.17:

(a) An ordinary real number 1.3
(b) A fuzzy number expressing "close to 1.3"
(c) A crisp closed interval [1.25, 1.35]
(d) A fuzzy interval (a fuzzy number with a flat region)

The bounded support of a fuzzy number and all its α-cuts for $\alpha \neq 0$ must be closed intervals to allow us to define meaningful arithmetic operations on fuzzy numbers in terms of standard arithmetic operations on closed intervals, well established in classical interval analysis [30].

For any fuzzy interval A, the α-cuts of A are expressed for all $\alpha \in (0, 1]$ by the formula:

$$^{\alpha}A = \begin{cases} f_A^{-1}(\alpha), g_A^{-1}(\alpha) & \text{when } \alpha \in (0, 1) \\ [b, c] & \text{when } \alpha = 1 \end{cases}$$

where f_A^{-1} and g_A^{-1} are the inverse functions of f_A and g_A, respectively.

Employing the α-cut representation, arithmetic operations on fuzzy intervals are defined in terms of the well-established arithmetic operations on closed intervals of real numbers. Given any pair of fuzzy intervals, A and B, the four basic arithmetic operations on the α-cuts of A and B are defined for all $\alpha \in (0, 1]$ by the general formula:

$$^{\alpha}(A * B) = \{a * b | \langle a, b \rangle \in {}^{\alpha}A \times {}^{\alpha}B\},$$

where $*$ denotes any of the four basic arithmetic operations; when the operation is division of A by B, it is required that $0 \neq {}^{\alpha}B$ for any $\alpha \in (0, 1]$. Let $\underline{a}$:

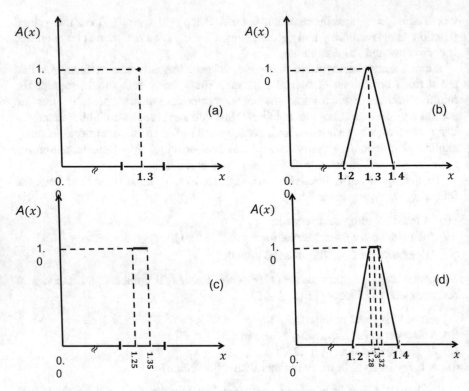

Fig. 4.17 A comparison of (**a**) an ordinary real number 1.3, (**b**) a fuzzy number expressing "close to 1.3," (**c**) a crisp closed interval [1.25, 1.35], and (**d**) a fuzzy interval (a fuzzy number with a flat region). (Adapted from [7])

$$^{\alpha}A = \left[\underline{a}(\alpha), \overline{a}(\alpha)\right]$$
$$^{\alpha}B = \left[\underline{b}(\alpha), \overline{b}(\alpha)\right].$$

Standard fuzzy arithmetic does not take into account constraints among fuzzy numbers that exist in various applications leading to a kind a deficiency of information. To remedy to this deficiency, *constrained fuzzy arithmetic* has been introduced to take all existing constraints among fuzzy numbers in each application into account [31, 32]. Fuzzy arithmetic is essential for evaluating algebraic expressions formed by operations of fuzzy arithmetic and fuzzy algebraic equations in which coefficients and unknowns are fuzzy numbers. Furthermore, fuzzy arithmetic is a basis for developing fuzzy calculus and for fuzzifying any area of mathematics that involves numbers.

Table 4.12 Definitions of diverse terms related to operations with lattices

Conjunctor	$C : L \times L \to L$	$C(0,1) = 0$ *and* $C(1,1) = 1$
Disjunctor	$D : L \times L \to L$	$D(0,1) = D(1,0) = 1$ *and* $D(0,0) = 0$
Implicator	$I : L \times L \to L$	$I(0,1) = I(1,1) = 1$ *and* $I(1,0) = 0$
Negator	$N : L \to L$	$N(0) = 1$ *and* $N(1) = 0$

4.7 Basics of Fuzzy Relational Calculus

The material of this section is mainly from Beg [6], Zimmerman [33], and Peeva and Kyosev [5]. As already mentioned, Zadeh [1, 2, 20] introduced the concept of fuzzy relation, defined the notion of equivalence, and gave the concept of fuzzy ordering. Fuzzy relations have broad utility. Compared with crisp relations, they have greater expressive power for expressing the strength of links between elements.

The classical concepts for Boolean relational equations have been generalized in fuzzy relational calculus. First the notions were extended for fuzzy relations and their compositions on [0, 1] and after that for lattices [3, 7]. *L*-fuzzy sets were introduced and studied in [12]. Note that *L*-fuzzy relations are considered because the lattice theory provides the most suitable tool for solving problems in fuzzy relational calculus. For a complete lattice $\mathbb{L} = (L, \sqcup, \sqcap, 0, 1)$, the Boolean logical operations $\wedge, \vee, \Rightarrow$, are extended as in Table 4.12.

Definition 4.20 A *fuzzy relation* from a universe X to a universe Y is a fuzzy set in $X \times Y$. $R(x, y)$ is called the degree of relationship between x and y.

In this case:

1. The *afterset xR* of x is the fuzzy set in Y defined by $xR(y) = R(x, y)$.
2. The *foreset Ry* of y is the fuzzy set in X defined by $Ry(x) = R(x, y)$.
3. The *domain* dom(R) of R is the fuzzy set in X defined by: $\text{dom}(R)(x) = \text{hgt}(xR)$.
4. The *range* rng(R) of R is the fuzzy set in Y defined by: $\text{rng}(R)(y) = \text{hgt}(Ry))$.
5. The *converse* fuzzy relation R^{-1} of R is the fuzzy relation from Y to X is defined as:

$$R^{-1}(y, x) = R(x, y).$$

6. The α-cuts of a relation $^{\alpha}R$ are defined as follows:

$$^{\alpha}R = \{(x, y) \mid (y, x) \in X \times Y : R(x, y) \geq \alpha\}.$$

7. The *resolution form* of a relation:

$$R = \bigvee_{\alpha \in [0,1]} {}^{\alpha}R$$

8. The *complement* relation:

$$R^c(x,y) = 1 - R(x,y)$$

Definition 4.20 Fuzzy relation If A and B are two fuzzy subsets of a given universe X, then we can define a fuzzy relation R on X, a conjunction operator such as "min" so that:

$$R = \{(x,y), \min(A(x), B(y))\}$$

This concept can also be extended to fuzzy sets in different universes. These definitions, so obtained, reduce to their crisp counterparts if the sets are crisp sets, and they encompass many interesting properties (for details, see [33]).

Beg and Ashraf [6] present different forms of fuzzy equivalence relations as similarity relations, likeness relations, probabilistic relations, and equivalence relations (Fig. 4.18). All these relations claim to model approximate equality. All these relations were also generalized under the name indistinguishability operators. Also in Beg and Ashraf [6] is presented a discussion of different types of fuzzy orderings and of fuzzy inclusion, an important concept that provides a basis for fuzzy similarity and measures of similarity.

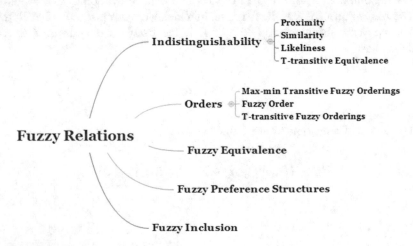

Fig. 4.18 An arborescence of fuzzy relations

Fuzzy preference structures that play an important role while working on fuzzy relations are also discussed. We let the reader consulting the numerous references cited in Beg and Ashraf [6] to get a deeper look at fuzzy relations depicted in Fig. 4.18.

4.7.1 *Images and Compositions for Fuzzy Relations*

Images and compositions are among the basic operations on crisp or fuzzy relations. The crisp case has been presented in Sect. 4.4. This section presents the fuzzy one.

Let $R \subseteq X \times Y$ be a fuzzy relation.

The *R-afterset xR* of $x \in X$ is the fuzzy set $xR(y)$ in Y defined by:

$$xR(y) = \{(\, y, \mu_R(x, y)) \mid y \in Y \wedge ((x, y), \mu_R(x, y) \in R)\}$$

The *R-foreset Ry* of $y \in Y$ is the fuzzy set $(x)Ry$ in X defined by:

$$(x)Ry = \{(\, x, \mu_R(x, y)) \mid x \in X \wedge ((x, y), \mu_R(x, y) \in R)\}$$

Let C be a *conjunctor* and I be an *implicator*. Let $A \subseteq X$ be a fuzzy set in X. *Direct image R(A)* of A under R is the fuzzy set in Y defined as follows:

$$R(A)(y) = \sup_{x \in X} C(A(x), R(x, y)).$$

Subdirect image $R_{\triangleleft}(A)$ of A under R is the fuzzy set in Y defined as follows:

$$R_{\triangleleft}(A)(y) = \inf_{x \in X} I(A(x), R(x, y))$$

Superdirect image $R_{\triangleright}(A)$ of A under R is the fuzzy set in Y defined as follows:

$$R_{\triangleright}(A)(y) = \inf_{x \in X} I(R(x, y), A(x))$$

Let $R \subseteq X \times Y$ and $C \subseteq Y \times Z$ be composable fuzzy relations. The following three kinds of binary compositions are defined:

Round composition R $\circ$ S:

$$R \circ S(x, z) = \sup_{y \in Y} C(R(x, y), S(y, z))$$

Subcomposition R $\triangleleft$ S:

$$R \triangleleft S(x, z) = \inf_{y \in Y} I(R(x, y), S(y, z))$$

Supercomposition $R \rhd S$:

$$R \rhd S(x, z) = \inf_{y \in Y} I(S(y, z), R(x, y))$$

4.7.2 Fuzzy Relations: Matrix Representation

When dealing with finite universes, fuzzy relations and their compositions can be represented by means of a matrix. A fuzzy relation R from $X = \{x_1, x_2, \cdots, x_l\}$ to $Y = \{y_1, y_2, y_3, \cdots, y_m\}$ can be represented by means of an $l \times m$ matrix, as follows:

$$R = \begin{bmatrix} R_{11} & \cdots & R_{1m} \\ \vdots & \ddots & \vdots \\ R_{l1} & \cdots & R_{lm} \end{bmatrix},$$

where R_{ij} stands for $R(x_i, y_j)$. A fuzzy set A in X can be represented by means of a row vector l with entries:

$$A = (A_1, A_1, \cdots, A_l),$$

where A_j stands for $A(x_i)$. The direct image of A under R can be written as follows:

$$R^T(A) = (A_1, A_1, \cdots, A_l) \begin{bmatrix} R_{11} & \cdots & R_{1m} \\ \vdots & \ddots & \vdots \\ R_{l1} & \cdots & R_{lm} \end{bmatrix},$$

where the matrix product is calculated using the triangular norm, T, as multiplication and the maximum operator as addition. Now consider a fuzzy relation S from Y to $Z = \{z_1, z_2, \cdots, z_n\}$. The *max-T* composition of R and S can be written as follows:

$$R \circ^T S = \begin{bmatrix} R_{11} & \cdots & R_{1m} \\ \vdots & \ddots & \vdots \\ R_{l1} & \cdots & R_{lm} \end{bmatrix} \begin{bmatrix} S_{11} & \cdots & S_{1n} \\ \vdots & \ddots & \vdots \\ S_{m1} & \cdots & S_{mn} \end{bmatrix}$$

The *max-T* composition is similar to the well-known matrix product, again by using the triangular norm as multiplication and the maximum operator as addition. Numerical examples are provided in Beg [6], p.30.

4.7.3 Fuzzy Relations and Membership Matrices

Peeva and Kyosev [5] (Chap. 2) consider membership matrices as suitable representatives for fuzzy relations that permit to apply the matrix calculus instead of computing with relations. They linked membership matrices on L to a bounded chain where L is the underlying totally ordered set of the bounded chain $\mathbb{L} = (L, \sqcup, \sqcap, 0, 1)$, 0 and 1 are the universal bounds and the operations are join $\sqcup$ and meet $\sqcap$.

Let $I \neq \varnothing$ and $J \neq \varnothing$ be index sets and $a : I \times J \to L$ be a map. The set of all maps from $I \times J$ to L is denoted by $L^{I \times J}$:

$$L^{I \times J} = \{a | a : I \times J \to L\}.$$

We write I_m and J_n for finite index sets $I = \{1, \dots, m\}$ and $J = \{1, \dots, n\}$, respectively, with cardinalities m and n, i.e., $|I| = m$ and $|J| = n$.

Definition 4.21 Membership matrix The matrix $A = [a_{ij}]_{m \times n}$ of type $m \times n$ is called a *membership matrix* on L if there exists a map $a : I_m \times J_n \to L$ such that for the elements a_{ij} of A and for the images $a(i, j) \in L$ by the map $a : I_m \times J_n \to L$ for each i, $1 \leq i \leq m$, and for each j, $1 \leq j \leq n$, we have $a_{ij} = a(i, j)$. The bounded chain provides operating on membership matrices. Membership matrices give a convenient representation of fuzzy relations.

Let $A = [a_{ij}]_{m \times n}$ and $B = [b_{ij}]_{m \times n}$ be matrices on L of the same type. Then:

(i) $A \leq B \Leftrightarrow a_{ij} \leq b_{ij}$ for each i, $1 \leq i \leq m$ and for each j, $1 \leq j \leq n$.
(ii) $A = B \Leftrightarrow a_{ij} = b_{ij}$ for each i, $1 \leq i \leq m$ and for each j, $1 \leq j \leq n$.

Consider a membership matrix $R = [r_{xy}]$ associated to any fuzzy relation $R \subseteq X \times Y$, with elements $r_{xy} = \mu_R(x, y)$ for each $(x, y) \in X \times Y$. If the relation is on a finite support, we represent it by a matrix of finite type. The relationship between membership matrices and fuzzy relations permits easily to perform operations on membership matrices instead of on fuzzy relations. Unary and binary operations can be performed:

Unary operations:—**inverse**[9] or **transpose** of a matrix is the first unary operation that is specific for membership matrices. The second unary operation is **complement** of a matrix. It makes sense when the matrix is on $\mathbb{I}$ or on a complemented lattice.

Simple binary operations: For matrices of the same type, two simple binary operations are being considered: the standard fuzzy union and the standard fuzzy intersection. When compatible fuzzy relations on finite support $R \subseteq X \times Y$ and $S \subseteq X \times Y$ are represented by matrices $R = [r_{xy}]$ and $S = [s_{xy}]$, respectively, we obtain the matrix representation of their standard fuzzy union by taking the

---

[9] . . .in traditional linear algebra, inverse of a matrix is not equal to its transpose.

maxima between the elements with the same indices and their standard fuzzy intersection by taking the minima between the elements with the same indices.

These operations are exemplified below as in the resolution of a direct problem (Example 4.16).

Example 4.16 Direct problem resolution with membership matrices Solve the direct problem if A and B are the following membership matrices on $[0, 1]$ and $\mathbb{I} = ([0, 1], \max, \min, 0, 1)$:

$$A = \begin{bmatrix} 0.2 & 0.5 & 0.1 \\ 0.7 & 0.6 & 0.9 \end{bmatrix}, B = \begin{bmatrix} 0.7 & 0.1 \\ 0.5 & 0.8 \\ 0.9 & 0.4 \end{bmatrix},$$

(i). for the standard product, $A \cdot B$, defined as:

$$c_{ij} = \bigvee_{k=1}^{p} \left(a_{ik} \wedge b_{kj} \right)$$

we obtain:

$$C = A \cdot B = \begin{bmatrix} 0.5 & 0.5 \\ 0.9 & 0.6 \end{bmatrix}$$

As it is obvious from the following computations:

$$C = \begin{bmatrix} (0.2 \wedge 0.7) \vee (0.5 \wedge 0.5) \vee (0.1 \wedge 0.9) & (0.2 \wedge 0.1) \vee (0.5 \wedge 0.8) \vee (0.1 \wedge 0.4) \\ (0.7 \wedge 0.7) \vee (0.6 \wedge 0.5) \vee (0.9 \wedge 0.9) & (0.7 \wedge 0.1) \vee (0.6 \wedge 0.8) \vee (0.9 \wedge 0.4) \end{bmatrix}$$

(ii). for the co-standard product, $A \circ B$, defined as:

$$c_{ij} = \bigwedge_{k=1}^{p} \left(a_{ik} \wedge b_{kj} \right)$$

we obtain:

$$A \circ B = \begin{bmatrix} 0.5 & 0.2 \\ 0.6 & 0.7 \end{bmatrix}$$

because:

$$\begin{bmatrix} (0.2 \vee 0.7) \wedge (0.5 \vee 0.5) \wedge (0.1 \vee 0.9) & (0.2 \vee 0.1) \wedge (0.5 \vee 0.8) \wedge (0.1 \vee 0.4) \\ (0.7 \vee 0.7) \wedge (0.6 \vee 0.5) \wedge (0.9 \vee 0.9) & (0.7 \vee 0.1) \wedge (0.6 \vee 0.8) \wedge (0.9 \vee 0.4) \end{bmatrix}$$

Other numerical examples are given by Peeva and Kyosev [5] (p.45). Direct problem is solvable in **polynomial time**. Solving direct problem requires the verification whether A and B are conformable matrices, according to Definition 4.22.

Definition 4.22 Conformal matrix The matrices A and B are called *conformable* (in this order), if the number of columns in A equals the number of rows in B.

4.8 Fuzzy Relational Equations: Direct and Inverse Problems

In this section, we discuss direct and inverse problem resolution in fuzzy relational calculus. We acknowledge the work of Peeva and Kyosev [5] who propose a unified methodology and procedures and software for solving fuzzy linear systems of equations and fuzzy linear systems of inequalities, as well as fuzzy relational equations and fuzzy relational inclusions. This is an excellent starting point for the practitioner engineer. They look at the *basic questions* such as:

- Is the system (equation, inequality, inclusion) solvable?
- If the system (equation, inequality, inclusion) is inconsistent, which connections could not be satisfied simultaneously with the other connections?
- If the system (equation, inequality, inclusion) is solvable, what is its complete solution set?

4.8.1 Direct and Inverse Problems in Fuzzy Relational Calculus

Let $R \subseteq X \times Y$ be a fuzzy relation and $A \subseteq X$ be a fuzzy set in X. The following types of problems are subject of the fuzzy relational calculus:

1. Computing $R(A)$, $R_\triangleleft(A)$ or $R_\triangleright(A)$ if R and A are given is called *direct problem resolution* for direct image, subdirect image, or superdirect image, respectively.
2. Equations of the type $R(A) = B$, $R_\triangleleft(A) = B$ or $R_\triangleright(A) = B$ when the unknown fuzzy set is A or the unknown fuzzy relation is R are called *image equations*. Solving them is called *inverse problem resolution* for the direct, subdirect, or superdirect image equation, respectively.
3. Computing composition $R \circ S$, $R \triangleleft S$ or $R \triangleright S$, if the fuzzy relations R and S are given, is called *direct problem resolution* for composition, subcomposition, or supercomposition, respectively.
4. Equations of the type $R \circ S = V$, $R \triangleleft S = V$ or $R \triangleright S = V$ when the unknown fuzzy relation is S (or R) are called *composition fuzzy relational equations*. Solving them in the unknown fuzzy relation on the left is called *inverse problem resolution* for the corresponding composition fuzzy relational equations.

4.8.2   The Role of Fuzzy Relational Equations

Any conditional *(if-then)* fuzzy proposition can be expressed in terms of a fuzzy relation R between the two variables involved. How do we determine the fuzzy relation R?

- To express R in terms of this fuzzy implication (require that the various generalized rules of inference coincide with their classical counterparts).
- The problem of determining R for a given conditional fuzzy proposition may be detached from fuzzy implications and viewed solely as a problem of solving the fuzzy relational equation for R.
- Since fuzzy relations are special fuzzy sets (subsets of Cartesian product), all operations on fuzzy sets are applicable to fuzzy relations as well.

4.8.3   Operations on Fuzzy Relations: Inverses, Compositions, and Joins

The *inverse* of a binary fuzzy relation R on $\times Y$, denoted by R^{-1}, is a relation on $Y \times X$ such that:

$$R^{-1}(y,x) = R(x,y)$$

For all pairs $\langle y, x \rangle \in Y \times X$. When R is represented by a matrix, R^{-1} is represented by the transpose of this matrix. This means that rows are replaced with columns and vice versa. Clearly:

$$\left(R^{-1} \right)^{-1} = R$$

holds for any binary relation.

Consider now two binary fuzzy relations P and S that are defined on set $X \times Y$ and $Y \times Z$, respectively. Any such relations, which are *connected* via the common set Y, can be composed to yield a relation on $Y \times Z$. The *standard composition* of these relations, which is denoted by $P \circ S$, produces a relation R on $Y \times Z$ defined by the formula:

$$R(x,z) = (P \circ S)(x,z) = \max_{y \in Y} \min \left[P(x,y), S(y,z) \right]$$

for all pairs $\langle x, z \rangle \in X \times Z$.

Other definitions of a composition of fuzzy relations, in which the min and max operations are replaced with other t-norms and t-conorms, respectively, are possible and useful in some applications. All compositions possess the following important properties:

$$(P \circ S) \circ Q = P \circ (S \circ Q)$$

$$(P \circ S)^{-1} = S^{-1} \circ P^{-1}$$

However, the standard fuzzy composition is the only one that is *cutworthy*.

4.8.4   Solving Fuzzy Relation Equations

The problem of solving fuzzy relation equations is any problem in which two of the relations are given and the third is to be determined via the equations.

– When P and Q are given, the problem of determining R is trivial (*direct problem*). It is solved by performing the composition $P \circ Q$, usually in terms of the matrix representations of P and Q.
– When R and Q (or R and P) are given, the problem of determining P (or Q) is considerably more difficult (*inverse problem*). This problem can be viewed as a *decomposition* of R with respect to Q or P. However, it is very important for many applications.

Several efficient methods have been developed for solving the *decomposition problem* of fuzzy relation equations. While these methods are rather tedious, they are highly suitable for parallel processing. Details of these methods are beyond the scope of this book. We present only the basic characteristics of the solutions obtained by some efficient methods.

Let $S(Q, R)$ denote the *solution set* obtained by solving the problem of decomposing R with respect to Q. That is, members of the solution set are all versions of the relation P for which the fuzzy relation equations are satisfied, given relations Q and R. Any member $\bar{P}$ of the solution set $S(Q, R)$ is called a *maximal solution* if, for all:

$$\forall P \in S(Q, R), \bar{P} \subseteq P \Longrightarrow P = \bar{P}.$$

Similarly, any member $\underline{P}$ of $S(Q, R)$ is called a *minimal solution* if:

$$\forall P \in S(Q, R), P \subseteq \underline{P} \Longrightarrow P = \underline{P}$$

Whenever the equations are solvable, the solution set always contains a *unique maximum solution*, $\bar{P}$, and it may contain several minimal solutions, $^1\underline{P}, ^2\underline{P}, \ldots, ^n\underline{P}$. Moreover, the solution set can be fully characterized by its maximum and minimal solutions. To describe this characterization, let:

Fig. 4.19 Structure of the solution set $S(Q,R)$. (Adapted from Klir & Demico [8], Chap. 2)

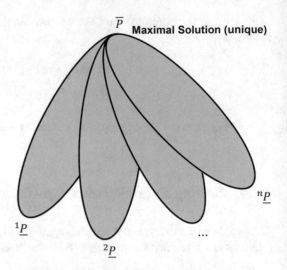

$$^i\mathcal{P} = \left\{ P \mid {}^i\underline{P} \subseteq P \subseteq \bar{P} \right\}$$

denote the family of relations that are between the maximum solution $\bar{P}$ and the minimal $^i\underline{P}$ for each $i \in \mathbb{N}_n$. The solution set is then described by taking the union of these families $^i\mathcal{P}$ for all the minimal solutions $^i\underline{P}$. That is:

$$S(Q,R) = \bigcup_{i=1}^{n} {}^i\mathcal{P}$$

This convenient way of characterizing the solution set is illustrated visually in Fig. 4.19.

Computationally speaking, it is very simple to determine the maximum solution in all the proposed methods for solving fuzzy relation equations. When the maximum solution is sufficient, this is of great advantage. When the solution to fuzzy relation equations is unique, it corresponds to the maximum solution. This is a rather rare case. When the given fuzzy relation equations are not solvable, we need to find a reasonable approximate solution. Finding approximate solutions of fuzzy relation equations is still a big problem and a subject of active research.

It is well-known that many problems coming from diverse applications of fuzzy set theory can be formulated in terms of fuzzy relation equations. Methods for solving these equations have great interest. Constructing rules of inference in fuzzy knowledge-based systems, knowledge acquisition, the problem of identifying fuzzy systems from input-output observations, and the problem of decomposing fuzzy systems are just a few examples illustrating this interest. A main source for fuzzy relation equations is the book by Di Nola et al. [3].

4.9 Conclusion

In this chapter, we presented preliminaries of crisp and fuzzy relational calculus that must be tailored to develop applications supporting the processing in a knowledge chain (described in previous chapters). We have described the basic elements allowing to perform different operations on the calculation of the mathematical relations and also to examine the interest of their properties according to their respective contexts of use. Obviously, a much deeper look (e.g., the following sources [5–10]) is required to fully appreciate the potential of fuzzy relations in the process of creation of actionable knowledge from an imperfect perception of the world. The following chapters attempt to apply these notions to the various processes of a knowledge chain.

References

1. L. A. Zadeh, "Fuzzy sets," *Information and control,* vol. 8, pp. 338-353, 1965.
2. L. A. Zadeh, "Similarity relations and fuzzy orderings," *Information sciences,* vol. 3, pp. 177-200, 1971.
3. A. Di Nola, S. Sessa, W. Pedrycz, and E. Sanchez, *Fuzzy relation equations and their applications to knowledge engineering*: Springer Science & Business Media, 1989.
4. L. Zadeh and C. Desoer, *Linear system theory: the state space approach*: Courier Dover Publications, 2008.
5. K. Peeva and Y. Kyosev, *Fuzzy relational calculus: theory, applications and software (with CD-ROM)* vol. 22: World Scientific, 2004.
6. I. Beg and S. Ashraf, "Fuzzy relational calculus," *Bulletin of the Malaysian Mathematical Sciences Society (2),* vol. 37, pp. 203-237, 2014.
7. G. J. Klir and B. Yuan, "Fuzzy sets and fuzzy logic: theory and applications," *Upper Saddle River,* p. 563, 1995.
8. G. J. Klir and R. V. Demicco, *Fuzzy logic in Geology*: Elsevier academic press, 2004.
9. L. Seymour and L. Marc, "Schaum's Outline of Discrete Mathematics, Revised," ed: McGraw-Hill, 2009.
10. B. Bede, *Mathematics of Fuzzy sets and Fuzzy logic*. New York: Springer, 2013.
11. V. K. Garg, *Introduction to lattice theory with computer science applications*: Wiley Online Library, 2015.
12. J. A. Goguen, "L-fuzzy sets," *Journal of mathematical analysis and applications,* vol. 18, pp. 145-174, 1967.
13. B. De Baets and E. Kerre, "Fuzzy relations and applications," in *Advances in Electronics and Electron Physics.* vol. 89, ed: Elsevier, 1994, pp. 255-324.
14. B. De Baets and E. Kerre, "A revision of Bandler-Kohout compositions of relations," *Mathematica Pannonica,* vol. 4, pp. 59-78, 1993.
15. D. Dubois and H. Prade, *Fundamentals of fuzzy sets* vol. 7: Springer Science & Business Media, 2000.
16. K. T. Atanassov, "Type-1 fuzzy sets and intuitionistic fuzzy sets," *Algorithms,* vol. 10, p. 106, 2017.
17. H. Bustince, F. Herrera, and J. Montero, *Fuzzy Sets and Their Extensions: Representation, Aggregation and Models: Intelligent Systems from Decision Making to Data Mining, Web Intelligence and Computer Vision* vol. 220: Springer, 2007.
18. Z. Xu, *Hesitant fuzzy sets theory* vol. 314: Springer, 2014.

19. K. T. Atanassov, "Intuitionistic fuzzy sets," in *Intuitionistic fuzzy sets*, ed: Springer, 1999, pp. 1-137.
20. L. A. Zadeh, "The concept of a linguistic variable and its application to approximate reasoning—I," *Information sciences,* vol. 8, pp. 199-249, 1975.
21. Z. Pawlak, "Rough sets," *International journal of computer & information sciences,* vol. 11, pp. 341-356, 1982.
22. K. Atanassov, "Intuitionistic fuzzy sets, 1983," *VII ITKR's Session, Sofia (deposed in Central Sci.-Technical Library of Bulg. Acad. of Sci., 1697/84)(in Bulgarian).*
23. K. T. Atanassov, *On intuitionistic fuzzy sets theory* vol. 283: Springer, 2012.
24. C. V. Negoiṭă and D. A. Ralescu, *Applications of fuzzy sets to systems analysis*: Springer, 1975.
25. B. Bede, *Mathematics of Fuzzy Sets and Fuzzy Logic,*Berlin: Springer, 2013
26. R. R. Yager, "On a general class of fuzzy connectives," *Fuzzy sets and Systems,* vol. 4, pp. 235-242, 1980.
27. V. V. Cross and T. A. Sudkamp, *Similarity and compatibility in fuzzy set theory: assessment and applications* vol. 93: Springer Science & Business Media, 2002.
28. G. Beliakov, A. Pradera, and T. Calvo, *Aggregation functions: A guide for practitioners* vol. 221: Springer, 2007.
29. G. Beliakov, H. B. Sola, and T. C. Sánchez, *A practical guide to averaging functions*: Springer, 2016.
30. R. E. Moore, R. B. Kearfott, and M. J. Cloud, *Introduction to interval analysis*: SIAM, 2009.
31. G. J. Klir, "The role of constrained fuzzy arithmetic in engineering," in *Uncertainty analysis in engineering and sciences: Fuzzy logic, statistics, and neural network approach,* ed: Springer, 1998, pp. 1-19.
32. J. Krejčí, *Pairwise Comparison Matrices and their Fuzzy Extension*: Springer, 2018.
33. H.-J. Zimmermann, *Fuzzy set theory—and its applications*: Springer Science & Business Media, 2011.

Chapter 5
Actionable Knowledge for Efficient Actions

Abstract This chapter looks at what facilitates the relevant decision-making and the modalities that can make the action (effect) more efficient. The aim is to provide a better understanding of the couple (knowledge, action). There is a strong dependency between the notion of knowing about a given world and the decisions that can be made and consecutively the potential actions that can be undertaken. We bring the notion of "mastering knowledge" for efficient actions. Mastering knowledge amounts to having a coherent set of means to represent the most useful knowledge in the context of the action, to know how to resort as necessary to the appropriate formalizations to model the situations.

5.1 Introduction

After discussing the transformation of the informational world in the previous chapters, we will now examine what facilitates the relevant decision-making and the modalities that can make the action (effect) more efficient [1]. There is a strong dependency between the notion of knowing about a given world and the decisions that can be made and consecutively the potential actions that can be undertaken. For the sake of efficiency, one cannot act on the constitutive objects of the world without first gathering useful knowledge related to this world, in other words, to acquire knowledge on it. If the disposition of a piece of knowledge favors the conduct of an action, it does not constitute a sufficient condition. A book or a knowledge base, although it contains knowledge, cannot act accordingly to the information content it conveys. This knowledge will only be used appropriately by the actor if, after a proper perception of the world, it allows him, then, to have both a good interpretation of the situations encountered and to consciously take the attitudes driving to the good conduct (effecting) of an action. In the absence of such dispositions, inherent to an actor endowed with intelligence, knowledge cannot be "mastered."

Let us be specific here what is meant, respectively, by mastering knowledge and efficiency of the action. Knowledge, in a broad sense, covers the notion of a set of more or less systematic knowledge acquired by a mental activity. Mastering knowledge amounts to having a coherent set of means to represent the most useful

M. Barès, É. Bossé, *Relational Calculus for Actionable Knowledge*, Information Fusion and Data Science, https://doi.org/10.1007/978-3-030-92430-0_5

knowledge in the context of the action, to know how to resort as necessary to the appropriate formalizations to model the situations. By questioning the deeper meaning of the notion of efficiency attached to an action, we will be led to propose an approach organized around the following points:

- To partition the universe in line with the aims of the projected actions, which leads to the proper discrimination of exogenous and endogenous data by proposing a bipartition of the universe of action
- To establish a typology of knowledge useful for action and their concomitant attributes
- To examine the impact that the "quality" of information can have on the conduct of an action and seek to improve it, particularly through the so-called concept of enrichment of semantics
- To master imperfect knowledge, by looking for the models most adapted to the situations in which their control can have a positive impact on the effective action
- To situate the notion of relevant information in relation to the efficiency of the action: the search for efficiency will in fact, in many circumstances, lead to partitioning the exogenous universe or refining its partitioning to consider, for example, certain criteria or categorization of elements

5.2   The Couple (Knowledge, Action)

In the context of a decision for a consecutive action, several situations are possible. Different relations exist between the decision-maker and the actor: the knowing agent and the perception that he has of the real-world objects via the objects of knowledge (i.e., his mental models of the physical objects as they are perceived). Before giving an illustration, let us first postulate that:

- All action is preceded by thoughtful decision-making.
- An agent, actor or decision-maker, is a conscious being able to know, to assimilate, and to interpret in a circumstantial way a piece of knowledge in view of a decision-making process.

The notion of actor must here be distinguished from that of an acting automaton (machine-agent), in this sense that he acts with conscience unlike the automaton. A conscious actor effectively retains the ability to freely control the circumstance, unlike an automaton, which acts only in function of a program of actions that ultimately reflects predefined objectives. It is now possible to define different situations that consider the "potentialities" of action according to the mastery of the knowing using the following three (3) parameters: **K**, knowledge (the knowing); **M**, mastery (the knowing); and **A**, action. Their role is specified as follows:

- **K**: the agent possesses the (a) knowledge on one (at least one) object of the real world.

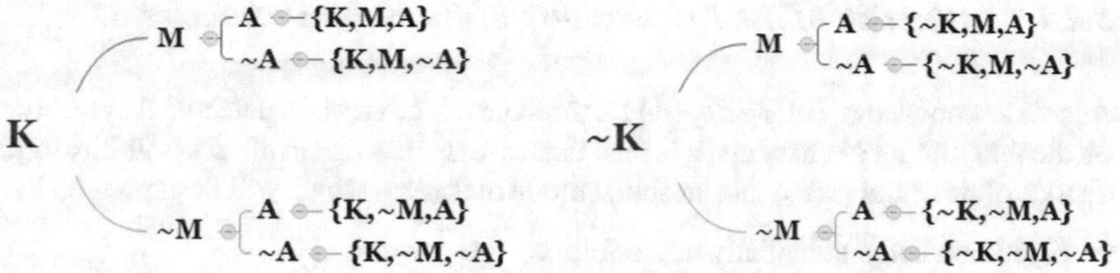

$$
K \quad
\begin{cases}
M
\begin{cases}
A \leftarrow \{K,M,A\} \\
{\sim}A \leftarrow \{K,M,{\sim}A\}
\end{cases} \\[2em]
{\sim}M
\begin{cases}
A \leftarrow \{K,{\sim}M,A\} \\
{\sim}A \leftarrow \{K,{\sim}M,{\sim}A\}
\end{cases}
\end{cases}
\qquad
{\sim}K \quad
\begin{cases}
M
\begin{cases}
A \leftarrow \{{\sim}K,M,A\} \\
{\sim}A \leftarrow \{{\sim}K,M,{\sim}A\}
\end{cases} \\[2em]
{\sim}M
\begin{cases}
A \leftarrow \{{\sim}K,{\sim}M,A\} \\
{\sim}A \leftarrow \{{\sim}K,{\sim}M,{\sim}A\}
\end{cases}
\end{cases}
$$

Fig. 5.1 The couple (knowledge, action) and the notion of mastering knowledge

- **M**: the knowing agent masters the object of knowledge, which gives him latitude, to decide an action with respect to that object.
- **A**: an action becomes possible from the known object because of the knowledge **K** that one has acquired about it.

With these three parameters, we can establish the following triplets with the negation operator represented by the symbol "~." By means of these triplets (Fig. 5.1), one can express conditions in which the action is rendered possible.

- **{K, M, A}**: The subject possesses knowledge about the object; he masters it, for example, an endogenous entity in a decision-action universe; and action is possible.
- **{K, M, ~A}**: The subject possesses knowledge of the object, he masters it, but action is impossible because, for example, imperfections of knowledge remain difficult to represent.
- **{~K, M, ~A}**: The subject does not possess knowledge of the object, although he has mastery of it. No action is therefore possible, except "to act blindly," but it would be an action insubordinate to a deliberate decision-making.
- **{~K, ~M, ~A}**: The subject has neither the knowledge nor he has mastery of it; obviously no action is possible.
- **{K, ~M, A}**: The subject possesses the knowledge about the object; although he does not have the mastery of it, for example, an endogenous entity in a universe of action-decision, an action remains possible from this object.
- **{K, ~M, ~A}**: The subject possesses knowledge about the object, he has no mastery over it, and action is impossible.

Note that if **K** is available, this does not give information on its level of imperfection, which can have consequences on the conduct of the action. The absence of knowledge "~ K" leads to the impossibility of acting. This derives from the starting postulate of an action subordinate to a decision based on a knowledge item. It must also be considered that mastering knowledge presupposes that it is relatively accessible and, above all, available at the time of decision; the availability of knowledge **K** does not mean that it is perfect. This remark leads us to introduce two characteristics belonging to **K** which will have an influence on the parameter **M**, respectively: accessibility and availability.

5.2.1   Influence of the Accessibility of K on the Parameter M

Access to knowledge **K** may depend on the state of activity of a perceptual system or on the functioning of channels. It is then useful to distinguish in the accessibility to K a mode of potential access and an actual mode of access, which will be expressed by:

- $\mathbf{K}_{pa}$: knowledge potentially accessible
- $\mathbf{K}_{aa}$: knowledge actually accessible

In the case of a "$\sim\mathbf{K}_{pa}$" access mode, this would mean that knowledge cannot be accessed, which can occur when the perceptual system does not properly perform its role due to any failure. Conversely, the perceptual system may very well be in perfect working order: $\mathbf{K}_{pa}$ :: true and be in different situations:

$$(\mathbf{K}_{pa} :: \text{true}) \wedge (\mathbf{K}_{aa} :: \text{true}) \Rightarrow \mathbf{K}$$

$$(\mathbf{K}_{pa} :: \text{true}) \wedge (\mathbf{K}_{aa} :: \text{false}) \Rightarrow \tilde{\mathbf{K}}$$

$$(\mathbf{K}_{pa} :: \text{true}) \wedge (\mathbf{K}_{aa} :: \text{true}) \Rightarrow \tilde{\mathbf{K}}$$

In the presence of $\sim \mathbf{K}$, we are brought back to the situation where it is not possible to act because there is no actually available and sufficient knowledge. The introduction of these new parameters leads us to propose a definition of a quadruplet giving more extensive relations than previously for the mastery of the knowing subject on the known object and consequently on the action that may follow. Possible changes in the previous relationships are as follows:

$$\{\mathbf{K}_{pa}, \mathbf{K}_{aa}, \mathbf{M}, \mathbf{A}\},$$

$$\{\mathbf{K}_{pa}, \mathbf{K}_{aa}, \mathbf{M}, \tilde{\mathbf{A}}\},$$

$$\{\mathbf{K}_{pa}, \tilde{\mathbf{K}}_{aa}, \mathbf{M}, \tilde{\mathbf{A}}\},$$

$$\{\mathbf{K}_{pa}, \mathbf{K}_{aa}, \tilde{\mathbf{M}}, \mathbf{A}\},$$

$$\{\mathbf{K}_{pa}, \mathbf{K}_{aa}, \tilde{\mathbf{M}}, \tilde{\mathbf{A}}\},$$

$$\{\mathbf{K}_{pa}, \tilde{\mathbf{K}}_{aa}, \tilde{\mathbf{M}}, \tilde{\mathbf{A}}\},$$

$$\{\tilde{\mathbf{K}}_{pa}, \tilde{\mathbf{K}}_{aa}, \mathbf{M}, \tilde{\mathbf{A}}\},$$

$$\{\tilde{\mathbf{K}}_{pa}, \tilde{\mathbf{K}}_{aa}, \tilde{\mathbf{M}}, \tilde{\mathbf{A}}\}.$$

Note that: $\sim\mathbf{K}_{pa}$ :: true $\Rightarrow \sim\mathbf{K}$ :: true translates the fact that it is impossible to act. When one is in the presence of **K**, the question remains to know whether one possesses all the knowledge useful and of sufficient quality for action: is it complete, reliable, perfect, etc.?

Table 5.1 Possible representation theories for imperfect knowledge

Type of imperfection	(~K, M)	(~K, ~M)
Imprecise: ~P	Fuzzy, probability, possibility	Not applicable
Incomplete: ~C	Probability, possibility	Not applicable
Uncertain: ~I	Hypothesis	Hypothesis

5.2.2   Availability of Parameter K

By availability, we try to locate the level of imperfection of **K** concerning mainly its completeness. Low availability may be an indication that **K** refers to fragmentary knowledge (incompleteness) that may have other imperfection indicia such as uncertainty or imprecision. Parameters **K**, $\mathbf{K}_{pa}$, $\mathbf{K}_{aa}$ provide information on the knowledge available and its modes of access. They do not specify, however, whether one possesses all the knowledge required and whether it is presented in the proper forms; in other words, one does not know if one possesses the necessary knowledge in quality and quantity: complete and free from uncertainty. The definition of the parameter **K**, as used up to now, is a broad definition in the sense that it is not considered that **K** can also refer to:

- A complete knowledge free of imperfections, an ideal condition
- Fragmentary elements of knowledge which, moreover, may prove to be imperfect

    To take these two remarks into account, we must redefine **K** in two ways:

- A parameter K^- corresponding to a restrictive use of knowledge that represents an ideal but unrealistic case.
- A parameter K^+ for an extended definition corresponding to a normal condition of knowledge: knowledge is considered with its imperfections. It requires the use of theories for the treatment of knowledge imperfections.

    The definition of **K** is completed with the operation:

$$\mathbf{K} \Leftrightarrow \mathbf{K}^+ \vee \mathbf{K}^-$$

    It is necessary to examine the theories that can be used to treat the doublets (~**K**, **M**) according to the formal value of the parameter **M**. Table 5.1 shows the possible theories according to three main classes of imperfections: precision (P), completeness (C), and certainty (I).

5.3   The Universe of Decision/Action

Let us now examine what facilitates the relevant decision-making and the modalities that can make the action more efficient. This will lead us to take a close look at the relationships between knowledge and action. There is obviously a strong link between the notions of knowing about a given world and the decisions that can be made consecutively and especially the potential actions that may arise. One cannot, for the sake of efficiency, act on the constitutive objects of the world, without first gathering a certain amount of knowledge precisely related to this world, in other words, acquiring knowledge over it. If the disposition of a knowing makes possible the conduct of an action, it does not constitute its sufficient condition. A knowledge base, although containing knowledge, is not able to act in function of the informative content that it conveys.

This knowing will be used wisely only if, preceded by a good perception of the world by the actor, it then allows him to have a proper interpretation of the situations encountered. Attributes referred to here as "mastery," which are inherent to an intelligent actor. A mastery that can of course, under certain technical conditions, be devolved to an automaton, by adaptation of the forms of an artificial reasoning or by symbolic learning. In the following, we will specify and illustrate what is meant by mastery and what role can be assigned to it in the knowledge-decision-action trilogy.

Intuitively, any action becomes as effective if it is previously supported by sufficient and well-structured knowledge. The choices involved in a decision, itself preceding the action, prove more judicious since they have been informed by the most relevant knowledge. Professional quality is thus measured by the acquisition and mastery of knowledge and especially the judicious use of a good chain of "professional" knowledge. The quality of the action carried out in a field of competence is judged by the yardstick of the know-how, in fact, of the experimental knowledge acquired throughout the practices respectively attested by the quality (and often the promptness) of the performed services.

For all decision-makers and professional actors (whatever their level of intervention), the linkage between decision-making and action takes on a greater importance as problems become more complex over time. The question is to know if the different parts of the universe in which one seeks to decide and to act can all be perceived, observed, and analyzed from the same angle. In view of the analysis, these parts of the universe are not sufficiently homogeneous in nature, in that sense that the degrees of freedom of action abound more, in a part of "friendly world," than in that which confronts us to the presence of numerous and uncontrollable hazards. This thus precludes any comprehensive and rational approach to their apprehension and, above all, a proper formal representation, which is a sine qua non condition of a good interpretation necessary for the foundation of subsequent decisions. It then becomes reasonable to distinguish between its own universe (or friend, endogenous) and an exogenous universe to consider differentiations in the modes of apprehension of reality and its formal representation. These are indeed very dependent on the control or not of the known object in the precise context where the action will be located. Failing to continuously be able to master the knowledge of the exogenous universe, we are led to use prediction processes (to compute probabilities), to hypothetical knowledge, even to consider stereotypes of behavior that make sense in possible worlds. The architecture of an information system, whatever its destination, can only gain coherence by considering the obligation:

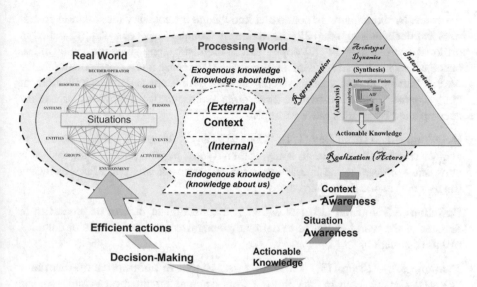

Fig. 5.2 Global context (big picture) of actionable knowledge for efficient actions

- Of a bipartite separation in the real world where the decision is made and where the action takes place
- To agree on the modes and modeling of knowledge representation that are separated according to whether one is in one or the other part of the universe

This results in a distinction between two classes of knowledge, endogenous and exogenous, as illustrated in Fig. 5.2. These two classes of knowledge have some equivalence with the notions of context used in several publications [2–11]. Figure 5.2 illustrates the main concepts that must be considered for the generation of actionable knowledge for efficient actions. Below is a brief discussion on these high-level concepts.

5.3.1 Context Awareness

The success of humans at conveying ideas to each other and reacting appropriately is due to many factors [10]: the richness of the language they share, the common understanding of how the world works, and an implicit understanding of everyday situations. The humans can use implicit *situational information*, or *context*, to increase the conversation semantics. However, this ability is hardly transferable to human-machine interactions. By improving the machine representation of context, we increase the richness of human-machine interactions and make them possible to produce more useful and efficient actions and services.

Data acquire meaning through context or contextual information and knowledge. Context establishes the basis for discerning meaning of its subjects and may occur at

many levels. Exploitation of contextual knowledge is necessary for situation awareness and decision support [12]. The notion of "context" has been widely studied in
different areas such as in philosophy and in linguistics, and since the 1980s, it has
become very popular in artificial intelligence (AI) because it is a key element of what
we call "smart" in devices and systems. Several definitions and characterizations of
context exist. Below are two or three quite representative definitions of context
accompanied by an illustrative example.

Definition 5.1 – (from Dey [10]): "Context is any information that can be used to
characterize the situation of an entity. An entity is a person, place, or object that is
considered relevant to the interaction between a user and an application, including
the user and applications themselves."

Definition 5.2 – (from [13]): "Context is a collection of objects or properties or
features of the world exploited to define or recognize and label simple or complex
events or situations."

Example 5.1 – (from [13]) – **Situation assessment in automotive applications**:
"Context consists both in data coming from cameras mounted on vehicles and in
other data form sensors measuring steering angle, speed, brakes, etc. The joint
analysis of on board/off board car context can be used to derive considerations on
driver's behaviour and then to detect possible dangerous complex situations
(driver's sleep, lane changes, etc.)."

Definition 5.3 – (from [13]): "Context is a collection of ranging data, sensed in a
subset of the world, exploited to build a correct or reliable perception of objects or
events."

Example 5.2 – (from [13]) – **Human or machine perception tasks**: "Any human
as well as machine perception task where properties owned by space-time proximal
or causally related objects provide the sensing agents with sufficient information to
drop out illusory experiences or, at least, to increase their believes on surrounding
world assets."

Transforming data into actionable knowledge is supported by technologies
regrouped under the appellation of Analytics and Information Fusion (AIF) as
illustrated in Fig. 5.2. The overall objective of the AIF processes is to support the
analysis and synthesis of information to generate actionable knowledge for decision-
making. Archetypal Dynamics [12, 14] (Fig. 5.2, *representation-interpretation-
realization*) is an integrating framework to guide the implementation of the numerous heterogeneous methods and techniques of the AIF processes. Context is necessary for the AIF processes to make and increase sense along the data-information-
knowledge (DIK) transformation chain. It is a key element in refining ambiguous
estimates, explaining observations, constraining processing, and therefore dealing
with all possible aspects of context. Both representation and exploitation of contexts

are underlying the current effervescence even the explosion of smart ICT[1] applications [4, 6, 11, 15, 16].

In the AI domain, *formalization* of context is required as discussed in Chap. 2 of this book as well as in several publications, namely, in Rogova et al. [2, 5] or in Steinberg et al. [17–21]. For instance, modern AIF systems [11, 15] designers must consider the specific characteristics of the application domain in which they have to operate, showing robust and context-sensitive behavior as well as considering different sources of contextual knowledge in addition to instantaneous sensory data to develop efficient AIF situation assessment processes.

For applications in a wide variety of domains, contextual knowledge may include structural models of the scene, known a priori relationships between the entities and the surrounding environment, user preferences, social norms, and cultures when estimating the situations of interest for the domain. Context includes conditions which augment and enhanced meaning in the DIK [22] cognitive knowledge pyramid. Context is the main component of the knowledge reference dimension discussed in Chap. 2 as cited in [13] and [23] "Context is a package of whatever parameters are needed to determine the referent [...] of the directly referential expressions . . . each parameter has an interpretation as a natural feature of a certain region of the world."

Example 5.3 – **External/internal contexts** (from [24]) – *sensor acquiring data*: the actual setting values of the sensor are part of the *internal context* (these settings intrinsically control the content outputs). However, the reasons (conditions, etc.) why the sensor has been set up with these settings belong to the *external context* which then includes the context of observation. The context of observation is an important part of the perception process as it is all that can have an influence on the perception of an event and all that is needed to understand an observation (or a message). Therefore, different situations may be perceived, relying on the same set of sensory information items, if they are interpreted within different contexts of observation. To let computers be somehow intelligent, one must provide them with this context, for a specific domain and a specific aim. Part of this context is the domain knowledge that every human uses to interpret and understand any perception.

Contextual information (CI) is crucial to support the AIF processes. Llinas [25] classifies the heterogeneity of the AIF processes supporting information under five classes described in Table 5.2. More importantly, he discusses several issues and challenges regarding the integration of contextual information within the AIF processes (detailed in next chapter):

- Full characterization and specification of CI at system/algorithm design time are not possible except in very closed worlds.

---

[1] Information and Communications Technology

Table 5.2 Five categories to classify the heterogeneity of supporting information for the AIF processes

Observational data	Data from sensors systems Data from intelligence organizations and systems (e.g., military domain) Human observers reports	Structured and unstructured digitized texts Hard data from sensors Soft data from reports and messages
Open source and social media data	Monitored open-source and social media feeds such as newswire feeds, twitter, and blog sources	Outputs of automated web crawlers and related capabilities Outputs of natural language processors
Contextual data	Information that can be said to "surround" a situation interest in the world Information that aids in understanding the (estimated) situation and also in reacting to the situation Big challenges: the relevance and system consistency of contextual information with respect to situational hypotheses	Relatively or fully static or can be dynamic, possibly changing along the same timeline as the situation (e.g., weather) Characterization might be incomplete Must consider the quality of contextual information and its impact on the AIF processes and architecture Effects of quality of context on situational estimates
Ontological data	Procedural and dynamic knowledge about the problem domain Tactics, techniques, and procedures Declarative knowledge: Temporal primitives along with structural/syntactic relations among entities	Represented in language and is available as digital text, in the same way as data from messages, documents, twitter, and so forth Specified ontologies that serve as providing consistent and grounded semantic terminology for any given system
Learned information	The class of information that could be learned (online) from all of the above sources with machine learning	If meaningful patterns of behavior can be learned, such patterns could be incorporated in a dynamically modifiable knowledge base

- Define an "a priori" framework for exploitation of CI that attempts to account for the effects on situational estimation that are known at design time (this is related to the endogenous knowledge discussed in next section).
- The issue related to the nature of the CI: for example, cases can arise that involve integrating symbolic CI into a numeric algorithm.
- Quality of contextual information, like observational data, has errors and inconsistencies itself, which has to be considered for hybrid (symbolic/numeric) algorithms of AIF processes [26].
- Some CI may have to be searched for and discovered at runtime, and a verification must be done on the consistency of a current situational hypothesis with the newly discovered (and situationally relevant) CI.
- Finally, system dependability dimensions such as accessibility, availability, reliability, etc. must also be considered with respect to CI.

As said in Steinberg [18, 27], exploiting context in AIF requires to

- Clarify concepts such as situation and context
- Properly account for the dimensions of quality of information (uncertainty, relevance, reliability, etc.) inherent in AIF processes as they apply to contextual reasoning

The concepts of situation, situation analysis, and situation awareness have been introduced in Chap. 1 and described lengthily in books related to information fusion [12, 28, 29]. For the sake of our discussion here, consider only the following short definitions.

Definition 5.4 – **Situation** – (from [27]): "Situation is defined as a set of relationships where a relationship is an instantiated relation" (e.g., marriage is a relation, but Mark Antony's marriage with Cleopatra is a relationship). In addition from [18], we cite: "Contexts are situations. They are situations that are differentiated by a person (or other agent) for some purpose."

Definition 5.5 – **Situation and situation analysis** – (from Roy [28, 30]). A **situation** is: "A specific combination of circumstances, i.e., conditions, facts, or states of affairs, at a certain moment." *Situation Analysis* (**SA**) as: "a process, the examination of a situation, its elements, and their relations, to provide and maintain a product, i.e., a state of situation awareness, for the decision maker."

The product of situation analysis is achieved by the AIF processes. "A state of situation awareness" is attained by the decision-maker when his *state of knowledge* (actionable) is enough for efficient actions. For AIF, context is used to refine ambiguous estimates, explain available data, and constrain processing, whether in cueing/tipping or managing the AIF processes. Context can be assessed either from the outside-in or from the inside-out [20]. These uses correspond to the notions of "context-of" and "context-for" introduced in [31]. These notions are in some sort linked to the endogenous and exogenous knowledge described below.

5.3.2   Class of Endogenous Knowledge

The endogenous universe characterizes the part of the world where it is possible to conduct highly deterministic actions, at least in their preliminary phases. The objects of knowledge of this universe are well-known, if not easily identifiable [1]. Of course, they are easily identifiable if they do not interact strongly with those of the exogenous world. Transformations as procedures, regulations, various protocols, etc. operating on these objects are also well-known. This universe is characterized by a large deterministic imprint that forms an informational standpoint: those objects of knowledge normally have cognitive imperfections clearly identifiable. At the event level, the mastery of the events of this universe when they are easily apprehensible, representable, and manageable remains quite large. This is of course justified; since the organizational structure is appropriate, the play of actors considered rational is properly regulated by acts and well-posed decisions.

5.3.3   Class of Exogenous Knowledge

The exogenous universe represents the part of the information system characterizing a world over which we do not have a priori control. The reason is either it is extremely complex or it appears highly reactive because of issues of domination and various antagonisms (case of an economic war). These can be distinguished by disorder arrivals of events whose meaning it is important to understand, regardless of the nature and weakness of the observation signals received. We must be able to know better to interpret their meaning, in other words, to agree on a semantics to attach to them, hence the need to introduce a stage of interpretation to make the link between *signifiers* and *signified* knowledge in the system. Compared to the endogenous universe, the latter is characteristic of a highly random world and thus has great "cognitive imperfections."

From an informational point of view, the formalisms used to represent knowledge must possess a richness of expressivity to translate at the same time also its quality: the veracity of the data collected, the uncertainty, the inaccuracy, but also, to include time intervals during which these characteristics can be considered valid. To this end, it is useful to develop the means and mathematical tools:

*   To formalize elements of processes that can be represented by probabilities, the existence of possible relations between these processes, and their dependencies according to contexts, etc.
*   To model random, parallel, concurrent, and asynchronous processes
*   To correctly represent the characteristics of any object of knowledge that can account for certain forms of knowledge: incompleteness, plausibility, etc.

5.3.4   Typology of Knowledge for Decision-Making and Actions

The usefulness of the knowledge required for decision-making or the conduct of an action is very dependent on the situation in which the decision-maker finds himself, as well as on the operating conditions in which he performs his actions. The conditions under which he requires and uses knowledge greatly differ. For example, if he must deal with an emergency, or to react to unforeseen events, or to anticipate a decision, or to operate in adverse conditions, or under a significant stress, all will not have the same degree of utility before and during the action. One can, therefore, legitimately question the existence and the nature of elements establishing a relation between knowledge and decision, but also ask the question: What does "useful or actionable knowledge" mean with respect to the conduct of an action?

At first, consider that a piece of knowledge is useful for a decision-maker only if it proceeds from the necessity to make him acquire a mental model of the world, at the very least, to give him a representation as faithful as possible of its objective reality. This must be done, moreover, at the sufficient level, i.e., corresponding to the provision of convincing elements of reflection to undertake and decide. The effectiveness of an action is often determined by the correct identification and then the consideration of criteria that characterize its operating context. These can directly influence the modalities of its conduct. A context can indeed be distinguished by circumstantial criteria but also by criteria of opportunity, those whose consideration must allow to know how to react at the right time and respond in "time."

Knowledge regardless of the destination of the service or application that requires it needs to be previously capitalized and organized in ways that allow a decision-maker to solicit it at once easily and especially opportunely: it is useless to have knowledge if we cannot use it at the opportune moment. It is also preferable to favor the search for knowledge, however fragmentary, but apparently more in relation to the context of the planned action, than to want to focus on the access to a global knowledge without any instructing criteria that make relation to this same context.

We also observe, and this is not the least difficult, that knowledge to be collected is proving to be more useful if they conform to the cultural level, even to a certain experience of the supposed users. Moreover, knowledge cannot be considered *actionable* if it does not allow the decision-maker:

- To properly locate the objects of the world on which he wants to act
- To determine properly the situations he seeks to evolve, in space and in time
- To give him the faculty to reason on or with these different elements
- To anticipate or to project its action into the future

The decision-maker is then confronted with a situation which, beyond his present horizon, forces him to imagine the means to reach a desired /desirable future. This situation can correspond to two different motivations according to the envisaged horizon:

- *Prospective horizon:* Starting from a present situation, it is a question of opening a range of possibilities with the aim of enlightening the future.
- *Strategic horizon:* It is necessary to define what actions are necessary to reach the goal envisaged in the future in a relatively long term.

The knowledge that is useful for this purpose must be used as a support to substantiate hypotheses taking their true meaning in different possible worlds [1]. By possible world, one understands here the action which accompanying the fact of making a hypothesis on a part of set of universe confers on it a certain semantics, either by making possible a certain application or by allowing the functioning of a mechanics of reasoning generally abductive. Let us remember that **abduction** can be understood from two aspects:

(a) It is the reasoning that makes it possible to add knowledge to complete a proof as in the example of an abductive correction of errors. In the case where an indisputable fact cannot be understood in the existing theory, one can complete it by a more global theory.

(b) In the other case, it is the opposite use of the modus ponens, i.e., we abduct A from $[(A \Rightarrow B) \wedge B]$, whereas in the strict application of the modus ponens, we deduce B of $[(A \Rightarrow B) \wedge A]$ (Fig. 5.3).

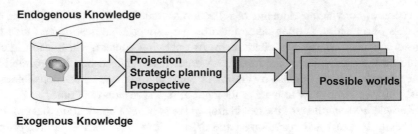

Fig. 5.3   Decision-maker prospective and strategic horizons

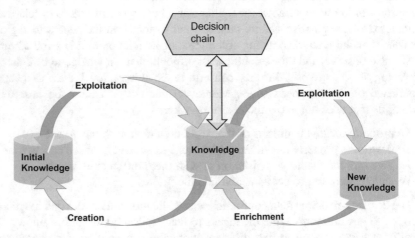

Fig. 5.4   Usefulness of knowledge in the decision-making process

Example 5.4 – Abduction – If, in the following case: $\forall x$ [fire(s) $\Rightarrow$ smoke(x)]. A fire is predicted by noting the presence of smoke (assertion of smoke(x)). An *abduction* is done while saying that one will have smoke if one observes a fire, amounts to making a deduction. An abduction can be described formally according to the inference diagram: (x) [$P(x) \Rightarrow Q(x)$], $Q(a)$, $P(a)$.

In the context of a chain of decision, or even of a chain of command, the usefulness of the knowledge must also make it possible to improve the knowledge, which can be observed upstream and downstream (Fig. 5.4).

- *Upstream*: creation of new knowledge that improves initial knowledge (frequent case of feedback).
- *Downstream*: the creation then provokes an effective enrichment by participating in the new knowledge.

Useful knowledge for the decision/action belongs to the classification that usually distinguishes:

- *Declarative knowledge*: it concerns the descriptive elements covering the knowledge that one has on a particular field. These elements can be varied in nature as simple facts concerning objects of the world but also relations that they maintain between them. Declarative knowledge contributes to the representation of a universe, without any visible concern, as to how it will be implemented later.

- *Procedural knowledge*: it determines the modalities of the "how to." To be useful, it is not enough to have "declared" knowledge; it is still necessary to have rules by specifying how they are used. They therefore concern procedures and, more generally, algorithmic approaches as soon as they are part of organized structures. They are the backbone of the inference drivers of knowledge-based systems.

- *Meta knowledge*: it constitutes an "over knowledge" or additional knowledge on how to use, for example, the initial knowledge and can represent an (expert) knowledge acquired over time. Meta knowledge, although specific, possesses all the attributes of knowledge. It can also concern the knowledge that it is useful to have on what others know. This may help to better cooperate with them to the success of a task.

- *Knowledge of common sense* (or common sense): it characterizes the knowledge that frequently occurs in common sense reasoning based often on unstated knowledge from habitual behavior and experiences. It poses a problem to the specialists (cognition) in charge of the constitution of bases because they are difficult to represent, by the preponderance of the implicit. They can singularize an individual by giving him either a cultural pre-eminence or a certain subtlety (or vice versa).

- *Heuristic knowledge*: it represents knowledge related to the means and processes that must make it possible to find a quick, if not effective, solution based on a human know-how which, if not rationally justified and formalized, proves effective in the resolution of a class of problems. The heuristic by its nature and approach is thus in opposition with the algorithm.

- *Deep and surface knowledge*: it is also customary to make a distinction between surface knowledge and deep knowledge. The former is mostly related to declarative knowledge, that of the expert, and represents a large part of the operational knowledge and will generally cover much of the need for the knowledge-based system. On the other hand, the so-called deep knowledge is more structured and will seek to explain why by providing, as and when necessary, justifying elements of theory that can form the reasoning basis.

To see where the different levels of utility of these different types of knowledge are, we make an analogy with the operation of an "intelligent" system organized around a knowledge-based system illustrated in Fig. 5.5.

In addition to the typology described above, psychologists identify two other large classes of knowledge.

- *Semantic knowledge*: they mainly concern the cognitive structure specific to the expert and reflect the verbal expressions, the meanings of the words/symbols, and the rules of use, references, and relations of the words/symbols.

- *Episodic knowledge*: they are specific to the experiences of events and relations in time and space between them.

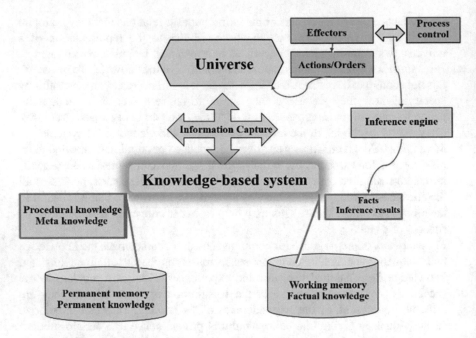

Fig. 5.5 Different types of knowledge in knowledge-based system

5.3.5 The Utility Field of Knowledge

The utility field of knowledge can be expanded or restricted under certain conditions of use or operational contexts, which leads to giving some indicative criteria of the conditions of utility of knowledge.

- *Volatility*: The "force of use" is variable according to time, its value of truth decreasing as time goes by, or its semantic content loses its interest, a frequent problem in the context of intelligence.
- *Hardening*: In a way similar to what is done for certain infrastructures, it is possible to reinforce the confinement of a knowledge to hinder access to users with malicious designs.
- *Convergence*: A knowledge item, a priori shareable, allows by its signified the consensual approach between several of its users.
- *Ownership*: In the case of apprenticeship, this is a remarkably interesting criterion, but one that should not obscure the fact that its application may have a counterpart, for example, diminishing the strategic interest.
- *Innovation/creativity:* Knowledge that is a support for creativity can be a source of added value for an organization that knows how to use it intelligently.
- *Strategic*: According to a deliberate choice (political reason), a sensitivity level (due to the nature), or a decision component (decision-maker), a piece of knowledge can be declared as strategic.

- *Globality*: It is a knowledge that identifies with the quantitative; its meaning is conferred by the multiplicity: statistical series, repetition of measures, and cloud of points.
- *Qualitative dimension*: It is necessary to be able to establish order relations on knowledge related to the same subject to indicate the gradual importance either relative to a subject or in relation to an area of interest.
- *Dissemination*: Knowledge can be disseminated rapidly and systematically in case of need: reaction to an attack or dissemination of an alert. This criterion, to be closer to the notion of broadcast, is not applicable to all knowledge without precautions.
- *Contextual*: Knowledge does not have the same weight placed in a context dominated by decision, prevention, or anticipation. It can play a different role depending on the operating context.
- *Shareability*: The idea underlying this criterion is that information is less the property of its holder than that of the one who needs it or to know it. This is an important criterion especially in distributed structures or coalitions.
- *Expressiveness*: A knowledge item must be able to express itself and to represent easily if not with simplicity in order to facilitate its communication and allow incidentally to make quickly converge its users on its semantics.
- *Intrinsic nature*: The manipulation of knowledge according to its logical nature, pure temporal, probabilistic, etc. will require adapted instruments and procedures.
- *Imperfection*: It must be postulated that any knowledge, no matter where it is captured, is a priori imperfect. An imperfection can be manifested in different ways as the following attributes attest: incomplete, uncertain, plausible, etc. The problematic and the control of these imperfections must be carefully studied.

5.3.6   A Closed World Assumption

It is necessary to specify the limits of the domain where the knowledge is useful in the same way that one specifies the field of definition of a variable in mathematics. The context in which knowledge is used is generally referred to as the universe of discourse. It is useful to make here a simplifying assumption like that often used in artificial intelligence (AI) known as the closed world hypothesis. It stipulates that knowledge that is not declared true is considered to be false or non-existent in relation to that which belongs to the world of action. This is to hide the fact that useful knowledge in the universe of discourse is necessarily incomplete. Proceeding here, we consider that a knowledge item is useful in an action only to the extent that we can assert its usefulness.

To be able to use knowledge, it is important to know exactly what one knows about it, as it is the same in a process of understanding, which requires knowing exactly what the other person knows if one wants to be able to understand oneself absolutely. We will consider here that a knowledge that is not known cannot be useful. In a classical logic, for an argument, the truth of the premises intervenes on the truth of the conclusion. The knowledge is essentially true if φ is known then φ is true, but the reciprocal does not always exist, which leads to say that it will not be possible to elaborate a logic of useful knowledge obeying the same schema. One can find oneself in the situation where one is no more able to know if φ is a useful knowledge as much as ~φ. An attitude which, in the context of a bivalent classical logic, contradicts the principle of non-contradiction since one cannot have both φ and ~ φ, (φ ∧ ~ φ = 0). We must therefore seek to extend the basic language of classical logic by defining an operator called epistemic **K** under the following conditions: **K**φ:: ~φ is a useful knowledge. The excluded third, for example, may be expressed as follows: **K**φ ∨ **K**~φ, φ is either known to be useful or known to be useless.

5.4   The Knowledge Relevance and Its Impact on the Couple (Decision, Action)

The word relevance derives from the Latin word "pertinens," a present participle of the verb "pertinere," referring to the expressions "which relates to the question," "which relates to the very essence of the cause," and "which is exactly appropriate to the object in question," which throw light on the strong coupling existing between a fact and a statement referring to it, without being the mark of a causality. Language habits such as "appropriate remark, quoted at the right time, arrived at the right time" allude (often without the knowledge of their authors) to something factual that is:

- Appeared when it proved most useful, in the case, for example, of an achievement taking place in the appropriate circumstances
- Manifested when it was most necessary with proper alignment as needed
- Arrived at the right moment, that is to say, opportunely

Moreover, the notion of relevance intuitively carries with it a character of opportunity or a situation relating to a useful time: "it is timely, etc." The term circumstantial is often the other aspect attached to the evocation of relevance: do we have adequate information sufficient and necessary to act in a given context, or said in a familiar way, under the right conditions? It is observed that the relevance is inherent to any technical or scientific environment as soon as its control is essential, as, for example, in:

- The foundation of a theory: notion of certainty
- The result of an experimentation: notion of credibility
- The modeling of a process: notion of constraints and reduction of approximations
- The representation of a situation: notion of vague and imprecise knowledge

In the military operational domain, the notion of relevance is highlighted with attributes such as certainty, likelihood, contradiction, precision, and completeness.

How can we identify and then report on the impact, the relevance to the decision, and the action that follows? Let us mention some singular traits that are more to the attention.

- *Logical level*: degree of veracity that can be asserted
- *The imperfection* of which the different facets characterizing the improper world account
- *The circumstantial aspect*: is the knowledge well in relation to the need to know it to act adequately in such context
- *Coherence*: does knowledge correspond to the problem posed, to the questions that arise before undertaking a given action
- *The opportunity*: often assimilated in practice to the useful time, is the information necessary for the action available at the required time, i.e., when it is most useful to dispose it of

It should be noted that the impact of the relevance of a knowledge item will not be the same depending on whether one is interested solely in the intrinsic qualities of that knowledge item or that it is simply referred to in the case of a given context or to respond to a particular need: issue a query on a knowledge base. We can indeed use knowledge:

- "Deferred" previously acquired and organized in a memory
- "On immediate time" depending on the desirability or the need to respond to a need, through the expression of a specific request or usual transactions

In the first case, the relevance is mainly dependent on the defects accumulated by the perceptual systems and modalities of the knowledge modeling which can account for the dimensions of imperfection, logic, and coherence. In the second case, there is an external aspect to consider the relevance depending partly on the context in which one operates. The way to ask the queries, beyond the only aspect of syntax, will play an important role. For the sake of simplification, and to consider the different points of view, let us distinguish several levels of relevance as indicated in Fig. 5.6.

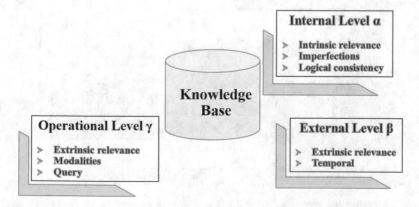

Fig. 5.6 Relevance at different levels

The three relevance levels of Fig. 5.6 have a greater or lesser impact on the efficiency of the action. The efficiency of the action is intricately linked to the dimensions of relevance attached to these levels:

$$\text{Efficient action} :: f(\alpha\ (..,..,\ ),\beta\ (..,..,..),\gamma\ (..,..,\ ...))$$

5.4.1 Intrinsic Relevance

Imperfection, consistency, and logic are the three essential dimensions for intrinsic relevance. Any knowledge item stored in a knowledge-based system has passed through:

- A phase of apprehension with different modalities
- A capture by a perceptual system
- A formalization and so possesses a formal representation

It is from these three stages that the interpretation necessary for decision-making takes place as illustrated in Fig. 5.7. The strong coupling existing between these stages, having at each stage their own imperfections, or deformations, directly affects the knowledge relevance and ultimately alter its quality. Figure 5.7 puts

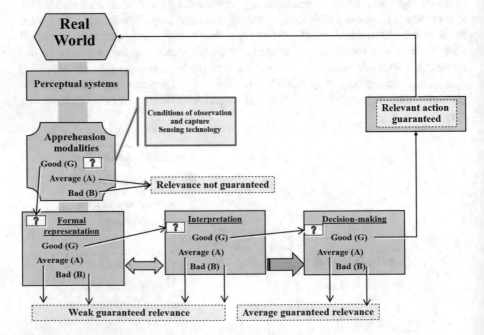

Fig. 5.7 Apprehension, representation, and interpretation stages of relevance levels

into perspective the level of quality to be attained at the different stages of the processes, for the sole purpose of limiting the accumulation of harmful consequences on the final stage determining the relevance of the action.

- *Consistency dimension* – It is based on reference to something that is pre-established and stable. These references can be of several types:

  - Criterion of belonging to a domain not respected, this domain being able to have possibly vague borders
  - Exceeding standards
  - Application of a rule: Boolean (15 meters) $\in[\min, \max]$ :: V/F, if F, then Boolean is inconsistent, unrealistic factual information

  Consistency can also be placed in relation to the trust that a user may place in the knowledge produced by a sending source, in the sense that if this trust is not within a bounded confidence interval, then it must be considered as inconsistent, in relation to the purpose or the agreed strategy. The lack of trust leads to considering this knowledge as incoherent since it prevents any decision-making and therefore prohibits any relevant action. On the other hand, a degree of confidence that is close to the confidence thresholds may lead to temporary indecision that does not necessarily imply a lack of relevance to the action. Consistency can also be examined from the perspective of consistency with the knowledge base. In other words, if at time t:

  - The knowledge represented by $K(t)$ is consistent with the other knowledge of the basis $K[t, \ldots$ that can be expressed by $\sum K[t, \ldots$
  - Does the addition of other knowledge to $t + \Delta t$ that impact the initial knowledge base in $K[t, \ldots + \sum K[t, \ldots$ preclude $K(t)$ to remain coherent with $K[t, \ldots + \sum K[t, \ldots$?

- *Logical relevance dimension* – The logical dimension of relevance concerns its degree of veracity, i.e., whether information or knowledge is assumed to be true enough to be used in a decision-making process. This logical dimension is based on two criteria: vagueness and uncertainty. The relevance dimension of an information cannot alone be considered sufficient apart from a use of this information making it interesting for a given problem or in a particular context of use. Indeed, information that is true but does not correspond to any use cannot be considered relevant. The logical relevance dimension can be expressed in two different ways in a time window determined by the dates $[d_1, d_2]$:

  - The knowledge is true at any time indicated by: $K(t)[d_1, d_2]$.
  - The knowledge is only true at certain instants mentioned by: $K(t)\langle d_1, d_2\rangle[d_1, d_2]$.

5.4.2   Temporal Relevance Dimension

Take the example of a competent authority that decides to apprehend a suspect. To carry out the action under circumstantial and timely conditions, the mandated police force must ensure that it has the "best" knowledge of the planned arrest at time t. The question asked is whether the knowledge $K(t)$ on the operation to be carried out is:

- Valid in the window $[d_n, d_0]$ and will be still at the future date $t_n > t$
- Valid in a projected interval and known $[d_n, d_0] \neq [d_1, d_2]$

It is then necessary to have two-time windows in the formalism required for exploitation:

- An interest validity window V_i.

  - V_i, being specific to the missions requested in general, which can be defined, compared to the current date t
  - V_b, a validity window of need, V_b corresponding specifically to the requested mission.

The two windows, V_i, V_b, defined by intervals will be in different situations as:

- In inclusion: $V_b \subseteq V_i$; the knowledge K(t) must have in this case a **certain relevance.**
- In intersection: $V_b \cap V_i \neq 0$, in which case K(t) can only have a **relative relevance.**
- In empty intersection: $V_b \cap V_i :: 0$; it is difficult to evaluate any level of relevance, according to the knowledge previously acquired.

One can also be in the case where: $V_i \cap V_b$, which implies that the need to know it can only be partially satisfied, and because of this, $K(t)$ will only be relatively relevant (Fig. 5.8).

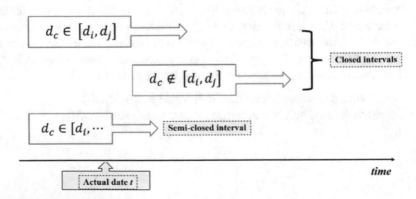

Fig. 5.8 Temporal relevance dimension

Example 5.5 – Extrinsic relevance – arrest of a suspect (1)

The police force, to make the arrest of a suspect, will at times seek to improve their knowledge by questioning their own knowledge base. The knowledge resulting from these interrogations cannot be relevant if it does not belong to at least a partial recovery of the windows V_i and V_b: $K(t) \in (V_b \subseteq V_i)$. This assumes that the queries issued are based on a certain formalism and, on the other hand, that one is sure that the requested knowledge base has been updated. In this specific context, matches can only be made if we agree on a certain formalism to describe a knowledge item or a request.

5.4.3 Validity of a Relevant Piece of Knowledge

A knowledge item is valid either all the time or intermittently for a given period, bounded by dates, and also depends on its date of entry into a knowledge base:

$$K(d_m) :: \left(d_m, [d_i, d_j] \vee \langle d_i, d_j \rangle, M(K) \right)$$

where

d_m: file date (of suspects)
$[d_i, d_j] \vee \langle d_i, d_j \rangle$: intervals of validity of interest (temporal window), with $d_m \geq d_j$
$M(K)$: modeling (logical formulation)

Example 5.6 – arrest of a suspect (2)

$K(t) :: (15/07/05, [04/07/05 - 10/07/05],$ *localization* $\langle Pr, Paris \rangle, (a))$ which will be interpreted by the fact that it was mentioned on July 15 that the suspect Pr was seen in Paris in the week of July 4 to July 10, 2005. If we want to know now, where was the Pr person on July 20, it is necessary to specify the need with a request, R, of the type: $V_b[20/07/05, 20/07/05]:R(t) :: (\ldots V_b[20/07/05, 20/07/05],$ *localization* $\langle Pr, x \rangle, (b))$.

Note that (a) is relevant to (b). For the arrest to be carried out in the best conditions, police officers will want to be able to access knowledge under certain dimensions of knowledge relevance. The previous formalism must be extended to answer these new concerns:

$$K(d_m) :: \left(d_m, V_i[d_i, d_j] \{REL_{dim}\}, M(K) \right)$$

where

d_m: file date (of suspects)
$V_i[d_i, d_j]$: intervals of interest validity (temporal window)
$\{REL_{dim}\}$: relevance dimensions associated with $K(d_m)$
$M(K)$: modeling (logical formulation)

Example 5.7 – arrest of a suspect (3)

$$(15/07/05, [04/07/05 - 10/07/05], [20/07/05 \ 21/07/05]$$
$$localization \ \langle Pr, Paris \rangle, source(5), intersected(3), spatial(2))$$

which is interpreted by the fact that what was mentioned at the date of July 15 is confirmed by a trusted observer as of 20/07/05, but that the localization of suspect Pr remains at this date imprecise. Improving knowledge before the intervention involves interrogating the knowledge base. The requests made will have to respect a certain formalism to be able to be matched with the available knowledge base. Formal definition of a query at time t:

$$R(t) :: \left(V_i[d_i, d_j], V_b[d_b, d_c], Q, M\{REL_{dim}\} \right)$$

where

$V_i[d_i, d_j]$: interval of interest validity for the request
$V_b[d_b, d_c]$: interval of the requirement validity (expressed in the request)
Q: logical formula expressing the knowledge sought
$M\{REL_{dim}\}$: modeling of dimension (s) of relevance to be considered

Example 5.8 – arrest of a suspect (4). If one wants to know by the 25th of July where the suspect was seen in Paris in the period from 4/07/05 to 10/07/05, the request can be made as follows:

$$R(t) :: ([04/07/05, 10/07/05], \langle 18/07/05, 20/07/05 \rangle, Paris \ (Pr, x))$$

To define the opportunity, it is necessary to reconcile the request with the knowledge that it has produced at the end of the interrogation. Consider the closed request:

$$R(t) :: \left(V_i[d_i, d_j], V_b[d_b, d_c], Q, M\{REL_{dim}\} \right)$$

The knowledge that results from the query is:

$$K(t) :: \left(d_m, V_i[d_i, d_j], M(K) \right)$$

$K(t)$ will be considered relevant if:

$$\left(d_m, V_i[d_i, d_j], M(K) \right) \cup K[t_k \ldots \models (V_b[d_b, d_c], Q)$$

where ($\models$ is the symbol for $'deduction'$)

$K[t_k$, represents the validated knowledge base from date t_k

$d_m < d_c$: indicates that $K(t)$ is relevant with respect to $R(t)$, defined in the previous conditions if, using a temporal reasoning on the intervals, as described above, one can deduce that Q was true in the window:

- $V_b[d_b, d_c]$ from the fact that $M(K)$ was itself true in the window $V_i[d_i, d_j]$; this correlates with the pre-established knowledge of the base $K[t_k \ldots$

The update date is before the end date of the requirement validity window expressed in the request, which is to say that the information about the defendant on the d_m date has been reported before the need deadline. In the case of an open query, we must give ourselves the latitude to enter free variables in (a) to give (b):

$$(d_m, V_i[d_i, d_j], M(K)) \cup K[t_k \ldots \models (V_b[d_b, d_c], Q) \qquad (a)$$

$$(d_m, V_i[d_i, d_j], M(K)) \cup K[t_k \ldots \models (V_b[d_b, d_c], Q(x)) \qquad (b)$$

The knowledge $(d_m, V_i[d_i, d_j], M(K))$ is relevant to the query:

$$R(t) :: (V_b[d_b, d_c], Q(x))$$

if and only if:

$$\exists A \, | \, (V_i[d_i, d_j], M(K)) \cup K[t_k \ldots \models (V_b[d_b, d_c], Q(A)), d_m \leq d_c \qquad (c)$$

In (b), we indicate that it is possible by temporal reasoning to say that $Q(A)$ is true or has been true over the interval V_b because $M(K)$ has been true in the interval V_i in association with all the pre-established knowledge in the database $[t_k. \ldots$ In (c), we express the fact that the date of entry into the database must be prior to the date of expiry of the need window; in other words, the knowledge was taken in useful time, before the need reaches deadline.

- *Dimension of circumstantial relevance* – It is obvious that a decision taken or the conduct of an action in an emergency or in a planning context does have different or special requirements in terms of actionable knowledge. Some aspects of the temporal dimension of relevance may become dominant in a particular context. The constants of time, the dating, the cycles of their own, the treatments that can be applied to them, differ in many ways. A piece of knowledge may also be of a circumstantial nature in a given context in that, having a critical role, it is always necessary and must be readily available. The circumstantial dimension involves indexing the time windows by these contexts:

$$R(t) :: (V_i(t)[d_i, d_j], V_b(t)[d_b, d_c], Q, M(t)\{REL_{dim}\})$$

5.5   Notion of Semantic Enrichment

The notion of semantic enrichment is associated here with the idea of seeing if there exist modalities allowing to increase the *signified* conveyed by the knowledge to which one has access. That amounts to looking for ways to improve the semantic dimension of "knowing." Semantic is often used in combination with terms such as *enrichment, tagging, markup, indexing, fingerprinting, classification, and categorization.* In this chapter, the term semantic enrichment refers to the following definition.

Definition 5.6 *Semantic enrichment* is the process of adding *signified* via symbolic fusion along the knowledge processing chain (data-information-knowledge-decision-action).

Remark 5.1 Semantic enrichment can refer to the various technologies and practices used to add semantic metadata to content. Given the explosion of available information, people have become reliant on computers to find the information that they need. Semantic metadata provide the answer to an important question, "What is the meaning of this content?" in a way that computers can process so that they can find, filter, and connect information.

Let us remember that any efficient action, like any rational decision, cannot be reasonably considered in the absence of a "well-established knowledge" or a "situation awareness" properly accumulated and enriched along a knowledge chain. This implies a proper and well-grounded grasping of knowledge that sheds light on the contexts in which "deciding" and "acting" must make full sense. That knowledge mastery must assure that we have the appropriate means to:

- Determine what pieces of knowledge have the most meaning (most *signified*)\
- Consider how it should be possible to have increased *signified* (semantic enrichment) by aggregating, combining, or fusing them

Three modalities are proposed below to achieve the implementation of a semantic enrichment concept using techniques such as graphs, partially ordered set, lattices, etc.:

- Development of concepts of symbolic fusion
- Usage of models and relations of compatibility
- Implementation of joint possibility distributions

5.5.1   Semantic Enrichment by "Symbolic Fusion"

We will try to show, via a mathematical formalism, how information, most often of a factual nature, applied to an already established knowledge (case of a situation) generates another, informally enriched, compared to the previous one.

Fig. 5.9 Notion of semantic
enrichment

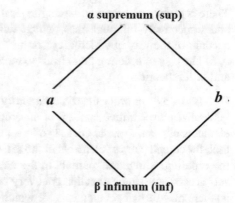

α supremum (sup)

a b

β infimum (inf)

Definition 5.7 – Symbolic fusion: What is called here symbolic fusion of knowledge is distinguished from data fusion, in that we seek to understand how such an application can lead to the improvement of the *signified* of a knowledge item.

- *Symbolic fusion (merging):* knowledge-applied φ (situation) ⇒ enriched situation (at the informational level), different from a usual aggregation merger where:

  – *Aggregation fusion::* [{aggregate data} {data, information}] ⇒ new situation

The purpose of the symbolic fusion is, by means of permanent knowledge of a situation, to enrich it so that its level of relevance can increase. By the means of knowledge modeling and its symbolic treatments, the idea of a semantic enrichment is more to intrinsically increase the informative content of a situation as opposed to a quantification in the form of criteria to account for that increase (as in the case of numerical criteria of likelihood oscillating over an interval $[-a, +a]$).

In this case, we will seek to build a relation to compare elements between them but also to see if it is possible to get semantic growth and respectively semantic decay. This operation is justified by the fact that the *maximal* of a pair (a, b) has intuitively more *signified* than the pair itself. At first, we will limit ourselves to the simple case involving two elements (a, b) as illustrated in Fig. 5.9 where:

- a, b, are the initial elements to be compared
- α the maximal element and β the minimal element that can be inferred

If we can show that with respect to (a, b):

$$(\alpha > a) \wedge (\alpha > b) \Rightarrow \alpha \rightarrow \text{maximal element,}$$

$$(\beta < a) \wedge (\beta < b) \Rightarrow \beta \rightarrow \text{minimal element,}$$

then the knowledge of α, β allows to see if there is an enrichment or not. The case where: $a = b$ means that α must not be greater than or β less than a, b. By the way, this highlights the important fact that a piece of knowledge cannot enrich itself.

There is a risk, of course, of having to deal with several maximal elements at once, and respectively minimal ones, which will have to be ordered, to give a sort of account of the way in which they "frame" the initial elements, in this case the pair *a*, *b*. The way of ordering them leads to recalling some basic definitions of the lattice and order theories.

Definition 5.8 In order theory, a *partially ordered set* (also **poset**) formalizes and generalizes the intuitive concept of an ordering, sequencing, or arrangement of the elements of a set. A *poset* consists of a set together with a binary relation indicating that, for certain pairs of elements in the set, one of the elements precedes the other in the ordering. The word "partial" in the names "partial order" or "partially ordered set" is used as an indication that not every pair of elements need be comparable. That is, there may be pairs of elements for which neither element precedes the other in the *poset*. Partial orders thus generalize total orders, in which every pair is comparable.

Example 5.9 The real numbers ordered by the standard *less-than-or-equal* relation ≤ (a totally ordered set as well). The set of subsets of a given set (its power set) ordered by inclusion. Similarly, the set of sequences ordered by subsequence, and the set of strings ordered by substring. The set of natural numbers equipped with the relation of divisibility. The vertex set of a directed acyclic graph ordered by reachability. The set of subspaces of a vector space ordered by inclusion, etc.

Definition 5.9 The *supremum* (abbreviated *sup*) of a subset *E* of a *poset T* is the least element in *T* that is greater than or equal to all elements of *S*, if such an element exists. Consequently, the supremum is also referred to as the *least upper bound* (or *LUB*).

Example 5.10 A set M of real numbers (blue balls), illustrated in Fig. 5.10, a set of upper bounds of M (red diamond and balls), and the smallest such upper bound, that is, the supremum of M (red diamond).

Definition 5.10 The *infimum* (abbreviated *inf*) of a subset *S* of a *poset T* is the greatest element in *T* that is less than or equal to all elements of *S*, if such an element exists. Consequently, the term *greatest lower bound* (abbreviated as *GLB*) is also commonly used.

Example 5.11 A set *T* of real numbers (red and green balls), a subset *S* of *T* (green balls), and the infimum of *S*. Note that for finite, totally ordered sets, the infimum and the minimum are equal (Fig. 5.11).

Fig. 5.10 Illustration of upper bounds for set M

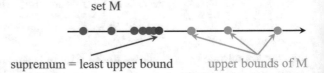

set M

supremum = least upper bound   upper bounds of M

Fig. 5.11 Illustration of infimum and supremum

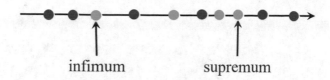

infimum   supremum

Definition 5.11 An element *g* in *poset P* is a *greatest element* if for every element *a* in *P*, $a \leq g$. An element *m* in *P* is a *least element* if for every element *a* in *P*, $a \geq m$. A *poset* can only have one greatest or least element.

Example 5.12 The subset of integers has no upper bound in the set $\mathbb{R}$ of real numbers. Let the relation "$\leq$" on $\{a, b, c, d\}$ be given by $a \leq c, a \leq d, b \leq c, b \leq d$. The set $\{a, b\}$ has upper bounds *c* and *d*, but no least upper bound, and no greatest element. In the rational numbers, the set of numbers with their square less than 2 has upper bounds but no greatest element and no least upper bound.

Definition 5.12 An element g in P is a *maximal element* if there is no element a in P such that a > g. Similarly, an element m in P is a *minimal element* if there is no element a in P such that a < m. If a *poset* has a greatest element, it must be the unique maximal element, but otherwise there can be more than one maximal element, and similarly for least elements and minimal elements.

Example 5.13 In the collection $S = \{\{d, o\}, \{d, o, g\}, \{g, o, a, d\}, \{o, a, f\}\}$
ordered by containment, the element $\{d, o\}$ is minimal as it contains no sets in the collection, the element $\{g, o, a, d\}$ is maximal as there are no sets in the collection which contain it, the element $\{d, o, g\}$ is neither, and the element $\{o, a, f\}$ is both minimal and maximal. By contrast, neither a maximum nor a minimal exists for *S*.

Definition 5.13 Upper bound and **lower bound**
For a subset A of P, an element x in P is an upper bound of A if $a \leq x$, for each element a in A. In particular, x need not be in A to be an upper bound of A. Similarly, an element x in P is a lower bound of A if $a \geq x$, for each element a in A. A greatest element of P is an upper bound of P itself, and a least element is a lower bound of P.

Example 5.14 5 is a lower bound for the set $\{5, 8, 42, 34, 13{,}934\}$; so is 4; but 6 is not. For the set $\{42\}$, the number 42 is both an upper bound and a lower bound; all other real numbers are either an upper bound or a lower bound for that set. Every finite subset of a non-empty totally ordered set has both upper and lower bounds

The smallest element, when it exists, of all the majorants, is called least upper bound (LUB) or supremum. Let us designate it by:

$$x \vee y \text{ the upper bound of the elements } x \text{ and } y,$$

and by:

$$\sup X \text{ the LUB of the subset X.}$$

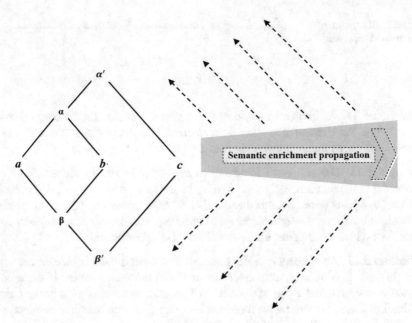

Fig. 5.12 Propagation of the semantic enrichment

We note in the same way the greatest element, when it exists, of the set of lower bounds, greatest lower bound (GLB) or infimum. We denote by:

$$x \wedge y \text{ the lower bound of } x \text{ and } y,$$

and by:

$$\inf X \text{ the GLB of the subset X.}$$

All the work of the symbolic fusion here will then consist, starting from a given pair $\{a, b\}$, of:

- Stage 1: searching for its infimum and supremum
- Stage 2: repeating the process with another element c
- Stage 3: propagating the enrichment, if possible, with other elements

These three stages must lead to increase the enrichment and thereby to continuously increase the level of *signified* as illustrated in Fig. 5.12.

Example 5.15 Let us illustrate the symbolic fusion through the following scenario. For political and religious reasons, a nation *Na* is preparing to attack the neighboring nation *Nb*. For this purpose, *Na* begins to deploy on the borders of *Nb* its arsenal according to precise rules and organization, related to a certain ORder of BATtle (ORBAT) [32].

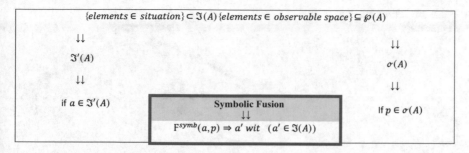

Fig. 5.13 The symbolic fusion, F^{symb}

These are represented by an arborescence A (tree structure). In preliminary, we observe that the perception system of Nb can capture only fragmentary elements, knowing that an ORBAT never unfolds completely and, secondly, that its sensors are necessarily limited by technology and also that there are no ideal observation conditions. This leads to defining the set of all the related parts of the tree A of Na noted:

$$\Im(A), (\Im, \subset) \text{ is a } \textbf{poset},$$

$\Im$ is stable by $\cap$ and unstable by $\cup$ except in the case where $\Im$ is a chain.[2]

What is engaged on Nb border corresponds to a certain subset: $\Im'(A)$ such that $\Im'(A) \subseteq \Im(A)$, in the observable space of the perception system of Nb; only some parts of Na(ORBAT) will be observed, and make a note of: $o(A) \subseteq \wp(A)$, ($\wp$ for part of sets), since a succession of observations will most certainly concern non-related elements of Na(ORBAT).

From the different points discussed above, it is possible to clarify what *symbolic fusion* means when applied to a situation (summarized in Fig. 5.13). It is necessary to be able to have a relation allowing to build and compare:

* The unit elements a, p,
* The possibility to induce the *maximal element*
* The *minimal element*.

It must indeed be possible to say whether a' is a maximal or minimal element with respect to a and p, in order to determine whether or not there has been an enrichment. The case where $a = p$ implies that a' must not be greater than a, which underlines an important fact: a knowledge item cannot be enriched from itself. In practice, under a symbolic fusion operation, F^{symb}, the following points are considered:

* A set $\Im(A)$ (ensemble of connected parts of the tree A)
* A set $\wp(A)$ (set of unconnected parts of A, each of them can be observations)

---

[2] A chain is a totally ordered set or a totally ordered subset of a poset.

Fig. 5.14 Illustration of a
regiment R

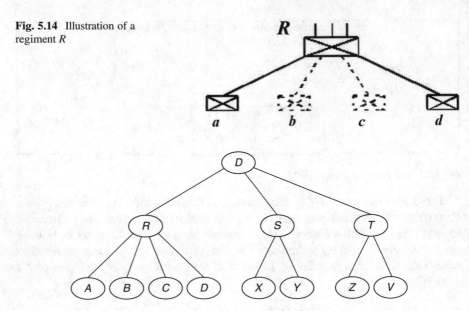

Fig. 5.15 An arborescence of an ORBAT

- An order relation: $\subset$ (inclusion) which makes it possible to order $\mathfrak{I}(A), \wp(\mathbf{A})$
  (although they do not concern the same objects)
- A universal element: A, and a null element: $\varnothing$

Referring to the example 5.15, we can imagine that Nb after observation, identification, cross-checking, and reconciliation with Na(ORBAT) does recognize a
regiment of Na, denoted R, with the identification of two of its subordinate elements,
noted (a, d). The organization, S, of the regiment R can be represented by a subtree
of A and with S that can be a assimilated to a connected component of A.

The dotted elements in Fig. 5.14 are missing for the moment since they have not
been identified as elements of the observation space o and relatively to what is
known as Na(ORBAT). The organization S represents then an incomplete knowledge of R. It should also be noted that the tree R is part of a more general set:
Na(ORBAT), which can be represented by $\mathfrak{I}(A)$. Figure 5.15 shows an arborescence
of an ORBAT.

The goal of all operators of the command system of Nb is to show now that:

- The root of R is an element of A: $A : root(R) \in A$
- To ease the matching of a tree from the observation space with that of
  Na(ORBAT):

$$R' \subset o(A) \to R \subset \mathfrak{I}(A)$$

- To complete R with leaves $x_i \in A$ such that from the tuples, we have: $(R, x_1, x_2,$
  $x_3, \ldots, x_n) \subset o(A)$

- To show, for example, that: $R(x_1, x_2) \subset \mathfrak{I}(A)$, $R(x_1, x_2, x_3, \ldots, x_n) \subset \mathfrak{I}(A)$
- Or what amounts to the same, to show that: $R(x_1, x_2, x_3, \ldots, x_n) \subset O(A)$ represent connected components of the tree

Definition 5.14 Let us define two operators, $\cap$, $\cup$. Note: not to be confused with $\cup$, $\cap$, set union and intersection, and two applications of $A \times A$ in A:

$$\cap / \ \forall \, a, a' \in A \quad a \cap a' \longrightarrow a \cap a'$$
$$\cup / \ \forall \, a, a' \in A \quad a \cup a' \longrightarrow a' \ si \ a \subset a'$$
$$a \cup a' \longrightarrow a \ si \ a' \subset a$$

It must be possible to constitute the smallest tree structure strictly containing a and a'. The triplet $\{\mathfrak{I}(A), \cup, \cap\}$ constitutes a lattice structure. The order relation induced on this structure is the *inclusion*. The symbolic fusion, $\mathsf{F}^{symb}\{a, a'\}$ of $\mathfrak{I}(A)$ in the case illustrated in Fig. 5.16, is described below.

In practical terms, we must find the *LUB* or supremum as well as the *GLB* or infimum for a and a' such as defined in Definitions 5.9 and 5.10. The tree structure of the resulting fusion is only constructed at the level of the lowest leaves of the two trees a and a'. That makes sense in practice since to fuse subordinate units at a higher level, the lower level brings nothing to the fusion represented by the rectangle in Fig. 5.16. Let us take a subgraph R from tree A as in Fig. 5.17.

The use of the two applications (see Definition 5.14) allows to write:

$$\{A, R\} \cap \{D, R\} = \{R\} \tag{5.1}$$

$$\{A, R\} \cup \{D, R\} = \{A, B, C, D, R\} \tag{5.2}$$

$$\{A, R\} \cap \{D\} = \varnothing \tag{5.3}$$

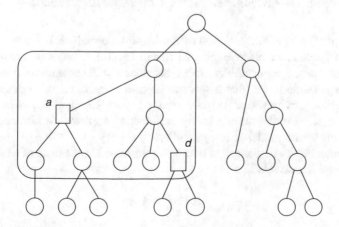

Fig. 5.16 Symbolic fusion: finding the smallest tree structure

Fig. 5.17 Symbolic fusion:
a subgraph R (regiment)

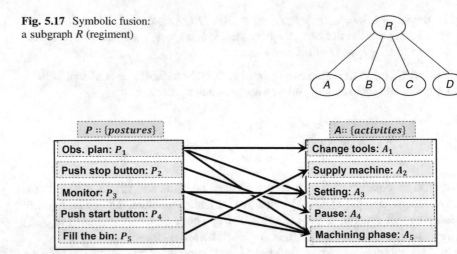

Fig. 5.18 Activities of an operator in a workstation

$${A, R} \cup {D} = {A, B, C, D, R} \tag{5.4}$$

Comparing (5.2) and (5.4), we observe that to enrich ${A, R}$ by the application, $\cup$, it suffices to use ${D}$, since ${D, R}$ that normally contains more knowledge does not bring more. If we observe (5.1) and (5.3), we note that any cross-checking or confirmation gives nothing for (5.3) but only to ${D}$ for (5.1). This would suggest that, for there to have enrichment relative to $\cap$, it is necessary to turn to structured objects.

5.5.2   Semantic Enrichment with Compatibility Model

The concept of compatibility relation offers a simple formalism, to put in correspondence descriptive elements belonging to two or several domains of knowledge. To illustrate the use of a compatibility model, consider first the example of a workstation where it is possible to say that a specific posture of an operator corresponds to a specific type of activity with: $P :: {postures}, A :: {activities}$, as shown in Fig. 5.18. An expert in work organization can, by establishing correspondences between sets P and A, construct a model of compatibility.

This compatibility model can of course be expressed in the form of a table which highlights the formal relation:

$$R \subset P \times A$$

($\times$-marks in Table 5.3).

Table 5.3 Highlights of the formal relation

	A_1	A_2	A_3	A_4	A_5
P_1	×		×		×
P_2				×	
P_3			×		×
P_4					×
P_5		×			

Most of the time, the planner cannot establish precisely these correspondences, but he is more able to say "how" it is possible that such attitude is compatible with such activity. It is therefore interesting to have a way to "graduate" these possibilities of assignments. This leads to the extension of the notion of normal relation, considering that they can be expressed vaguely, imprecisely, or incomplete as the case may be. To do this, we will define a fuzzy relation R_f different from the usual relation $R \subset P \times A$. The fact of giving a degree of possibility to each arrow of Fig. 5.18 amounts in fact to defining a fuzzy subset:

$$R_f \subset P \times A \neq R \subset P \times A.$$

Example 5.16 A fuzzy relation

$$R_f :: \{0.2/(P_1, A_1); 0.4/(P_1, A_3); 0.1/(P_1, A_5); 0.4/(P_2, A_4); 0.5/(P_3, A_3) \cup 0.8/$$
$$(P_3, A_5); 0.8/(P_4, A_5); 0.8/(P_5, A_2)\},$$

with its matrix form:

$$R_f \subset P \times A = \begin{bmatrix} 0.2 & 0.0 & 0.4 & 0.0 & 0.1 \\ 0.0 & 0.0 & 0.0 & 0.4 & 0.0 \\ 0.0 & 0.0 & 0.5 & 0.0 & 0.8 \\ 0.0 & 0.0 & 0.0 & 0.0 & 0.8 \\ 0.0 & 0.8 & 0.0 & 0.0 & 0.0 \end{bmatrix}$$

Incomplete or vague knowledge about the sets P and A would amount to assimilating them to fuzzy sets P_f and A_f, and in doing so, to defining another fuzzy relationship R'_f taking meaning in a fuzzy Cartesian product. Since $P_f \times A_f$ is a fuzzy set, it is known only by its membership function, evaluated as follows:

$$\mu_{P_f \times A_f}(x, y) = \min [\mu_P(x), \mu_A(y)]$$

In the case of the fuzzy Cartesian product, the "vague" characterizing P_f and A_f will influence any relation that is to be defined. If we consider in the preceding

Table 5.4 Compatibility relations

	P_1, A_1	P_1, A_2	P_1, A_5	P_2, A_1	P_2, A_2	P_2, A_5
$A \times B(x, y)$	0.8	1	0.8	0.8	1	0.8
	P_4, A_1	P_4, A_2	P_4, A_5	P_5, A_1	P_5, A_2	P_5, A_5
$\mu_{A \times B}(x, y)$	0.2	0.2	0.2	0.8	1	1

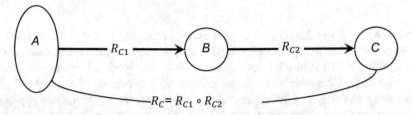

Fig. 5.19 Enrichment by transitivity through compositions of relations

example that P and A are imperfectly known then we obtain the following using Table 5.4:

$$P :: \{1/P_1, 1/P_2, 0.2/P_3, 0/P_4, 1/P_5\}$$
$$A :: \{0.8/A_1, 1/A_2, 0/A_3, 0/A_4, 0.8/A_5\}$$

which modifies the previous matrix of $R_f \subset P \times A$ to:

$$R_f \subset P_f \times A_f = \begin{bmatrix} 0.8 & 1 & 0 & 0 & 0.8 \\ 0.8 & 1 & 0 & 0 & 0.8 \\ 0 & 0 & 0 & 0 & 0 \\ 0 & 0 & 0 & 0 & 0 \\ 0.8 & 0 & 0 & 1 & 1 \end{bmatrix}$$

Enrichment by compatibility model can be generalized to many domains by transitivity. Since we know how to build several compatibility models after each other, it becomes possible to conduct enrichment on several levels as illustrated in Fig. 5.19.

$$R_c(A, C) = R_{c1}(A, B) \& R_{c2}(B, C)$$

When sets are crisp, this amounts to making relations compositions in the classical sense: $R_c = R_{c2} \odot R_{c1}$ and so on. If the sets are fuzzy, a membership function is usually defined by:

Fig. 5.20 Zadeh's
extension principle of fuzzy
sets

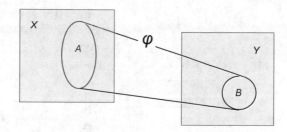

$$\mu_{R_{c2} \odot R_{c1}}(a, c) = \max_{b \in B} \left[\min \left(\mu_A(a, b), \mu_B(b, c) \right) \right]$$

This expression results from the application of a fundamental principle of fuzzy set theory: the *extension principle* (illustrated in Fig. 5.20). This one was introduced by Zadeh to give the possibility of defining the image of a fuzzy subset by an application. Concretely, this allows to induce, from a given universe, fuzzy subsets in another universe via a given application. From X and Y universes, consider an application φ from X to Y represented by the fuzzy subsets A and B,

$$\forall y \in Y, \mu_B(y) = \text{Sup}_{x \in X | y = \varphi(X)} \mu_A(X), \text{if } \varphi^{-1}(\{y\}) \neq \emptyset, \text{otherwise } \mu_B(y) = 0$$

Note that the determination of B becomes very simple if φ is bijective since $\forall y \in Y, \varphi_B(y) = \varphi^{-1}(\{y\})$.

If A itself results from a Cartesian product, the extension principle and the definition of the fuzzy Cartesian product make it possible to associate a fuzzy subset B of the universe Y with the respective fuzzy subsets P, A of the "operator at workstation" example. This is characterized by the membership function defined as follows:

$$\forall y \in Y, \mu_B(y) = \text{Sup}_{[x = (x_1, x_2, \dots, x_n) \in X | y = \varphi(X)]} \min \left[\mu_P(x_1), \dots, \mu_A(x_1), \dots \right]$$

5.5.3 Semantic Enrichment with Distribution of Possibilities

Let us now take the case where an observer is not able to say with certainty in which postures the operator is, but, on the other hand, he is only capable to estimate a certain degree of possibility from his observations. This is illustrated in Fig. 5.21. This amounts to saying that:

- $x(posture) :: p_i$, with a certain degree of possibility, it is then possible to deduce
- $y(attitude) :: a_i$, with a certain degree of possibility, considering the fact that there are relations between postures and activities defined in the model of compatibility established by the expert

(Compatibility model)

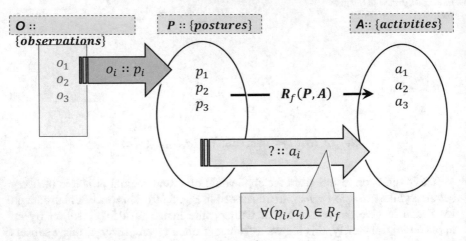

Fig. 5.21 Enrichment by distributions of possibilities

The implementation of the preceding scheme is made possible only by using several elements of the theory of possibilities recalled briefly below.

- *Notion of measure of possibility*

Let R be a reference set, and we define the measure of possibility as an application Π:

$$\Pi : P(R) \rightarrow [0,1] \text{ with } \Pi(\varnothing) = 0 \text{ and } \Pi(R) = 1$$

where

$$\forall E_i \in P(E_i) \rightarrow \Pi\left(\underset{i=1,\dots,n}{\cup} E_i\right) = \text{Sup}_{i=1,\dots,n}\Pi(E_i)$$

It should be noted that by making $i = 2$, we find the logical sum of two terms (axioms of a fuzzy measure).

$$\Pi(E_1 \cup E_2) = \max\left[\Pi(E_1), \Pi(E_2)\right]$$
$$\Pi(E_1 \cap E_2) = \min\left[\Pi(E_1), \Pi(E_2)\right]$$

The measure of possibility must have the characteristics of a fuzzy measure since the application,

$$\Pi : P(r) \longrightarrow [0, 1]$$

with the reference set R, satisfies the three following axioms:

1. Limit cases: $\Pi(0) = 0$, $\Pi(r) = 1$,
2. Monotonicity: $A, B \in P(r) | B \supseteq A \Rightarrow \Pi(B) \geq \Pi(A)$,
3. Continuity: $A_1 \subseteq A_2 \ldots \subseteq A_{n-1} \subseteq A_n$.

It is worth noting these two important relations:

$$\Pi(A \cup B) \geq \max[\Pi(A), \Pi(B)]$$

$$\Pi(A \cap B) \leq \min[\Pi(A), \Pi(B)]$$

- *Notion of measure of distribution of possibilities*

A measure of possibility $\Pi : P(r) \longrightarrow [0, 1]$ is completely defined in practice if we can assign a coefficient to each part $E \in P(r)$. If we now consider the elements of R (and not the parts of sets) and we assign them a coefficient, we then define a distribution of possibilities, in the form of a function π:

$$\pi : P(r) \longrightarrow [0, 1] \text{ with } \text{Sup}_{e \in R} \pi(e) = 1$$

(π is called normalized).

There is an interesting and useful relationship between measurement and distribution of possibilities, since:

- From the *distribution* π: it is possible to construct Π,

$$\forall E \in P(r), \text{ we have, } \text{Sup}_{e \in E} \pi(e) = 1$$

- From the *measure* Π: it is possible to construct π,

$$\forall e \in R \longrightarrow \pi(e) = \Pi(\{e\})$$

Note that $\Pi(\{e\})$ is a measure of possibility because $\Pi(r) = 1 \Rightarrow \exists e \in R | \pi(e) = \Pi(\{e\}) = 1$

- *Notion of joint distribution*

We will expand the previous definition to make it applicable to the Cartesian product: $X \times Y$, that is, to have a degree, which indicates how much each pair ($x \in X$, $y \in Y$) is possible. The distribution defined on the Cartesian product must indicate to what extent it is possible that $x = x_0 \in X$ appears at the same time as $y = y_0 \in Y$,

$$\forall x, \forall y\ \pi(x, y) : \text{Cartesian Product} = X \times Y \longrightarrow [0, 1]$$

The term $\pi(x, y)$ provides information on the X, Y reference sets represented individually by their marginal possibility distributions, which are obtained by keeping the largest value of π relative to the reference set.

- *Notion of distribution of marginal possibilities*

From a distribution of joint possibilities π, it is possible to define the marginal distributions on X and Y as follows:

$$x \in X \longrightarrow \pi_x(x) = \text{Sup}_{y \in Y} \pi(x, y)$$

$$y \in Y \longrightarrow \pi_y(y) = \text{Sup}_{x \in X} \pi(x, y)$$

These two distributions verify on X and Y:

$$\forall x, \forall y\ \pi(x, y) \leq\ \min\left[\pi_x(x), \pi_y(y)\right]$$

$$x \in X, \forall y =\ \min\left[\pi_x(x), \pi_y(y)\right]$$

All this supposes that the reference sets X, Y are *interacting*. They have a reciprocal influence on each other. The term $\pi_{Y/X}$ then indicates the coefficients of possibility of $y = y_0 \in Y$ knowing that $x = x_0 \in X$ is possible (analogy with the computation of the conditional probabilities). Conversely, X, Y are said to be *non-interactive* if their joint distribution is the largest of all those compatible with the marginal distributions $\pi_x(x)$ and $\pi_y(y)$. The notion of non-interactive sets is to be compared with that of independent events in probability.

- *Notion of distribution of conditional possibilities*

A distribution of conditional possibilities $\pi_{Y/X}$ is represented by the set of degrees of possibility of $y = y_0 \in Y$ knowing that the element $x = x_0 \in X$ is itself possible (to be compared with the approach of the calculation of the probabilities conditional):

$$\forall x \in X, \forall y \in Y, \pi_{Y/X} = \pi_{Y/X}(x, y)\ \varsigma\ \pi_x(x)$$

The combination operator ς generally corresponds to the minimum or the product. We will now show, from the example of Fig. 5.22, how one can concretely realize a semantic enrichment by exploiting the notions just described above.

A probability distribution projection associated with the variable $o \in O$ (observations) must be made, where o_i will be mixed in p_i using the following tools:

- Distribution of possibilities of O/P noted: $\pi_P(o_i \approx p_i)$
- Distribution of possibilities of A noted: $\pi_A(a_i)$
- Distribution of joint possibilities, $P \times A$ noted: $\pi_{P/A}(p_i, a_i)$

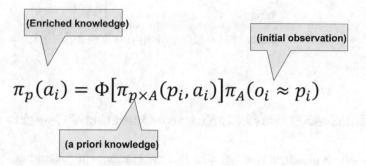

Fig. 5.22 Enrichment by distributions of possibilities

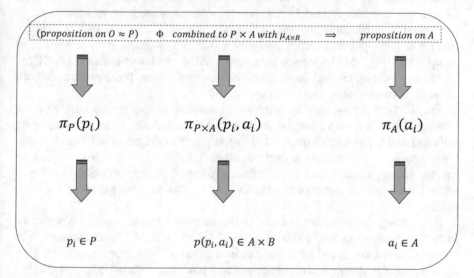

Fig. 5.23 Projection combined with the fuzzy Cartesian product

From a proposition expressed on the set O/P, the role of the operator Φ is in fact to be able to obtain a proposition expressed on the set of the activities A, thus summarized in Fig. 5.23 diagram.

Consider:

$$\pi_P(p_i) :: \{(p_i, n) | p_i \in P, n \in [0, 1]\}$$

$$\pi_{P \times A}(p_i, a_i) :: \{(p_i, a_i, n') | p_i \in P, a_i \in A, n' \in [0, 1]\}$$

$$\pi_A(a_i) :: \{(a_i, n'') | a_i \in A, n'' \in [0, 1]\}$$

and if we meld Φ in a minimum function (the most frequent case), we obtain:

$$\pi_A(a_i) = \Phi[\pi_{P \times A}(p_i, a_i)]\pi_A(o_i \approx p_i)$$

$$\pi_A(a_i) = \{(a_i, n'') | \exists(p_i, a_i, n') \in \pi_{P \times A}(p_i, a_i), \exists(p_i, n) \in \pi_P(p_i), n'' = \min(n; n')\}$$

5.6   Efficient Actions in Interactive Operational Contexts

The discussion presented here is to illustrate the semantics of "interactions" using a relatively simple context. We limit ourselves to a case of decision-making and actions within the framework of embedded systems not connected to Internet. Operational contexts created by the presence of a connectivity to Internet (e.g., networked systems such as cyber-physical systems (CPS), Internet of Things (IoT), and cyber-physical and social systems (CPSS)) possess a level of interaction complexity (e.g., H2S, human-to-system; M2M, machine-to-machine; H2M, human-to-machine) beyond the scope of our discussion here. There exists a plethora of recent books on those networked systems.

We choose a simple case: an embedded system participating in a mission (e.g., military). Such a system identifies itself with a technical device which, despite its technical and technological qualities giving it a certain degree of autonomy, needs the cooperation of a human to perform its task (or a mission). This will induce successive interactions, which involve the evolution of the processes of the system to achieve the sub-goals prescribed by the task and to consider the logic of the decision-maker.

Even though an embedded system is the simplest case, its level of complexity induces complications in the human system interaction mechanisms, through elaborate behaviors, combined with efficient interaction technologies (but implying on the other hand protocols that are ergonomically binding). With the human still in the loop, it is necessary to ask himself the question of what it means to interact on (or act in) a given system for any operator, and whatever the requested orders have specific terms of use, themselves subject to constraints, which make their interactions operative, or not at all in a particular employment context.

For example, for an aircraft pilot using the "voice command" interaction channel, it can be imagined that the "voice command" modality can be used (operationally) in flight up to the load factor 3 g. Beyond this threshold, under the effect of tight maneuvers (inducing strong accelerations), the performance of speech recognition algorithms deteriorating rapidly no longer allows its use in satisfactory operational conditions. Constraints can be associated with the modalities: the "voice command" modality remains applicable if the "recognition rate" constraint remains constant and does not fall below 0.98%. On a formal level, this amounts to asking the question of whether the representation of interactions in a space representative of the context of evolution of a process, under the effect of a command, has a meaning or not and of whether it remains applicable. In short, it would be necessary to be able to represent the semantics of the interactions in such a space, to know if, and how, it is possible to activate them.

Let's start by distinguishing two important notions:

- *An Interaction Space*: Existing space between the domain of the operator, identifying most often with the structure of a dialogue of type: voice, gestural, etc. and the action universe where the commands run as tasks. The interaction space is the seat of the transformations necessary to make operational orders in the form of basic actions to be carried out in the universe of actions. The semantics of a multimodal interaction in this space must consider all the transitions that take place there.
- *A Multimodal Grammar*: Consideration and simultaneous management of several modes of interaction, multi-modality involving constant changes of interaction channels depending on the state of the actions to be taken and the progress of the processes. In the interaction space, several types of interaction channels may be solicited sequentially or simultaneously to circulate a command from the decision-maker universe to that of the actions. Each interaction channel has its own modalities of use which must be identified and described with their specific constraints: physical limits of a modality for a given interaction allowing to express an operator's requirement.

The interaction space brings out three differentiated domains, from which the relations and their compositions will be established. This alone will make it possible to base the semantics of an interaction. To overcome certain difficulties of representation, we can look at the expression of relations between representative sets of the space of interaction, and to the compositions of these relations between them.

It can be observed in Fig. 5.24 that the sending of commands for the execution of a concrete act in the action universe must go through the intangible software layers by taking an interaction channel (associated with a physical privileged communication medium). The resulting interactions will undergo transformations induced by the specific modalities of each channel according to the circumstantial conditions of their use. These modalities are themselves subject to constraints attached to the interaction channel, taking into account both the technical and technological realities of their own but also the specific conditions of their use. Moreover, Fig. 5.24 suggests that an interaction will depend on the existence of an "interaction entity" only to the extent that one can specify its belonging to the Cartesian product:

Interact $\in$ [{interaction channels} $\times$ {interaction modalities} $\times$ {constraints on modalities}].

Functionally, to clarify Fig. 5.24, we must define a number of sets and specify their role.

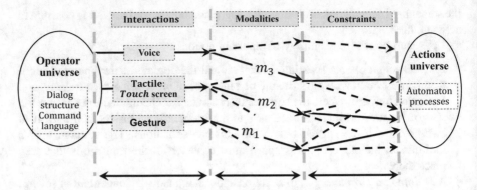

Fig. 5.24 Actions universe of human-machine (interaction space)

5.6.1 Set I of Interaction Channels

The set I can consist of several types of channels, for example: I_v, for voice interaction channel; I_t, for touch interaction channel; I_g, for gesture interaction channel; I_p, for pushbutton interaction channel; I_o, for oculometry interaction channel, etc.

$$CI :: \left\{ I_v, I_t, I_g, I_p, I_o \right\}$$

That set is a declarative list, where each element is associated with a sequence of attributes (response time, precision, etc.).

5.6.2 Set M of Modalities

For each interaction channel, there exists one or more characteristic modalities of use of this channel; M represents the set of all the unary modalities attached to the interaction channels of I being able to be expressed by propositions or propositional variables. Some of these unary modalities are obviously dedicated to a single interaction channel, others to multiple channels. Also, from these unary modalities, one can build "logical propositions" representing the modalities of use of the channels; some can be common to different channels. We will represent the modalities by:

$$F(M) :: \{\text{modalities formulas} \} \supset M$$

These can be understood as predicates whose evaluation, from a perspective of first-order logic, makes it possible to know if the employment of an interaction

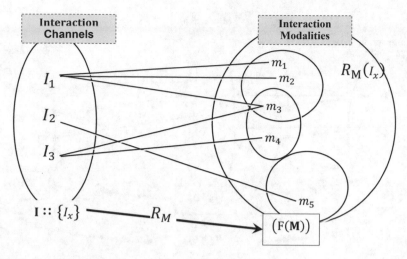

Fig. 5.25 Channels-modalities relations (interaction space)

channel remains or is not legitimate. It is then a question of grouping these modalities according to the interactions put into play, thanks to the establishment of a relation:

$$R_M \subset I \times F(M)$$

An interaction channel, $I_x \in I$, is associated with one or more modalities (Fig. 5.25), that is to say, one or more elements of $F(M)$ whose image is:

$$R_M(I_x) \subset F(M)$$

5.6.3   Set C of Constraints

This set includes the known constraints on the terms of use, those expressed by the orders, and those imported from the universe of actions (e.g., contextual, environmental). This is a declarative list of unary constraints likely to validate or invalidate the modalities for the usage of the interaction channels, provided that these constraints are grouped into "logical propositions." All these formulas will be noted $F(C)$; it is therefore essential to link these sets by adequate relations. This relation therefore relates to the Cartesian product: $I \times C$, being the set of parts of $F(M)$.

A second relation R_C (Fig. 5.26) consists of associating with any element $R_M(I)$, image by R_M of the set I, one or more constraints of $F(C)$, or none, if a modality undergoes no constraint. In general, one will be in the case where:

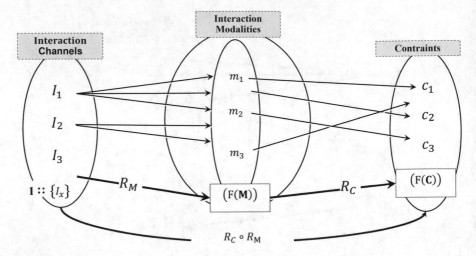

Fig. 5.26 Modalities-constraints relations (interaction space)

$$\forall m_i \in R_M(I), \exists c_j \in C$$

R_C relates to a;n extended Cartesian product:

$$R_C \subset F(M) \times \mathcal{P}(F(C))$$

where $\mathcal{P}(F(C))$ represents the set of parts of $F(C)$. The composition of the relations $R_C \circ R_M$ makes it possible to cross the space of interactions. The Cartesian product $I \times R_M \times (R_C \circ R_M)(I)$ is necessary to prove the semantics of an interaction.

The evolution of the constraints in the time circumstantially invalidates the use of certain interaction channels, which means that the terms of use of the channel are no longer respected. It is thus by asserting the modalities that the use of a channel is inhibited or not. The constraints are then as many factual events imported from the universe of actions, coming to change over time the evaluation of the predicates translating the modalities. These constraints can therefore be represented by predicates whose evaluation will play directly on the modality of use of the interaction channel. The interaction semantics is based on the knowledge of the relations R_M and R_C whose composition makes it possible to cross the interaction space of the operator universe to the universe of actions. A triplet $\{I_x, m_y, c_z\}$ where:

- I_x is a specific interaction channel
- m_y is one of the modalities associated with the ci_x channel
- c_z is a constraint related to the modality m_y

is semantically valid if it is possible to show:

$$\{I_x, m_y, c_z\} \in I \times R_M \times (R_C \circ R_M)(I)$$

All the previously described relationships will be useful for basing the operating principles of a multimodal interaction grammar. The question now is how will the interaction semantics allow the interaction engine to execute a command or not? For a channel, $I_x \in I$, the relation R_M associates with I_x the set $R_M(c_x) \times \{m_y\}$ of the possible modalities (R_M consists of associating with the channel the predicate that models its use); two cases may occur:

- $R_M(c_x) = \varnothing$; the role of the grammar is to invalidate the interaction
- $R_M(c_x) \neq \varnothing$; it is then a question of scanning the elements $m_y \in R_M(ci_x)$ and for each m_x, to consider all the constraints $R_C(m_y)$

At the level of the constraints, two cases can also be presented:

- $R_C(m_y) = \varnothing$; no constraint blocking the use of the modality, the interaction channel is valid; the semantics being then verified, the command is then executable
- $R_C(m_y) \neq \varnothing$; $R_C(m_y)$ is a set of constraints that, as factual events from the system's action universe, can invalidate the modality, m_y

If none of the constraints of $R'(m_x)$ comes to invalidate the modality, then this one is asserted as well as the interaction channel; the execution of the command can then take place. Otherwise, if a constraint invalidates the modality, then although the interaction has semantics, the command cannot be executed. The interaction engine must find another modality that can be associated with the interaction channel, whose interaction semantics are valid and for which the constraints from the action universe allow the execution of the command. If the interaction engine does not highlight such a modality, then the channel must in turn be invalidated and the system must propose other channels of interaction allowing the process to continue to evolve. A multimodal interaction can thus be described by a series of actions carried out by the interaction engine, by asserting certain modalities, and by prohibiting others, etc. On the other hand, the interaction engine follows an evolution over time. It then appears necessary to define a new notion, comparable to an "object of knowledge," and called "interact," whose objective is to describe the state and evolution of the multimodal interaction as a process that can be interrupted, paused, resumed, etc.

5.7 Efficient Actions in Cooperative Operational Contexts

We will now be interested in the action taken in contexts manifested by the obligation of making several systems to simultaneously work together. This singles out working contexts that imply for the systems to have some of the common qualities that can be reduced to the following two points:

- Interconnectivity: all hardware and software devices able to ensure the proper functioning of exchange mechanisms between communicating systems

- Interoperability: the development of the opening of systems to improve their interworking, its prerequisite being the interconnectivity

If we want to make this interoperability more efficient, in particular by exchanging knowledge and sharing the know-how in the execution of processes, in other words by putting intelligence into inter-system exchanges, we then define a new so-called cooperative context whose interoperability becomes the prerequisite. Barès [1, 32–35] proposes a formal modeling approach to the problem of interoperability in large information systems. His approach is relying on three main concepts: openness structure of a federation of systems, a definition of distinct domains for interoperability and cooperability, as well as mathematical tools in which parameters can be used to assess interoperability.

A great deal of important transformations generated by the evolution of geopolitical contexts implies the responsibility of the international community as soon as a crisis or a conflict is emerging. All nations in a position to share this "responsibility" are more and more often involved in an international coalition. The different technical systems, networks, and C4ISR[3] that are taking part of a coalition must cooperate for executing a common mission fixed by a coalition authority under specific conditions and temporal constraints. The term "cooperate" is intentionally used to highlight the type of communication required in a coalition. This term completely exceeds the simple exchange of messages as it will be shown later. For now, let us illustrate it by an agent who realizes, for the reason either by insufficient know-how or a lack of knowledge, that he cannot alone achieve an objective. This agent will request assistance from other agents. In explaining the considered case, the agent specifies a "context of openness" in a potential cooperative scheme for other agents to bring the requested assistance. When doing so, agents must position themselves in a cooperative scheme. Moreover, that hypothetical coalition cannot be successful if his members are not willing:

- To exchange their knowledge about situations that necessitate their intervention. This knowledge must be enriched as long as the process is going on (validity of an information may depend on time)
- To exchange the know-how about operating processes and methods of application
- To share, timely and under appropriate conditions, actionable knowledge with all agents of the cooperation (support to the evolution of actuation)
- To add "intelligence" at the different levels of the interoperability mechanisms (that may be of a great help to other members)

For the sake of this discussion, consider the organizational paradigm of a coalition. However, the concepts could be adapted to other organizational paradigms such as holarchies, teams, federations, etc. with the support of a substantial legacy of analyses from the multiagent community [36–43]. In a coalition, the first step to be

---

[3]C4ISR stands for Command, Control, Computers, Communications, Intelligence, Surveillance and Reconnaissance. This is an acronym used in defense and security domain.

considered is to make systems cooperate in achieving a mission. To that end, a major problem to face is the fact that these systems are heterogeneous: a heterogeneity, inherent to national design and employment concepts. As a result, they present enormous deficiencies at the interoperability level. One could argue that it is always possible to solve this question by making gateways, but that represents a temporary solution. Moreover, gateways solution cannot be easily and reasonably generalized. Such generalization is getting more difficult as the number of systems increases (combinatory). Therefore, what seems more reasonable is to design new concepts of interoperability embedded in each system.

5.7.1   The Concept of Cooperability

The cooperation between a system A and a system C presupposes that both systems are interoperable and the interoperability between A and C presupposes that they are connectable within each other. This is illustrated in Fig. 5.27. Cooperability between systems is only possible under certain conditions. In fact, systems must be:

- *Connectable* – systems be capable to exchange data according to known protocols
- Interoperable – systems designed to progressively open their structures to make possible richer exchanges to pursue and to meet common goals

    Interoperability requires the following:

- A common representation and understanding (common semantics) of the actuation world
- A shared knowledge about the required materials and tools for achieving the task

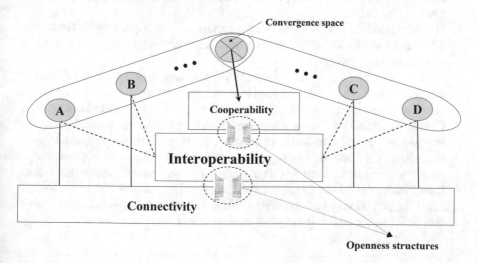

Fig. 5.27 Three distinct domains for interoperability

- A convergence with respect to viewpoints, processes, etc. (convergence space)
- A shareable knowledge environment
- Team-up to assemble all means necessary to establish a cooperation

As illustrated in Fig. 5.27, a space is defined in which distinction is made between three domains relevant to interoperability. These domains are structured; they have semantic links and their own different techniques and approaches. Interoperability is making the essential link between connectivity and cooperability. Interoperability must be understood more than just being an exchange of simple messages. Modeling and symbolic representations are being involved to carry and share knowledge. For instance, in a coalition, C4IRS systems bring each other a mutual assistance in their cooperative action to reach a common objective. C4IRS systems must possess a joint comprehension of what they are doing. That makes necessary to introduce semantic notions in the interoperability domain and to determine modalities that enable to add "intelligence" and to interpret that intelligence in the exchange mechanisms. At last, the cooperability domain represents the final level to reach, through the definition of a world, in which all (cooperative) systems can share all elements constituting their common activity in the coalition.

Cooperability implies other system notions such as concurrency, parallelism, communication, and synchronization. For instance, communication supports synchronization to get timely and useful information for proper system functioning and dependability. These system notions being mainly concerned with two cybernetics functions: coordination and control. A cooperative system has certain abilities when placed in a coalition framework. These abilities are:

1. *Openness ability*: Quality of a system, previously connected with others, to share a common understanding with them, relative to some themes of a coalition, for instance: ground evacuation and medical assistance. The openness of a system appears to be a subset of the openness structure of the coalition (illustrated by the gates in Fig. 5.27).
2. *Interoperability ability*: Capability of a system to (inter)operate with (interoperable) actions, relevant to the cooperation, more precisely orders and missions fixed within the coalition.
3. *Cooperability ability*: We will consider that a system is (inter) cooperable when this system is capable to share knowledge and the know-how with its neighboring systems. A system can be understood as a network of people and machines (e.g., CPSS). In addition to being interoperable, systems must cooperate by sharing professional experiences, skills, and knowledge (e.g. Skills-Rules-Knowledge of Rasmussen's model [44, 45]) to perform at best in doing a common task.
4. *Ability to conduct actions*: Let us assume a system owns all the abilities mentioned above 1), 2), and 3) to perform a mission in a coalition. The mission can fail if the conditions of temporal intervals are not strictly respected. An action is only valid within a precise temporal interval sometime referred to as opportunity.

5.7.2   The Concept of System Openness

The first and necessary step of interoperability is the concept of "openness." An openness context is ways and limits of opening systems for interoperability beyond the notion of connectivity. First, let us define some pertinent notions to a coalition.

- *System in a coalition*: a system i will be designated by: S^i where $i \in [1, n]$, $n =$ number of systems placed in the coalition. They are supposed to be able to share a minimum common knowledge and to have common comprehension of fundamental commands.
- *Theme in a coalition*: a theme is a set of knowledge required for the coalition and describing an ability: for instance, medical assistance or civil rescue. A theme t will be designated by T_t, with $t \in [1, q]$; q is the number of themes of the coalition. T_t encompasses a variable number of interoperable actions. An action j will be designated by A_j.

The context of openness is the network of existing relationships between systems and the coalition-themes described in a formal way: the formalization of which is based on some mathematical notions such as relations and Galois connections. The context is formally defined by a triplet (S, T, R):

$$S :: S^1, \ldots, S^n$$

$$T :: T_1, \ldots, T_q$$

$$R :: R \subset S \times T$$

where R is a binary relation.

The context may be given a priori when the coalition is defining the mission of each system. It can be also defined a posteriori when the coalition is running and situations are evolving.

Example 5.17 Three different nations are asked to interoperate within a coalition framework for rescuing civil people in an African state. The coalition is defined as follows: three nation-systems S^1, S^2, S^3 are concerned, and three coalition-themes are defined: ground evacuation operations (T_1), airborne transportation (T_2), and logistical medical aid (T_3).

Example 5.17 suggests that systems can interoperate on different actions relevant to the coalition-themes and secondly they exchange knowledge required to achieve their respective missions as in the following expressions:

$$R\left(S^1, T_1\right), R\left(S^1, T_2\right), R\left(S^1, T_3\right)$$
$$R\left(S^2, T_1\right), R\left(S^2, T_2\right), R\left(S^3, T_3\right)$$

Relation R	T_1	T_2	T_3
S^1	■	■	■
S^2	■	■	■
S^3	■	■	■

(Example 1)

Relation R	T_1	T_2	T_3
S^1	■	■	
S^2	■	■	■
S^3	■		■

(Example 2)

Fig. 5.28   Illustration of a context of openness

$$R(S^3, T_1), R(S^3, T_2), R(S^3, T_3) \subset S \times T$$

Example 1, in Fig. 5.28, is a particular case where relation R on $\{S^1, S^2, S^3\} \times \{T_1, T_2, T_3\}$ is total. Considering strictly the semantic point of view, systems are totally open to the themes involved in this coalition. This example describes an ideal situation that rarely takes place in reality. Consequently, we get a unique totally open couple in the sense of a Galois connection [46, 47]. The subset $\{S^1, S^2, S^3\}$ must be considered as totally open on the subset $\{T_1, T_2, T_3\}$. In the second example of Fig. 5.28 (Example 2), two couples are absent. The absence of certain couples, (S^1, T_3), (S^3, T^2), makes $R \subset \{S^1, S^2, S^3\} \times \{T_1, T_2, T_3\}$ not total. The coalition must reduce its openness context.

Example 1 describes an ideal case, since all systems of S are totally open to all themes of T. A condition of openness is defined as:

$$\exists i, t \mid S^i \in S \text{ and } T_t \in T, \text{ and } \mid \exists (S^i, T_t) \subset R \subset S \times T$$

$$i \in [1, 2, \ldots, n], t \in [1, 2, \ldots, q]$$

and a totally interoperable group (IG) is defined as:

$$< IG - (< S > \rho < T >) >$$

ρ means that R is a total relation on $S \times T$; in other words, there exists only one dependency between the subset S and the subset T. The subset S is totally open on the subset T. All IG must be numbered (IG-#) to construct a lattice for the openness structure of a coalition.

5.7.3   Openness Structure of a Coalition

The notion of interoperable groups (IG) supports a formal structure to represent a coalition openness: a lattice with its interesting properties. Let us illustrate that with an example. Fig. 5.29 presents a coalition C with an openness context composed of

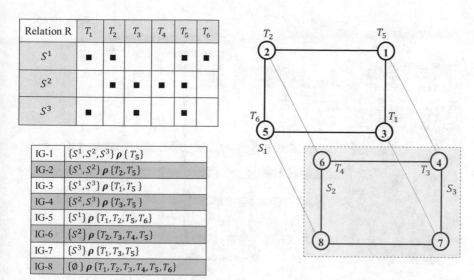

Relation R	T_1	T_2	T_3	T_4	T_5	T_6
S^1	■	■			■	■
S^2			■	■	■	■
S^3	■		■		■	

IG-1	$\{S^1, S^2, S^3\}\, \rho\, \{T_5\}$
IG-2	$\{S^1, S^2\}\, \rho\, \{T_2, T_5\}$
IG-3	$\{S^1, S^3\}\, \rho\, \{T_1, T_5\}$
IG-4	$\{S^2, S^3\}\, \rho\, \{T_3, T_5\}$
IG-5	$\{S^1\}\, \rho\, \{T_1, T_2, T_5, T_6\}$
IG-6	$\{S^2\}\, \rho\, \{T_2, T_3, T_4, T_5\}$
IG-7	$\{S^3\}\, \rho\, \{T_1, T_3, T_5\}$
IG-8	$\{\emptyset\}\, \rho\, \{T_1, T_2, T_3, T_4, T_5, T_6\}$

Fig. 5.29 Openness context and structure of a coalition C

eight subsets of systems. This context is translated through an IG structure. From those IG, a graph is built where its nodes correspond to numbered IG (IG-#).

Right part of Fig. 5.29 represents the only possible openness structure of the coalition C that, obviously, depends upon the way to fix the openness context. The graph in Fig. 5.29 allows the following remarks about the openness of the coalition C:

- Every IG-# inherits all coalition-themes linked up to it in the graph.
- Every node number is constituted of all systems that are linked down to it.
- The graph can be used as a tool to visualize consequences of actions from coalition C decision-makers upon basic interoperability: e.g., elimination of a link, assignment of a system to a coalition-theme, suppression of themes, restriction of a node for security reason, etc.

5.7.4 Definition of an Interoperability Space

Space interoperability relies upon different notions where some of them appear in Fig. 5.30. First, let us consider an action to be not interoperable with itself but with only systems that are capable to handle it. This designates an interoperable action to be represented by a couple: (S^i, A_j), where $S^i \in S$ and $A_j \in A$ (the set of all allowed actions in a coalition). This couple (S^i, A_j) must encompass temporal dimension since situations and systems are dynamics. The validity of (S^i, A_j) will depend on a temporal window, θ_M, or "opportunity window" denoted as follows: (S^i, A_j, θ_M). System S^i interoperates on action A_j in a temporal interval θ_M assigned to mission M. Time parameters are determined by coalition deciders.

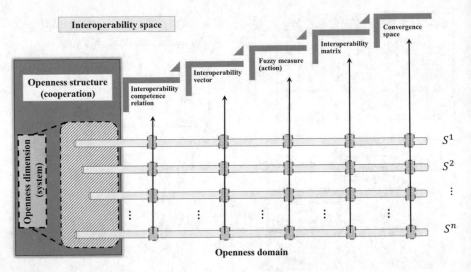

Fig. 5.30 Interoperability space

5.7.4.1 Interoperability Competence Relation

Consider a predicative interoperability relation, $\mathcal{R}$, in a propositional calculus view. The *arity* of that relation is 3, and by that relation, any system can evaluate its ability to operate on an action of a coalition. This relation must be applied by every system and to every action of the coalition. We form the following proposition:

$$\mathcal{R}\left(S^i, \{A_j\}, \theta_M\right), \forall i \in [1,n], \forall j \in [1,p]$$

System S^i considers itself competent to interoperate on actions $\{A_j\}$ where all A_j can be described in a formal way. Each system is bound to determine a first condition, necessary but not sufficient for its interoperability. According to its self-awareness of its neighboring world, a system can say if it has the required competence to interoperate on an action. In fact, the relation $\mathcal{R}$ allows to define an effective interoperability in the following fashion: *a system S^i can evaluate its competence to operate on any action, in a window time θ, for a specific mission M, under normal and usual conditions.* As $\mathcal{R}\left(S^i, \{A_j\}, \theta_M\right)$ is considered as a proposition, we can assign a truth value to it:

$$\mathrm{Val}\left[\mathcal{R}\left(S^i, \{A_j\}, \theta_M\right)\right] :: \mathrm{True}\left(\frac{T}{1}\right) \ \text{ or } \mathrm{False}(F/0)$$

Meaning:

- If True, S^i can interoperate on A_j, in time window θ, fixed by mission M. $\forall i \in [1, n], \forall j \in [1,p]$.
- If False, S^i is not capable to interoperate on A_j.

In practice, decision-makers who are responsible for S^i are entitled to apply this relation and, thus, to decide about the interoperability status of S^i with respect to their mission context. In a formal way, S^i is interpreting its own possible world.

5.7.4.2   Fuzzy Representation of an Interoperable Action

A fuzzy measure refers to a means of expressing uncertainty in the presence of incomplete information. Since we do not dispose of complete information, where probabilities could have been used, the suggestion here is to determine, in a subjective way, numerical coefficients or certainty degrees. These coefficients indicate how it is necessary for a system to interoperate on an action that has been beforehand declared possible. Let us assume that a system only executes one interoperable action at a time. We can therefore define a universe, W, from the following singletons:

$$W = \left\{ \left(S^i, A_1\right), \ldots, \left(S^q, A_p\right) \right\}$$

with $d\left(S^i, A_n\right)$ :: degree of possibility, and $d\left(S^i, A_n\right) \in [0, 1]$

where the value, $d(S^i, A_n)$, assesses the possibility that S^i executes the action A_n.

A fuzzy measure is completely defined as soon as a coefficient of possibility has been attached to every subset of a universal set U. If the cardinality of U is n, we must state 2^n coefficients, in order to specify a measure of possibility. Here, we will proceed by simply observing that each subset of U may be regarded as a union of singletons it encompasses. So, the determination of a measure of possibility can be done from only n elements. Fig. 5.31 presents an illustration of an interoperable action using the concept of a "fuzzy cube" where the dimensions are:

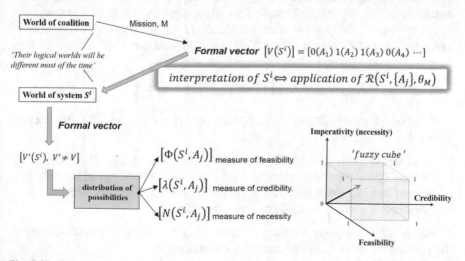

Fig. 5.31 Fuzzy representation of an interoperable action

(a) A *feasibility* measure comparable to a possibility
(b) An *imperativity* measure comparable to a necessity which will be dual of (a)
(c) A *credibility* measure to assess trust on systems in fulfillment of an action

Dimensions (a), (b) will be defined by distributions of possibility.

5.7.4.3   Fuzzy Matrices of Interoperability

For a given system S^i, if we successively apply the predicative relation of interoperability, $\mathcal{R}$, to couples (S^i, A_j), j varying from 1 to p, we obtain a binary vector $V(S^i)$. There are as many vectors as systems in the coalition: Val $\left[\mathcal{R}\left(S^i, A_1\right)\right] :: T$, Val $\left[\mathcal{R}\left(S^i, A_3\right)\right] :: F, \ldots\ldots\ldots\ldots,$ Val $\left[\mathcal{R}\left(S^i, A_p\right)\right] :: T$.
Let us valuate a component of vector $V(S^i)_j$ (row j), for instance:

$$\text{Val}\left[V\left(S^i\right)_j\right] :: F \Longrightarrow \neg \exists \mathcal{R}\left(S^i, A_j\right)$$

False (F), in equation, means that there is no openness structure for the coalition, and therefore, S^i has no semantics to evaluate. $V(S^i)_j$ is not supposed to exist. From the binary vector, or from the (logic) world resulting in the application of $\mathcal{R}$, it becomes possible to assign fuzzy measures to each vector component for values different of False (F). These fuzzy vectors are computed by taking:

- Couples of the world $\mathcal{R}$ such as: $V(S^i)_{j\,=\,1,\,2,\,\ldots,\,p} :: 1$, or.
- Vector's elements $V(S^i)$ such as: $\left[V\left(S^i\right)_{j=1,2,\ldots,p}\right](\mathcal{R}) :: 1$.

The evaluation of the semantics, Val $[V(S^i)_{j\,=\,1,\,p}] :: T$, has been made necessary because either unpredictable events happen in own system's world or changes in the mission could modify the world of S^i. This means a potential shift in the meaning of A_j for system S^i and possibly also for the coalition. In gathering all vectors of interoperability $V(S^i)$, we get what we call an interoperability matrix.

$$\left[I\left(S^i\right)_{i=1,2,\ldots,n}\right] = \left[V\left(S^1\right)\,V\left(S^2\right)\ldots V\left(S^n\right)\right]$$

This matrix represents only an apparent interoperability and can be used in different ways such as:

- To indicate what is theoretically the most interoperable system, S^i, for action, A_j.
- To identify the most adequate system, S^i, to operate under special conditions: a mission that imposes a temporal constraint to execute an action. Three kinds of fuzzy interoperability matrices can be constructed.

Matrix of feasible interoperability: This matrix gives a dimension of feasibility of the interoperability of $\{S^i\}$. The matrix is denoted by:

$$\left[\mathbf{I} - \Phi\left(S^i\right)_{i=1,2,...,n}\right]$$

Matrix of imperative interoperability: The matrix of necessary interoperability is built with fuzzy vectors of necessity as described above and in Fig. 5.31. It informs about the necessary interoperating conditions that are imposed to some systems. This matrix is denoted by:

$$\left[\mathbf{I} - N\left(S^i\right)_{i=1,2,...,n}\right]$$

Matrix of credible interoperability: This matrix brings visibility on systems that are in the best position to interoperate successfully. It will be denoted by:

$$\left[\mathbf{I} - \lambda\left(S^i\right)_{i=1,2,...,n}\right]$$

Barès et al. [33, 48] present other detailed examples of the usage of these three matrices.

5.7.5   Definition of a Cooperability Space

This section defines concepts of a cooperability space. The same methodology is being used as for the interoperability space above. For cooperability, in addition to system self-ability to execute any action, a system must interpret other systems' ability for interoperating on actions. This means that in addition to the predicative relation of interoperability of S^i, a predicative relation of cooperability of system S^i is also required.

Assume that a system S^i has a competence in cooperability when S^i can "judge" the ability of adjoining (cooperative) systems to interoperate on a set of actions $\{A_j\}$, in a time window θ_M. This competence is denoted by a quadruplet:

$$\left[S^i \because S^k, \{A_j\}, \theta_M\right] \forall i, k \in [1, n], j \in [1, p]$$

(the symbol $\because$ indicates the method of interpretation)

A predicative relation of cooperability, $\mathcal{R}'$, is defined according to the following conditions: $\mathcal{R}'$:: "is able to cooperate", and $\mathcal{R}'$:: "interpret the other systems' aptitude to interoperate on $\{A_j\}$." Then, the following predicate relation can be written:

$$\mathcal{R}'\left[S^i \because S^k, \{A_j\}, \theta_M\right] \forall i, k \in [1, n], j \in [1, p]$$

This means that: S^i judges that S^k can interoperate on $\{A_j\}$ in the time-window θ_M. This evaluation is made with a fuzzy measure of credibility, λ. In a predicate calculus view, the relation $\mathcal{R}'$ defined under these conditions is equivalent to a propositional function: representing the variables (S^i, S^k, A_j), (θ_M may be considered here as a constant). So, for a given S^i, the truth value of the predicate, S^k, is interoperable on each $A_{j\,=\,1,\,2,\,\ldots,\,p}$ and can be evaluated as:

$$\mathrm{Val}\left[\,\mathcal{R}'\left[S^i \,\therefore\, S^k, \{A_j\}, \theta_M\right]\right] \,::\, \text{True}$$

If "true" means: S^i interprets that S^k is able to interoperate on the actions $\{A_j\}_{j\,=\,1,\,p}$.

$$\mathrm{Val}\left[\,\mathcal{R}'\left[S^i \,\therefore\, S^k, \{A_j\}, \theta_M\right]\right] \,::\, \text{False}$$

If "false" means: S^i considers that S^k is unable to interoperate on the actions $\{A_j\}_{j\,=\,1,\,p}$.

A cooperability matrix is obtained by assembling the vectors of cooperability. Although the interoperability matrix is unique, it is necessary to establish two categories of matrices in the domain of cooperability. The first category, called *cooperability-system*, is going to indicate how the set of systems interpret their respective interoperability. The second one, called *cooperability-action*, is regarding actions, i.e., a matrix to comprehend the interoperability of the coalition from its elementary actions.

5.7.5.1 Matrix of Cooperability-System

The cooperability matrix of a system S^i is denoted by $[C(S^k)]$. Special computations can be made on rows and columns that provide useful insights to characterize the cooperability capacity of a coalition, i.e., the visibility about the quality (e.g., easiness) of system interoperation.

Properties of a column: Let $[C(S^k)]$ be the matrix of cooperability-system of S^k, and consider the mth column of that matrix. If we sum up all components of the vector-column m of the matrix $[C(S^k)]$, then we get a scalar, $\alpha_C(S^k)$, defined as:

$$\alpha_C\left(S^k\right) = \sum_{j=1}^{p} \left[C\left(S^k\right)\right]_{j,m}$$

The scalar $\alpha_C(S^k)$ indicates how S^m assesses the interoperability "strength" of S^k regarding actions. Below are two cases to illustrate the interpretation of that cooperability-system matrix.

Case 1 $\alpha_C(S^k) = 0$. According to its evaluation, S^m considers S^k being not interoperable. S^k must not play any role in the coalition since S^k is unable to execute any action. This does not mean that S^k has no interoperable capacity. At this point, the other systems assessment is not known, and several issues can be raised such as: "Is

$$[C(S^2)] = \begin{bmatrix} \mathcal{R}'(S^1 \because (S^2, A_1)) & \mathcal{R}'(S^2 \because (S^2, A_1)) & \mathcal{R}'(S^3 \because (S^2, A_1)) \\ \mathcal{R}'(S^1 \because (S^2, A_2)) & \mathcal{R}'(S^2 \because (S^2, A_2)) & \mathcal{R}'(S^3 \because (S^2, A_2)) \\ \mathcal{R}'(S^1 \because (S^2, A_3)) & \mathcal{R}'(S^2 \because (S^2, A_3)) & \mathcal{R}'(S^3 \because (S^2, A_3)) \\ \mathcal{R}'(S^1 \because (S^2, A_4)) & \mathcal{R}'(S^2 \because (S^2, A_4)) & \mathcal{R}'(S^3 \because (S^2, A_4)) \end{bmatrix}$$

Application of the relation $\mathcal{R}'$ **Measures of credibility**

$$[C(S^2)] = \begin{bmatrix} 1 & 1 & 0 \\ 0 & 0 & 1 \\ 0 & 0 & 0 \\ 1 & 1 & 0 \end{bmatrix} \qquad\qquad C[S^2] = \begin{bmatrix} 0.3 & 0.6 & 0.0 \\ 0.0 & 0.0 & 0.7 \\ 0.0 & 0.0 & 0.0 \\ 0.3 & 0.2 & 0.0 \end{bmatrix}$$

Fig. 5.32 Matrix of cooperability-system

there a misjudgment between S^m and S^k?"; "Is there anything shareable between S^m and S^k?"; etc.

Case 2 $\alpha_C(S^k) \neq 0$. This means S^m trusts more or less the ability of S^k to interoperate. If the value of α_C is close to 0, S^m trusts weakly S^k and thinks that S^k is likely to fail. If $\alpha_C(S^k)$ is close to 1, S^m trusts strongly S^k and thinks that S^k will be successful in its interoperations.

Figure 5.32 presents an example of a matrix of cooperability-system with system S^2 and 4 actions. From the columns, one can get a given system point of view: for example, on which action the system c is the most interoperable? From the rows, one can get the points of view of all systems (consensus notion – property of a line l): for example, what is the system which is in the best position to interoperate on action l?

5.7.5.2   Matrix of Cooperability-Action

We now define another kind of fuzzy matrix allowing to get the required visibility of the cooperability of all systems of the coalition. This special matrix is assumed to indicate the systems which are in the best conditions to interoperate on actions. For this reason, these matrices will be called matrices of cooperability-action. If the jth row in each of the previous matrices is considered, we form a new matrix that reports about the systems cooperability capacity relative to action A_j. This matrix is denoted by $[C(A_j)]$, and the process is illustrated in Fig. 5.33.

The matrix of cooperability-action, $[C(A_j)]$, presents interesting features: its structure is square; it allows to understand what are the systems of the coalition which are in the best position to interoperate on an action A_j; and, greatly important,

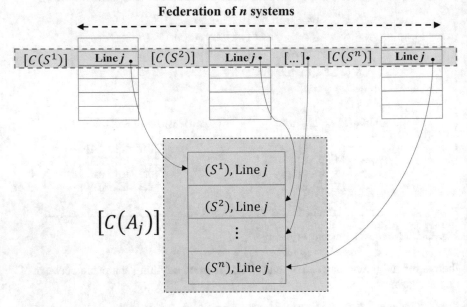

Fig. 5.33 Matrix of cooperability-action

it gives an assessment of the level of difficulty to interoperate on specific actions. For instance, the computation of the p matrices of all the actions

$$\{A_j\}_{j=1,p}$$

of the coalition gives visibility on what are the actions that are difficult to carry out and what are the ones that are likely either to get the coalition into trouble or to force the coalition to face difficult issues.

5.8 Conclusion

This chapter brought contributions for a better understanding of the couple (knowledge, action). They can be summarized by the following stretch objectives:

- To establish a typology of knowledge useful for actions and their concomitant attributes.
- To examine what facilitates the relevant decision-making and the modalities that can make the action more efficient.
- To partition the universe in line with the aims of the projected actions, which leads to the proper discrimination of exogenous and endogenous data by proposing a bipartition of the universe of action.

- To examine the impact that the "quality" of information can have on the conduct of an action and seek to improve it, particularly through the so-called concept of semantics enrichment.
- To master imperfect knowledge, by looking for the models most adapted to the situations in which their control can have a positive impact on the effective action.
- To situate the notion of relevant information in relation to the efficiency of the action.
- To examine the notions of cooperability and interoperability in a complex world of actions.

References

1. M. Barès, *Maîtrise du savoir et efficience de l'action*: Editions L'Harmattan, 2007.
2. G. L. Rogova and L. Snidaro, "Quality, context, and information fusion," in *Information Quality in Information Fusion and Decision Making*, ed.: Springer, 2019, pp. 219–242.
3. E. Blasch, É. Bossé, and D. A. Lambert, Eds., *High-Level Information Fusion Management and Systems Design*. Artech House, 2012, p.^pp. Pages.
4. J.-y. Hong, E.-h. Suh, and S.-J. Kim, "Context-aware systems: A literature review and classification," *Expert Systems with Applications,* vol. 36, pp. 8509–8522, 2009.
5. G. Rogova, M. Hadzagic, M. St-Hilaire, M. C. Florea, and P. Valin, "Context-based information quality for sequential decision making," in *Cognitive Methods in Situation Awareness and Decision Support (CogSIMA), 2013 IEEE International Multi-Disciplinary Conference on*, 2013, pp. 16–21.
6. K. Wan and V. Alagar, "Dependable Context-Sensitive Services in Cyber Physical Systems," in *Trust, Security and Privacy in Computing and Communications (TrustCom), 2011 IEEE tenth International Conference on*, 2011, pp. 687–694.
7. A. Bikakis and G. Antoniou, "Distributed Reasoning with Conflicts in a Multi-Context Framework," in *Twenty-Third AAAI Conference on Artificial Intelligence, AAAI 2008*, Chicago, Illinois, USA, 2008, pp. 1778–1779.
8. A. Zimmermann, A. Lorenz, and R. Oppermann, "An operational definition of context," in *Modeling and using context*, ed.: Springer, 2007, pp. 558–571.
9. C. Bettini, O. Brdiczka, K. Henricksen, J. Indulska, D. Nicklas, A. Ranganathan, et al., "A survey of context modelling and reasoning techniques," *Pervasive and Mobile Computing,* vol. 6, pp. 161–180, 2010.
10. A. Dey, "Understanding and Using Context," *Personal and Ubiquitous Computing,* vol. 5, pp. 4–7, 2001.
11. L. Snidaro, J. Garcia-Herrera, J. Llinas, and E. Blasch, Eds., *Context-enhanced information fusion* (Boosting Real-World Performance with Domain Knowledge. Springer, 2016, p.^pp. Pages.
12. É. Bossé and B. Solaiman, *Fusion of Information and Analytics for Big Data and IoT*: Artech House, Inc., 2016.
13. G. Ferrin, L. Snidaro, and G. L. Foresti, "Contexts, co-texts and situations in fusion domain," in *14th International Conference on Information Fusion*, 2011, pp. 1–6.
14. W. Sulis, "Archetypal dynamics, emergent situations, and the reality game," *Nonlinear dynamics, psychology, and life sciences,* vol. 14, pp. 209–238, 2010.
15. L. Snidaro, J. García, and J. Llinas, "Context-based information fusion: a survey and discussion," *Information Fusion,* vol. 25, pp. 16–31, 2015.
16. M. Beigl, H. Christiansen, T. R. Roth-Berghofer, A. Kofod-Petersen, K. R. Coventry, and H. R. Schmidtke, *Modeling and Using Context: seventh International and Interdisciplinary*

Conference, CONTEXT 2011, Karlsruhe, Germany, September 26–30, 2011, Proceedings vol. 6967: Springer, 2011.

17. A. N. Steinberg, "Situations and contexts," *ISIF Perspectives on Information Fusion (Perspectives),* vol. 1, 2015.

18. A. N. Steinberg, "Context-sensitive data fusion using structural equation modeling," in *2009 12th International Conference on Information Fusion,* 2009, pp. 725–731.

19. G. L. Rogova and A. N. Steinberg, "Formalization of "Context" for Information Fusion," in *Context-Enhanced Information Fusion,* ed.: Springer, 2016, pp. 27–43.

20. A. N. Steinberg and G. Rogova, "Situation and context in data fusion and natural language understanding," in *2008 11th International Conference on Information Fusion,* 2008, pp. 1–8.

21. A. N. Steinberg and G. L. Rogova, "System-level use of contextual information," in *Context-Enhanced Information Fusion,* ed.: Springer, 2016, pp. 157–183.

22. M. Frické, "The knowledge pyramid: a critique of the DIKW hierarchy," *Journal of information science,* vol. 35, pp. 131–142, 2009.

23. J. Almog, J. Perry, and H. Wettstein, *Themes from Kaplan*: Oxford University Press, 1989.

24. É. Bossé and B. Solaiman, *Information fusion and analytics for big data and IoT*: Artech House, 2016.

25. J. Llinas, "Challenges in Information Fusion Technology Capabilities for Modern Intelligence and Security Problems," in *Multisensor Data Fusion,* ed.: CRC Press, 2017, pp. 3–14.

26. J. Llinas, A.-L. Jousselme, and G. Gross, "Context as an uncertain source," in *Context-Enhanced Information Fusion,* ed.: Springer, 2016, pp. 45–72.

27. A. N. Steinberg, "Foundations of situation and threat assessment," in *Handbook of multisensor data fusion,* ed.: CRC Press, 2017, pp. 457–522.

28. É. Bossé, J. Roy, and S. Wark, *Concepts, models, and tools for information fusion*: Artech House, Inc., 2007.

29. E. Blasch, E. Bosse, and D. A. Lambert, *High-level Information Fusion Management and Systems Design.* Boston & London: Artech House, 2012.

30. J. Roy, "From data fusion to situation analysis," in *Information Fusion (FUSION), 2001 fourth Conference on,* 2001, pp. 1–8.

31. L. Gong, "Contextual modeling and applications," in *2005 IEEE International Conference on Systems, Man and Cybernetics,* 2005, pp. 381–386.

32. M. Barès, *Pour une prospective des systèmes de commandement*: Polytechnica, 1996.

33. M. Bares, "Formal approach of the interoperability of C4IRS operating within a coalition," in *Proceedings of the 5 th International Command and Control Research and Technology Symposium,* 2001.

34. M. Barès, "Proposal for Modeling a Coalition Interoperability," in *sixth International Command and Control Research and Technology Symposium,* 2001.

35. H. A. Handley, A. H. Levis, and M. Bares, "Levels of interoperability in coalition systems," DTIC Document2001.

36. A. M. Elmogy, F. Karray, and A. M. Khamis, "Auction-Based Consensus Mechanism for Cooperative Tracking in Multi-Sensor Surveillance Systems," *JACIII,* vol. 14, pp. 13–20, 2010.

37. A. Khamis, "Role of Cooperation in Multi-robot."

38. A. M. Khamis, M. S. Kamel, and M. Salichs, "Cooperation: concepts and general typology," in *Systems, Man and Cybernetics, 2006. SMC'06. IEEE International Conference on,* 2006, pp. 1499–1505.

39. B. Horling and V. Lesser, "A survey of multi-agent organizational paradigms," *The Knowledge Engineering Review,* vol. 19, pp. 281–316, 2004.

40. P. Xuan, V. Lesser, and S. Zilberstein, "Communication decisions in multi-agent cooperation: Model and experiments," in *Proceedings of the fifth international conference on Autonomous agents,* 2001, pp. 616–623.

41. H. Yu, Z. Shen, C. Leung, C. Miao, and V. R. Lesser, "A survey of multi-agent trust management systems," *Access, IEEE,* vol. 1, pp. 35–50, 2013.

42. S. Hart and A. Mas-Colell, *Cooperation: game-theoretic approaches* vol. 155: Springer Science & Business Media, 2012.

43. M. Yokoo, *Distributed constraint satisfaction: foundations of cooperation in multi-agent systems*: Springer Science & Business Media, 2012.

44. J. Rasmussen, A. M. Pejtersen, and L. P. Goodstein, *Cognitive systems engineering*: Wiley, 1994.

45. J. Rasmussen, "Skills, rules, and knowledge; signals, signs, and symbols, and other distinctions in human performance models," *Systems, Man and Cybernetics, IEEE Transactions on,* pp. 257–266, 1983.

46. L. Chaudron and N. Maille, "Generalized formal concept analysis," in *Conceptual Structures: Logical, Linguistic, and Computational Issues*, ed.: Springer, 2000, pp. 357–370.

47. A. Jaoua and S. Elloumi, "Galois connection, formal concepts and Galois lattice in real relations: application in a real classifier," *Journal of Systems and Software,* vol. 60, pp. 149–163, 2002.

48. M. Barès and É. Bossé, "Besoin de coopérabilité pour les C4ISR opérant dans le cadre d'une coalition multinationale "presented at the Symposium - Transformation: vers des capacités européennes en réseau Paris, France, 2008.

Chapter 6
Relational Calculus to Support Analytics and Information Fusion

Abstract This chapter proposes a discussion on the usage of relational calculus when applied to the analytics and information fusion (AIF) core processes that support the generation of actionable knowledge. This chapter discusses how relations and its calculus make AIF processes more capable technologically speaking to support the processing chain of transforming data into actionable knowledge.

6.1 Introduction

Analytics and information fusion (AIF) processes are what we can call the technological dimension, the technology, to support the generation of actionable knowledge as illustrated in Fig. 6.1. The previous chapter discussed the complexity of the universe decision-action and the efficiency of actions [1, 2]. This chapter proposes a discussion on the usage of relational calculus when applied to the AIF core processes. AIF processes are being implemented via an assemblage of techniques and methods that should ideally be guided by a sort of transcendence principle or an integrating framework (e.g., archetypal dynamics in Fig. 6.1). This integrating framework allows to coherently and semantically frame a complex processing chain that "transform data into actionable knowledge." Recall that actionable knowledge is a desired state of knowledge coming out of a situation awareness process (i.e., situation assessment). Figure 6.1 positions AIF in a global context as the technological support to get situation awareness [3–8].

6.1.1 Archetypal Dynamics: An AIF Integrating Framework

The assemblage of a plethora of techniques and methods into a processing chain that "transform data into actionable knowledge," i.e., to a specific goal (e.g., a computer-based support system), must be guided by an integrating framework. The development of a framework in which knowledge, information, and uncertainty can be represented, organized, structured, and processed to achieve a specific objective or

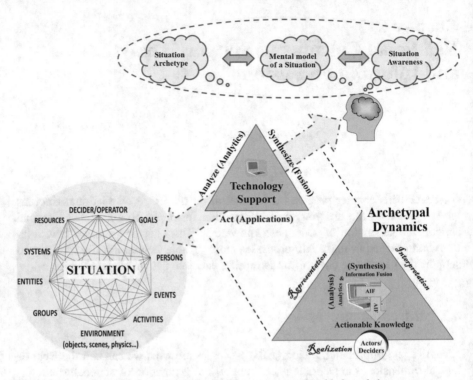

Fig. 6.1 Analytics and information fusion (AIF) to support actionable knowledge

multiple objectives is a core requirement to cope with the complications inherent with the design of such an AIF-based support system.

This integrating AIF framework should:

1. Provide means to represent knowledge through well-defined notions of situation and awareness
2. Support the modeling of uncertainty, belief, and belief update
3. Provide the key "computational model" for AIF and linkage to actions (users and machines)
4. Provide practical support for system design through modularization, refinement, validation, and verification
5. Offer a good compromise between operational and functional modeling in capturing systems behavior
6. Enable rapid prototyping and experimental validation of fairly abstract models
7. Support modeling of multi-agent systems

A very limited number of AIF frameworks have been proposed in the literature that offer partial fulfillment of the above requirements. A number of significant and powerful "Analytics" tools and techniques have also been offered to scientific and engineering communities for organizing, integrating, and visualizing large volumes of data [9]. In the domain of information fusion alone, noticeable efforts have recently been dedicated to define models and integrating frameworks [4, 5], namely, a promising one based on archetypal dynamics theory [5, 6].

Sulis [10] introduced archetypal dynamics as follows:

"Archetypal dynamics is a formal framework for dealing with the study of meaning laden information flows within complex systems."

"This is a formal framework for dealing with the study of the relationships between systems, frames and their representations and the flow of information among these different entities. The framework consists of a triad of: semantic frame (representation), realizations (system) and interpretation (agent/user). Real systems relate to semantic frames through one of the dimensions of that triad. The viewpoint of archetypal dynamics is that meaning is tight with actions. A semantic frame is an organizing principle that ascribes meaning in a coherent and consistent manner to phenomena that have been parsed into distinct entities, mode of being, modes of behaving, modes of acting and interacting."

The archetypal dynamics is a fundamental processing triad composed of the following axes: representation, interpretation, and realization. In archetypal dynamics, the way the information is understood is not in the sense of Shannon (i.e., the quantity of information), but in its active sense: "Information possesses content and elicits meaning." A holonic computational model has been proposed in [6] using the ideas from archetypal dynamics and "holon" from complex systems theory to progress towards a potential AIF computational model and an integrating framework.

Analytics and information fusion (AIF) is an assemblage of techniques and methods to analyze (analytics) and to synthesize (fusion) information from multiple sources to support a decision cycle. In Fig. 6.2 an assemblage of AIF is proposed

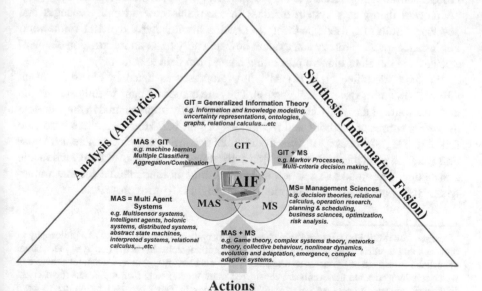

Fig. 6.2 Methods and techniques for analytics and information fusion (AIF)

around three main categories to be considered in a holistic fashion for the design of AIF-based support systems [5, 8]:

- Multi-Agent Systems (MAS) theories to formalize the distributed system aspect and the notion of autonomy
- Generalized Information Theory (GIT) for knowledge, information, and uncertainty representation
- Decision theories (represented by Management Sciences (MS) in Fig. 6.2) to explicitly account for actuation (e.g., decisions-actions and their impacts)

The challenge is to assemble an appropriate set of techniques and methods that will support measuring-organizing-reasoning-understanding-deciding and acting about/upon situations in complex environments. Note that the three categories in Fig. 6.2 follow the line of thought of the archetypal dynamics triad: representation (GIT) – interpretation (MS) – realization (distributed systems – MAS). Meaning transcends from actuation (contexts of actions). Semantic raises as the transformations (AIF processes) progress data towards actionable knowledge. Relations and relational calculus are foundational for the three categories of Fig. 6.2.

6.1.2 Generic AIF Core Processes

The following set of core processes are being proposed as a basic composition for an AIF-based processing chain [5, 6]. We use the term "process" here rather than function or relation to avoid confusion of terms with mathematics and computer sciences. Process is used with reference to the process theory[1] of Whitehead's philosophy [11]: "A process theory is a system of ideas that explains how an entity changes and develops." Sulis [12] used this theory to develop his archetypal dynamic framework. For instance, process theory can explain how an entity (e.g., an information element) changes and develops along a processing chain from data to actions.

We have identified eight (8) AIF core processes as listed in Fig. 6.3 that an AIF-based support system shall possess. They are named by pairs to fully capture the notions. Some of these AIF core processes are currently implemented in the literature through a subset of methods and techniques of Fig. 6.2. However, there is still no full-blown eight-process AIF system designed according to some coherent integration rules or principles which can frame the transformation of data to actionable knowledge. The notion of relation and its calculus is fundamental for that framework. Below are informal descriptions of the eight-AIF core processes.

---

[1] Excerpts from Sulis Ph.D. thesis: "Whitehead's philosophy has been described as a philosophy of organism and re-introduced a concept of subjectivity into metaphysics in the notion of prehension. Whitehead conceived of a process as a sequence of events having a coherent temporal structure in which relations between the events are considered more fundamental than the events themselves. Whitehead viewed process as being ontologically prior to substance and becoming to be a fundamental aspect of being."

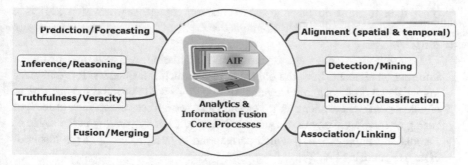

Fig. 6.3 Analytics and information fusion (AIF) core processes

1. *Alignment* (spatial and temporal)

   Description: World is observed in time and space. To make sense out of observations, one needs to understand the context and common referencing under time and space. Numerous examples can be provided. Registration [13–15] in time and space is well-known in image processing. Sensor alignment is crucial in multi-sensor multi-target tracking [16]. For instance, in bioinformatics [17], a sequence alignment is a way of arranging the sequences of DNA, RNA, or protein to identify regions of similarity that may be a consequence of functional, structural, or evolutionary relationships between the sequences.

2. *Detection/Mining*

   Description: Multi-sensor detection and data processing [18, 19] is well-known in engineering and physics. Applying the term "detection" to all levels of the Data-Information-Knowledge processing pyramid implies to also use the term "knowledge discovery" which is an equivalence in the data mining domain [20–22]. Knowledge discovery is the process of automatically searching large volumes of data to search for complex patterns. The term "data mining" is referred for methods and algorithms that allow extracting and analyzing data to find rules and patterns describing the characteristic properties of the information. Techniques of data mining can be applied to any type of data to learn more about hidden structures and connections. For instance, anomaly detection [23] techniques are used to detect surprising or unexpected behaviors (e.g., intrusions). Image mining, a data mining technique where images are used as data, is also an example of what that process can refer to.

3. *Partition/Classification*

   Description: Facing the complexity of the world, partitioning eases its apprehension and understanding. Partitioning is the act of dividing a unit into its components. Clustering and classification are examples of partitioning. Classification is the orderly, systematic arrangement of related things in accordance with a governing principle or basis. Clustering is the task of grouping a set of objects in such a way that objects in the same group are more similar (in some sense) to each other than to those in other groups (clusters). Image segmentation [24] is a representative example of partitioning.

4. *Fusion/Merging*

Description: Aggregation, integration, combination, merging, and fusion of information elements have the objective to gain better awareness. It is generally known that merge is a joining together of two flows while fusion is the merging of similar or different elements into a union. Here, we address the specific process of fusing or merging, i.e., the merging or fusing rules and operators in the sense of Dubois and Prade [25–27], which correspond to sub-processes of the inspired-JDL AIF models [4–6, 28].

5. *Truthfulness/Veracity*: (true, ¬true, degree of truth) Elements, objects, relations

Description: The action of putting forward some statement or proposition as true. The statement, or proposition, can also be put forward not as assertion, but as a supposition or hypothesis, as possibly true, and so on. Assertion using logical approaches is an example of that process. Reliability and relevance are also notions that can be assessed under that process. In fact, the impact of all dimensions of quality of information [29] on the AIF processes is an overall concern of that process.

6. *Inference/Reasoning*

Description: Inference is using observation and background to reach a logical conclusion. An inference is the process of drawing a conclusion from supporting evidence. This is the way to understand the world. Inferences are steps in reasoning. Inference is theoretically divided into three main categories: deduction, induction, and abduction. Deduction is inference deriving logical conclusion from premises known or assumed to be true. Induction is inference from particular premises to a universal conclusion. Abduction starts with an observation or set of observations and then seeks the simplest and most likely conclusion from the observations.

Statistical inference uses mathematics to draw conclusions in the presence of uncertainty. State estimation in multi-target tracking is an inference.

7. *Prediction/Forecasting*

Description: Prediction is concerned with estimating the outcomes for unseen data. Forecasting is a sub-discipline of prediction in which we are making predictions about the future, based on time-series data. Thus, the only difference between prediction and forecasting is that we consider the temporal dimension. One of the challenges of forecasting is finding the number of previous events that should be considered when making predictions about the future. This also depends on whether you are making about the immediate or the distant future. So, for a forecasting model with exogenous inputs (e.g., weather features), you basically need to model two things: (1) model the exogenous, non-temporal features (the feature model) and (2) model the historical, temporal data (the temporal model). Another example is tracking the dynamics and predicting the evolution of a system: the prediction (analysis) of an unknown true state by combining observations and system dynamics (model output). Simple object tracking as well as more complex objects like event tracking or group tracking or situation tracking can be seen as examples of this AIF process.

8. *Association/Linking*

Description: This important AIF process is about the identification and characterization of any link or relation between objects, knowledge items, and information elements considering all dimensions. Correlation between information elements, association between detections and object tracks, application such as link analysis, or relations in social networks are all examples of implementation of that process. The major difference between link and association is that link is a physical or theoretical connection between the objects, whereas association is a group of links with same structure and semantics. Associations are implemented in programming languages as a reference model in which one object is referenced from the other, while links cannot be referenced as these are not objects by itself but rely on the objects. The logical or physical connection among objects is referred to as link. These links are used to relate multiple objects and represent a relationship between objects. We cannot reference links, because a link is not a component of either object by its own but does rely on the objects. The link can be explained by the example such as students studying in university or universities in which there would be several numbers of students studying in one or more than one university. The links can be of three types – one-to-one, one-to-many, and many-to-many.

The following sections discuss how and where relational calculus can be of any use in the implementation of the AIF core processes. Note that the processing structure can be implemented in a holonic manner [6].

6.2   A Brief Recall of Relational Calculus

Under the term relational calculus, we group here the basic elements allowing to perform different operations on the calculation of the mathematical relations and also to examine the interest of their properties, according to their respective contexts of use. The notion of relation is ubiquitous in our human actions, as long as we establish links between objects, doing matching, making various analogies, and so on. It is also in the universe of our discourse, with frequent statements eliciting the dependencies and causal links, between ideas or concepts. So this is a concept that proves to be important in many areas, such as in analytics and information fusion (AIF), as it will be seen in what follows.

6.2.1   Relations Represented By Cuts

The representation "by cuts" has not been presented in Chap. 4. However, relations represented "by cuts" are very practical because it makes apparent the relation [1],

Fig. 6.4 Relation
R represented by five cuts

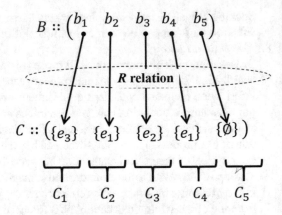

i.e., the linking. Consider Example 6.1 that illustrates the correspondence of a specific subset of nuts with each bolt (the list of the elements of E which have the same screw thread):

$$\{C_i\} \begin{pmatrix} b_1 & b_2 & b_3 & b_4 & b_5 \\ \{e_3\} & \{e_1\} & \{e_2\} & \{e_1\} & \varnothing \end{pmatrix}$$

Any cut C_i of R with $b_i \in B$ will be noted:

$$C_i \to R(b_i) :: \{e_k \in E | (b_i, e_k) \in R, b_i \in B\}, C_1 :: R(b_1), C_2 :: R(b_2), \ldots, C_5 :: R(b_5).$$

Example 6.1 Cartesian product: B :: {bolts of 5 different types}, E :: {nuts of 3 given types}, suppose that the ordered pair[2] according to the threads are: (b_1, e_3), (b_2, e_1), (b_3, e_2), (b_4, e_1) which corresponds to intersections in Fig. 6.4. In fact, they particularize a subset $R \subset B \times E$, defining a binary relation R (the binary term is often omitted).

The representation by cuts is interesting in several aspects:

- It can be empty, case of $R(b_5) \to \varnothing$ in Fig. 6.4.
- A possible interpretation of a binary relation between two sets as being a unary relation on the set of ordered pairs forming the Cartesian product, since for each element b_i, it is possible to say whether $R(b_i)$ is or is not verified.

Note that the set of cuts of a relation R is nothing else than the quotient of B by R (usually noted B/R or $B \div R$) and, interestingly, amounts to defining a partition on this set.

---

[2] *Ordered pairs* aka *couples* aka *2-tuples*

Fig. 6.5 Equivalence
classes induced by a relation

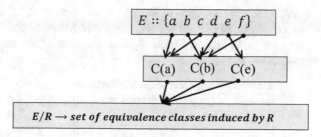

$$E :: \{a\ b\ c\ d\ e\ f\}$$

$$C(a)\quad C(b)\quad C(e)$$

$$E/R \rightarrow set\ of\ equivalence\ classes\ induced\ by\ R$$

Consider the set $E :: \{a, b, c, d, e, f\}$, and make the cuts:

$$C(a) :: \{a, c\}, C(b) :: \{b, d, f\}, C(e) :: \{c\}$$

according to the diagram of Fig. 6.5.

A difficulty remains in the sense that $C(x) \in (E \div R)$, while it is itself a set of elements: $a, b \in C(a)$. One observes that the three subsets $C(a), C(b), C(e)$ of E form three equivalence classes $C(x)$ (x being representative of the class), the elements of a class $C(x)$ being in relation by R with x. Two elements of the same class can be considered as equivalent: $a \equiv c$, $b \equiv d \equiv f$; the different classes are disjoint: $C(a) \cap C(b) \cap C(e) = 0$ and $C(a) \cup C(b) \cup C(e) \Leftrightarrow E$. The three classes $C(x)$ form a partition of E.

6.2.2   *Typology and Characterization of Relations*

Depending on how they are used, we can classify the binary relations into several categories, each of which has very interesting characteristics. For example, the equivalence relations which prove useful in the establishment of the classes/categories of elements are equivalent with respect to a given criterion (or to be more precise modulo something) to partitions of a set. Preorder relations are required when it is necessary to compare and order various elements/sensors or symbols. In computer science, they are constantly involved in sorting algorithms. Important preorder relations include preference relations such as that used in sequencing and scheduling problems. The functional relations $R \subset A \times B$ are a functional relation if:

$$x \in A \Rightarrow R(x) \in B$$

is unique; when its symmetric R^{-1} is also functional, R is said to be *bijective*.

Relations being sets $R_1, R_2 \subset E$, we can apply set operations to them and check their properties: commutativity, associativity, neutral element, and absorbing element. Other operations are applicable to them such as "symmetrization" and composition. The binary relations of the same referential can be considered as subsets of it; the set operations are therefore applicable to them. Sometimes it is convenient to combine several relations to form only one. Let $R_1 \subset E \subseteq E$ and $R_2 \subset E \subseteq E$; any couple $(a, b) \in R_1 \cup R_2$, if it verifies either R_1 or R_2. This does not exclude the fact that (a, b) satisfies both R_1 and R_2. The union of several relations can be represented in two ways:

- By superimposing their respective graphs
- By the sum of the representative Boolean matrices $M[R_1]$ of the relations R_i

$$M[R_1 \cup R_2] = M[R_1] + M[R_2].$$

The union is obviously commutative:

$$(a, b) \in R_1 \cup R_2 \leftrightarrow (a, b) \in R_2 \cup R_1$$

with $R_1 \subset E \times E, R_2 \subset E \times E, R_3 \subset E \times E$.

The union is distributive to the intersection if:

$$R_1 \cup (R_2 \cap R_3) \Leftrightarrow (R_1 \cup R_2) \cap (R_1 \cup R_2), (a, b) \in R_1 \cup (R_2 \cap R_3)$$

It means the following. Let $(a, b) \in R_1$, that is,

$$(a, b) \in (R_2 \cap R_3), (a, b) \in R_2 \cap R_3 \Rightarrow [(a, b) \in R_2], [(a, b) \in R_3],$$

so the couple (a, b) belongs either to R_1 or to

$$R_2 \ (\rightarrow (a, b) \in R_1 \cup R_2),$$

but (a, b) also belongs to either R_1 or to

$$R_3 \ (\rightarrow (a, b) \in R_1 \cup R_3),$$

when:

$$[(a, b) \in R_1 \cup R_2] \wedge [(a, b) \in R_1 \cup R_3] \Leftrightarrow (a, b) \in R_1 \cup (R_2 \cup R_3),$$
$$R_1 \cup (R_2 \cap R_3) \Leftrightarrow (R_1 \cup R_2) \cap (R_1 \cup R_3).$$

The intersection between relations corresponds to the operation that makes it possible to obtain all the couples common to relations appearing in the intersection. To define an intersection between two relations R_1 and R_2 of a referential E amounts in fact to partition this latter into four disjoint subsets:

$$R_1 \cap R_2,$$

$$C_E \, R_1 \cap R_2,$$

$$C_E \, R_2 \cap R_1,$$

$$C_E \, R_1 \cap C_E \, R_2.$$

6.2.3 Properties Common to Operations: Idempotence, Absorption, and Involution

This overview of set operations must be supplemented by a brief reminder of some general characteristics concerning them such as idempotence, absorption, involution, etc. In mathematics, an element $e \in E$ is called idempotent if, E being a set provided with an internal law of internal composition "$\circ$," we verify that:

$$e \circ e \to e;$$

in other words, e composing with itself gives as result e. If $\forall e \in E$ is idempotent, then the law "$\circ$" and the set E are also idempotent. The union is idempotent because the elements common or not to the relation R are elements of R; therefore: $R \cup R \Leftrightarrow R$.

The intersection is also idempotent: $R \cap R \Leftrightarrow R$, since the common elements are obviously those of R. Note that $R \cap \varnothing \Leftrightarrow \varnothing$; with E as a referential, we have: $R \cap E \Leftrightarrow R$, notation corresponding to the definition in comprehension of a set. Considering the distribution of the union with respect to the intersection:

$$R_1 \cup (R_1 \cap R_2) \Leftrightarrow (R_1 \cup R_1) \cap (R_1 \cup R_2),$$

and applying the idempotence:

$$R_1 \cap (R_1 \cup R_2),$$

such as:

$$(R_1 \cap R_2) \subset R_1,$$

we have:

$$R_1 \cup (R_1 \cap R_2) \Leftrightarrow R_1.$$

It follows the absorption property:

$$R_1 \cup (R_1 \cap R_2) \Leftrightarrow R_1 \cap (R_1 \cup R_2) \Leftrightarrow R_1.$$

If a set E with a law of internal composition "$\circ$" admits a neutral element e, $x \in E$ is said to be *involutory* if:

$$x \circ x = e,$$

the case of the "f" mapping matching a real "x" its opposite "$-x$" (*involutory* since the opposite of "$-x$" is "x"). The complementation of a relation is *involutory* because if one applies C_E to C_{ER}, one falls back on R since: $C_E(C_{ER}) \Leftrightarrow R$. This property presents a great interest for the simplification of certain expressions.

6.2.4   Usefulness of Morgan's Laws

The use of these laws is omnipresent in the manipulation of relational expressions. We will briefly recall the principles. Let us consider the two relations $R_1 \subset E$ and $R_2 \subset E$; we can express their union according to three intersections:

$$R_1 \cup R_2 \Leftrightarrow (R_1 \cap R_2) \cup (C_E R_1 \cap R_2) \cup (R_1 \cap C_E R_2).$$

The meeting $R_1 \cup R_2$ in the form of three intersections shows that its complement corresponds to all the couples that belong neither to R_1 nor to R_2. In other words:

$$C_E(R_1 \cup R_2) \rightarrow \{\text{couples} \notin (R_1 \cup R_2)\},$$
$$C_E R_1 \cap C_E R_2 \rightarrow \{\text{couples} \notin R_1\} \& \{\text{couples} \notin R_2\},$$

we can write:

$$C_E(R_1 \cup R_2) \Leftrightarrow C_E R_1 \cap C_E R_2 \tag{6.1}$$

If we replace R_1 and R_2 by their respective complement and then take the complement of the two members of Eq. (6.1), then:

$$C_E[C_E R_1 \cup C_E R_2] \Leftrightarrow C_E(C_E R_1) \cap C_E(C_E R_2), \tag{6.2}$$

with the involution of the complementation on the second member of Eq. (6.2) just above then:

$$C_E(C_E R_1) \rightarrow R_1, C_E(C_E R_2) \rightarrow R_2,$$

and Eq. (6.2) becomes:

$$C_E(C_E R_1 \cup C_E R_2) \Leftrightarrow R_1 \cap R_2,$$

taking the complement of both members,

$$C_E(R_1 \cap R_2) \Leftrightarrow C_E(C_E R_1 \cup C_E R_2), \tag{6.3}$$

taking into account the involution on the second member of (6.3),

$$C_E(R_1 \cap R_2) \Leftrightarrow C_E(C_E R_1 \cup C_E R_2) \tag{6.4}$$

Equations (6.1) and (6.4) constitute the so-called Morgan formulas, which of course are generalized to any number of subsets. To demonstrate them more quickly, we can use a Carroll diagram (correspondingly to make in a set E a vertical partition $A \subset E$ and $C_E A$ and a horizontal partition $B \subset E$ and $C_E B$ and thus obtain four disjoint squares). We thus find by direct reading of the Carroll diagrams the previous formulas (6.1) and (6.4).

6.2.5   Interest of Transitivity in AIF Processes

Transitive relations are quite common in everyday life:

R_1:: "*there is a shortcut between*"
R_2:: "*to be a brother of*"
R_3:: "*is a part of*"

A relation R on E is transitive when a first element of E is in relation to a second element of E, the latter being also an element of E; either by generalizing:

$$\forall x \forall y \forall z \, [(x, y) \in R \wedge (y, z) \in R] \Rightarrow (x, z),$$

which amounts to saying that R is transitive if its graph contains the one of its composed with itself:

$$G_{(R \circ R)} \subseteq G_R.$$

That is understandable moreover intuitively, if one takes the example of:

$$R :: \text{``}a \text{ } passage \text{ } exists \text{ } between\text{''};$$

starting from $R(a, b) \wedge R(b, c)$, we obtain $R(a, c)$; one is still in the relation, after having, however, "associated" (composition) the relation with itself. If we make, respectively, the cuts of R and $R \circ R$ as in Fig. 6.6:

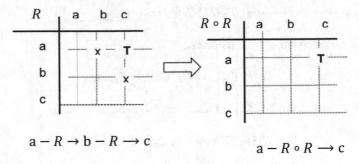

$$a - R \rightarrow b - R \rightarrow c \qquad\qquad a - R \circ R \rightarrow c$$

Fig. 6.6 Representation of transitivity

$$R \rightarrow \begin{pmatrix} a & b & c \\ \vdots & \vdots & \vdots \\ \downarrow & \downarrow & \downarrow \\ \{b,c\} & \{c\} & \{\emptyset\} \end{pmatrix} \qquad R \circ R \rightarrow \begin{pmatrix} a & b & c \\ \vdots & \vdots & \vdots \\ \downarrow & \downarrow & \downarrow \\ \{c\} & \{\emptyset\} & \{\emptyset\} \end{pmatrix}$$

Fig. 6.7 Representation by cuts of transitivity

we can see (Fig. 6.7) then that: $[R \circ R :: \{\{c\}, \emptyset, \emptyset\}] \subset [R :: \{\{b,c\}, \{c\}, \emptyset\}]$.

Example 6.2 R_1, R_2 are not transitive, whereas R_3 is
$$R_1 :: \{(1,1), (1,2), (2,1), (2,2), (3,3)\}$$

$$R_2 :: \{(1,2), (2,3), (4,4), (1,3)\}$$

$R_3 :: \{(1,3), (3,2), (2,4)\}$

If for a relation R in E, $\forall x \, \forall y \, \forall z \, [(x,y) \in R \wedge (y,z) \in R] \Rightarrow (x,z) \notin R$, R is called *intransitive*, which is equivalent to:

$$R \circ R \subset C_E \, R \text{ or } (R \circ R) \cap R \Leftrightarrow \emptyset.$$

The relation $R::$ "$\perp$" (to be perpendicular to) in the plane is intransitive: $[\perp(a, b) \wedge \perp (b,c)] \Rightarrow \parallel (a,c)$.

6.2.6 Importance of Relations "Closure"

The closure of a relation corresponds to the idea of conferring a certain "complete-ness" to the characteristics of reflexivity, symmetry, and transitivity of a relation. If one observes a relation R being 'a little reflexive' or "non-reflexive," the fact of completing the empty spaces of its diagonal makes R "completely" reflexive. One can make the same reasoning concerning the transitivity of R by completing the symmetrical elements of R absent in R^{-1}. A relation $R \subset A$ with a *property p* of R will be called *p-relation*, for example, *p-reflexive*, *p-symmetry*, and *p-transitivity*. The *p-closure* of any relation on A denoted $p(R)$ is a relation on A containing R and such that:

$$R \subseteq p(R) \subseteq S$$

for every *p-relation S* containing R. We will denote by *reflexive (R)*, *symmetric (R)*, and *transitive (R)* the different closures of R. Of course, $R \Leftrightarrow p(R)$ if R itself has the property p.

Transitive closure Verification of transitive closure requires, in addition to the union operation, the composition of relations. Recall that: $R^2 = R \circ R$, $R^3 = R^2 \circ R$, ..., $R^n = R^{n-1} \circ R$. Let us take the example of a set $E :: \{a, b, c\}$ and a relation $R \subset E$, where:

$$R :: \{(a, b), (b, c), (c, c)\};$$

the transitivity of R apparent on the representative graph and in the matrix calculus of $R^2 = R \circ R$ is translated by:

$$R^2 \rightarrow \{(a, c), (a, b), (b, c), (c, c)\}.$$
$$R^3 \rightarrow \{(a, c), (a, b), (b, c), (c, c)\},$$

as: $R^2 \Leftrightarrow R^3$, it follows that:

$$transitive\ (R) \rightarrow R \cup R^2 \cup R^3,$$

which generalizes to n unions:

$$transitive\ (R) \rightarrow R \cup R^2 \cup \ldots \cup R^n \rightarrow \{(a, c), (a, b), (b, c), (c, c)\}.$$

If E is a finite set of n elements, the transitive closure is denoted by

$$R^* : R^* \Leftrightarrow R \cup R^2 \cup \ldots \cup R^n,$$

generalizing:

$$R^* \Leftrightarrow \bigcup_{n=1}^{\infty} R^n$$

6.3   Various Notions of Order Relations in AIF Processes

In AIF processes, order relations occur under different facets and play an important role in most of the eight processes: alignment, detection, partition, fusion, etc. Also, it is appropriate to recall the generalities and to report the impact of their specificities. When we must compare the results produced by a multi-sensor system, the question is then to know to what extent they are comparable and, above all, to examine in which conditions they can be ordered. Likewise, seeking to satisfy the functionality of an alignment or being able to set up the necessary synchronization amounts to placing a total order on all the signals produced.

6.3.1   Notion of Order on an Ensemble of Signals

We are in the presence of a total order if the relation of order makes it possible to compare all the elements of a set as in the case of $R :: " \leq "$ applying on Z. Given two relative numbers, it is always possible to say, for a given ordered pair, if the first one is larger, smaller, or equal to the second: ... $-2 \leq -1 \leq 0 \leq 1 \leq 2 \leq 3$. The search for a total order on a set can be likened to the establishment of a relation of precedence or succession. Let an order be defined on E: $R \subset E$ and $(x, y) \in R$; we call y successor of x, if $R(x, y)$ being satisfied, $z \in E$, $\nexists z \mid R(x, z) \wedge R(z, y)$. One can define the same an immediate predecessor (Fig. 6.8). We usually represent a total order relation on a finite set E of n elements by a graph; if $i < j$, then x_i is predecessor of x_j; each arc is representative of $R(x_i, x_j)$. The numbering of the elements of $E :: \{x_i\}$, $i = 1, n$ makes graphing easier. The alphabetical order also gives a good illustration of what is a total order because, given two letters x, y, either x or y is placed in front of the other.

   When comparing signals one may be led to set apart some of them considered as equal or similar with respect to a criterion established by the operator. We must then introduce a restriction in some way on the total order; in this case we speak of strict order. Let R be a relation R' obtained from R; if: $\forall x \ \forall y \mid x \neq y$, we have $R(x, y) \Rightarrow R'(x, y)$; R' is called a strict order relation associated with the relation R. We observe that R' is a transitive and antisymmetric relation. Starting from the total order relation "$\leq$," if we define "$<$," to write $x < y$ means that although x is in relation with y, it cannot be equal to it. The relation "$<$" is no longer an order relation because it

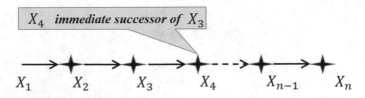

Fig. 6.8 Graph of successors

does not respect the property of reflexivity. It is nevertheless called a strict order relation. The word "strict" is commonly used now to clarify that we exclude equality in the use of inequality in the comparison.

Partial Order and Preorder in Comparison of Signals

Now if it becomes impossible to compare all the elements of a given set, in other words, to completely establish a relation of order on the set, then partial order is declared. A relation of order R is qualified as a partial order as soon as it does not make it possible to compare, according to the definition of R, all the elements of a set as in the case of:

$$R :: \text{``} \subset \text{''} \text{ on } \mathcal{P}(E).$$

Consider $E :: \{a, b, c\}$; R makes it possible:

$$\varnothing \subset \{b, c\} \subset \{a, b, c\},$$

but R does not make it possible to compare: $\{b, c\}$ and $\{a, b\}$. A relation is said of preorder if it satisfies at the same time the property of reflexivity:

$$R \subset E, \forall x \, (x, x) \in \Delta \subset R,$$

and the property of transitivity:

$$[(x, y) \in R] \wedge [(y, z) \in R] \Rightarrow (x, z) \in R,$$

if $(x, z) \in R \circ R$, also knowing that:

$$\Delta \subset R \Rightarrow R \subset R \circ R, \text{ so } [(R \subset R \circ R) \wedge (R \circ R \subset R] \leftrightarrow R,$$

hence the condition necessary and sufficient for R to be a preorder relation: $\Delta \subset R \circ R \leftrightarrow R$.

Relation of Large Order

We will examine under what conditions a relationship becomes a relationship of large order. Consider first a relation of preorder:

$$\Delta \subset R \circ R \leftrightarrow R;$$

if moreover it is reflexive, it entails:

$$[\Delta \subset R \leftrightarrow \Delta \subset R^{-1}] \Rightarrow [\Delta \subset R \cap R^{-1}].$$

If we assume that R is unsymmetrical, then:

$$R \cap R^{-1} \neq 0, C_E\, R \cap R^{-1} \neq 0;$$

this is verified because of its reflexivity, $\Delta \subset R \cap R^{-1}$, that requires the following condition:

$$\exists x \exists y \mid (x \neq y) \text{ and } [(x, y) \in R \wedge (y, x) \notin R].$$

In the particular case where $x = y$, R becomes antisymmetric which results in:

$$R \cap R^{-1} \cap \Delta,$$

but as R is also reflexive, we have $\Delta \subset R \cap R^{-1}$; in summary:

$$[R \cap R^{-1} \subset \Delta] \wedge [\Delta \subset R \cap R^{-1}] \Rightarrow \Delta \leftrightarrow R \cap R^{-1}.$$

An antisymmetric preorder relation R is characterized by

$$(R \cap R^{-1}) \leftrightarrow \Delta \leftrightarrow (R \circ R) \leftrightarrow R;$$

any relation that is at once reflexive, antisymmetric, or transitive is called a relation of large order (or relation of order):

$$\forall x \forall y\, [(x, y) \in R \wedge (y, x) \in R] \Leftrightarrow (x \leftrightarrow y);$$

in other words:

$$R \cap R^{-1} \leftrightarrow \Delta.$$

Example 6.3 Large order relation   Consider the subsets $A, B, C \subset E$; the relation of inclusion on $\mathcal{P}(E)$ is indeed a relation of large order because one can verify:

- The reflexivity: $\forall A,\ A \subset A$
- The transitivity: $\forall A\ \ \forall B\ \forall C,\ [(A \subset B) \wedge (B \subset A)]\ ssi\ (A \leftrightarrow B)$
- The antisymmetry: $\forall A\ \forall B,\ [(A \subset B) \wedge (B \subset C)]$

Note that in $\mathbb{R}$ (set of real numbers), the relations $R :: {}^{"} \leq {}^{"}$, $R' :: {}^{"} \div {}^{"}$ (x divide y) are also relations of large order.

6.3.2   Useful Relations for the AIF Alignment Process

The alignment is an essential AIF process since it allows to value the temporal and spatial contexts for the AIF's other processes. Contexts can be evaluated according to certain dimensions: situation in space-time space and vector of signified. At the relational level, the context understanding is greatly enlightened if one can locate the signals collected, according to the aforementioned criteria, with respect to each other. For example, if a preorder relation is needed in the alignment feature, then it is a total order relation.

6.3.3   Preorder Relations Applied to the AIF Detection Process

If after identification of signals, we seek to find out which ones are the most interesting in terms of qualifying criteria, for instance, which ones carry the most significance in the specific context of a detection operation, the task is often made difficult by the fact that it is challenging to distinguish them from an apparent similarity. If we cannot order or strictly classify these signals between them, one can be brought somehow to "position" them with respect to the others. Then, one will associate to them a notion of preorder. In mathematics, especially in order theory, a preorder or quasiorder is a binary relation that is reflexive and transitive. Preorders are more general than equivalence relations and (non-strict) partial orders, both of which are special cases of a preorder. An antisymmetric preorder is a partial order, and a symmetric preorder is an equivalence relation.

Let us examine two signals s_1, s_2; according to R:: ", s_1 has a mark greater than or equal to that of s_2. It is a preorder relation insofar as two signals having the same rating are not distinguished in the ranking and said, for this reason, ex aequo. In this case, R becomes a relation of order only on the condition that all the scores of the signals are all different from each other.

To illustrate the notion of preorder relation, we will place ourselves in the situation where we observe a certain situation with seven (7) sensors: A, B, C, D, E, F, and G. From the relation, R:: 'has a detection at least as good as', we want to interpret the results obtained. Some detections very close to each other can be considered as equivalent; we will try to "position" them in a relative way. The results obtained are given in Fig. 6.9.

The presence of the crosses (**x**), moving from left to right in Fig. 6.9, indicates which are the best detections checking the relation "x is at least as good as y." On the other hand, a large number of crosses in a column give an idea of the detection quality of this sensor. This is obtained by going from the best to the least (good) detections:

$$1^{st} : B, C \; (\text{ex} - \text{aequo}) \rightarrow 2^{nd} : A \rightarrow 3^{rd} : E \rightarrow 4^{th} : F, G \rightarrow 5^{th} : D$$

By direct reading of Fig. 6.9, we can see what are the detections: the worst, number of crosses in a column associated with a sensor, case of D, and the best,

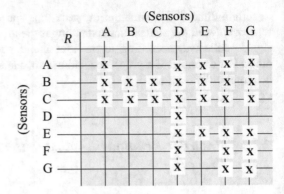

Fig. 6.9 Application of a preorder relation to an AIF signal detection

according to the number of crosses on a line to a given sensor. Moreover, we note that:

- R is reflexive: presence of crosses on the diagonal
- R is transitive because of its definition: $R(x, y) \wedge R(y, z) \Rightarrow R(x, z)$

From this example it is easy to establish the principles of a simple preorder algorithm for establishing a first classification of signals from an observed situation.

6.3.4   Relations and Equivalence Relations

Equivalence is a particularly important notion in mathematics. It can be approached from several angles, reflecting in a certain way ideas that are subordinate to it: partition, connectivity, equivalence classes, quotient sets, equipotent classes, etc. The notion of equivalence [2] which is intuitively apprehended will be of great importance in several AIF processes as soon as it becomes important to group observations in families showing one or more characteristics or to make partitions on a set of data or elements of an observed situation. When a person wanting to store washers in drawers, according to their respective diameters, carries out on the set "washers" a divider or partition of this set, each drawer, more precisely its content, constitutes a class of the whole. Each partition is therefore specific given the criterion used to establish this separation. It is appropriate to say that two washers of the same class are equivalent, relative to the relation:

$$R :: \text{ 'to have the same diameter as'}.$$

This amounts to evoking in an underlying way previously discussed properties: reflexivity, transitivity, and symmetry.

In mathematics, an *equivalence relation* is a binary relation that is reflexive, symmetric, and transitive. The relation "is equal to" is the canonical example of an equivalence relation, where for any objects a, b, and c:

- $a = a$ (reflexive property)
- if $a = b$ then $b = a$ (symmetric property)
- if $a = b$ and $b = c$ then $a = c$ (transitive property)

Because of the reflexive, symmetric, and transitive properties, any equivalence relation provides a partition of the underlying set into disjoint equivalence classes. Two elements of the given set are equivalent to each other if and only if they belong to the same equivalence class.

Notion of class for the AIF linking process   In a broad sense, the idea of class refers to the notion of a collection or group of elements or objects according to a common characteristic. This is the idea of classification or categorization. In set theory, the notion of class refers to the idea of classification of sets made possible by bijection between them. The notion of class generalizing that of a whole was introduced around 1937 by Von Neumann and Bernays in a perspective of axiomatization of mathematics. The sets appear to be specific classes; a class A is a set if:

$$\exists \text{class } B \mid A \subset B.$$

Equivalence class   Consider a set $E :: \{a, b, c, d, e, f\}$ and a relation $R \subset E \times E : R :: \{(a, a), (a, c), (b, b), (b, d), (b, e), (c, a), (c, c), (d, b), (d, d), (d, e), (e, b), (e, d), (e, e), (f, f)\}$. By examining R, we see that it is reflexive, symmetrical, and transitive. Let us highlight the transitivity by isolating some couples as shown in Fig. 6.10.

One can observe in Fig. 6.10 that the columns and lines identified by c and e play, in a way, a particular role in isolating specific squares as indicated in Fig. 6.11.

We note in Fig. 6.11 that all the elements of a given square are symmetrical with respect to its diagonal. Two elements x, y will be said to be equivalent to $R \subset E \times E$ if they appear on the diagonal of the same square:

$$\forall x \forall y \ (x, x) \text{ and } (y, y) \in \Delta.$$

Fig. 6.10 Equivalence relation, R transitivity

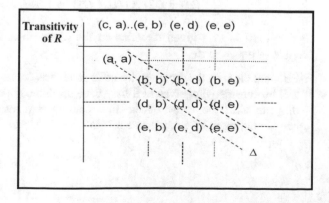

Fig. 6.11 Division into
equivalence classes

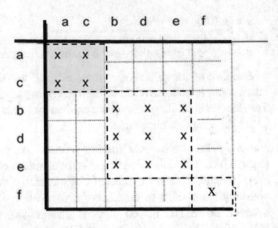

In a general way, we will say that x is equivalent to y modulo R, according to the relation R, if the existence of the couple (x, y) implies the existence of its symmetric (y, x), which is denoted by:

$$x \equiv y \bmod R \quad (x \text{ is equivalent to } y \text{ modulo } R).$$

We also observe in Fig. 6.11 that the relation R shares $E :: \{a, b, c, d, e, f\}$ in three classes of equivalence corresponding to the three disjoint subsets:

- $\{a, c\}$ where $a \equiv c \bmod R$, ("$\equiv$" used symbol for equivalence)
- $\{b, d, e\}$ where $(b \equiv d \equiv e \bmod R)$
- $\{f\}$

It is also interesting to note that the set $\{\{a, c\}, \{b, d, e\}, \{f\}\}$ is none other than the quotient $(E \div R)$ of the set E by the relation R.

Example 6.4 Equivalence classes
1. Relations of parallelism of lines in a plane. Such a relation is reflexive, symmetrical, and transitive:

$$(D1 \parallel D2) \wedge (D2 \parallel D3) \Rightarrow (D1 \parallel D3);$$

all the parallels to a given direction in the plane constitute as many classes as possible of equivalence.

2. Personal file with the same service code in the first column: two articles in this file will be considered equivalent if they have the same code in this column. Sorting this file with the value of the first column realizes a partition of this file corresponding to an equivalence relation.

6.3.5 Applying the Equivalence Relation to AIF Processes

To give an equivalence relation on a set of observed elements is in fact to make a partition on this set. The sharing criterion is indicated by the equivalence relation:

$$R :: \text{'} signals\ having\ the\ same\ template \text{'};$$

this amounts to giving, in practical terms, the rule which allows to classify in the same class or partition elements considered as equivalent regarding the same template. To generalize the use of such a partitioning rule is to define an algorithm. If from an enumerable set of E signals and two elements $a \in E$, $b \in E$ there exists a finite set of rules for deciding whether $a \equiv b$, then there is an algorithm that can:

- Perform a "classification" according to the classes of the partition, defined by an equivalence relation
- Name each class by one of its elements entitled its "representative"
- Enumerate the elements of each class e_i; otherwise e_{i+1} becomes the title or the "label"[3] of a new class

6.3.6 A Quotient Set of Sensors

A quotient set is a set derived from another by an equivalence relation. The notion of a quotient set will prove to be an interesting use to organize an observed situation but also to proceed to summary classifications of observations or signals relative to rules fixed a priori. The realization of a quotient set is often necessary when one must deal with large files, for instance, to facilitate the discrimination of their elements:

- To distinguish signals of equivalent size in the overall file of the signals collected
- To consider them as equivalent, according to a given analysis criterion

In the case of the example about the storage of washers in drawers according to their diameter, one can, for convenience of use, designate the drawers by a label specifying the diameters of the washers contained in this drawer. If we are interested only in the totality of labels appearing on the drawers, this amounts then to consider a new set compared to the relation R of comparison, in fact the set quotient designated by $B \div R$, or B/R. Consider the set $E :: \{a, b, c, d, e, f\}$ and an equivalence relation R which makes it possible to carry out the partition under the following conditions:

---

[3]The term *class label* is usually used in the context of supervised machine learning, and in classification in particular, where one is given a set of examples of the form (attribute values, class label) and the goal is to learn a rule that computes the label from the attribute values. The class label always takes on a *finite* (as opposed to infinite) number of different values.

$$A :: \{b, c\}, B :: \{a, d, f\}, C :: \{e\},$$
$$A, B, C \in (E), A \cap B \cap C \;\rightarrow\; \varnothing, A \cup B \cup C \rightarrow E.$$

A, B, C define three equivalence classes $Cl(x)$:

$$A \Leftrightarrow Cl\,(b) \Leftrightarrow \{x \in E \div R(x, b)\} \Leftrightarrow \{b, c\},$$
$$B \Leftrightarrow Cl\,(a) \Leftrightarrow \{x \in E \div R(x, a)\} \Leftrightarrow \{a, d, f\},$$
$$B \Leftrightarrow Cl\,(e) \Leftrightarrow \{e\} \Leftrightarrow \{x \in E \div R(x, e)\} \Leftrightarrow \{e\}.$$

If we join the three classes A, B, C, we form the new set $\{A, B, C\}$, which is entitled exactly quotient set $E \div R$; its elements are the classes of the partition made by $R: A \in E \div R,\; B \in E \div R,\; C \in E \div R$. Different levels of belonging are to be distinguished:

$$a \in E, b \in E, \dots, F \in E, A \in \mathfrak{P}(E), \dots, C \in \mathfrak{P}(E)$$

but also $A \in \mathfrak{P}(E \div R), \dots, C \in \mathfrak{P}(E \div R)$, because the subsets A, B, C in turn become elements with respect to $E \div R$. As equivalents we have:

- For $A \Leftrightarrow Cl\,(b): b \equiv c$
- For $B \Leftrightarrow Cl\,(a): a \equiv d \equiv f$

The following diagram summarizes the different levels of membership by scanning it from top to bottom (the arrows materialize a membership link) (Fig. 6.12).

The quotient set of a set E can be obtained by an equivalence relation R or a certain partition $\mathfrak{P}$ on E; it corresponds to the set of equivalence classes. Incidentally, we have with the notion of a quotient set a new way of defining a set. An element of a class is called a class representative. It should also be noted that it is possible to nest a quotient set with inclusions of their respective relation $R \subset R' \subset R''$, which makes it possible to refine discrimination or make it more precise.

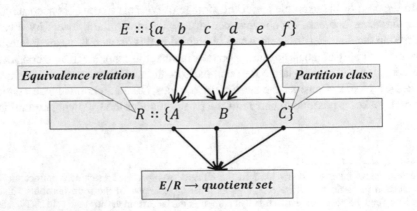

Fig. 6.12 Representation of a quotient set

6.3.7 Order Relation Induced on a Quotient Set

If we take again the example of the storage of bolts according to their length in suitable drawers:

$$B \; :: \; \{bolts\}, R \; :: \; `to\ be\ of\ the\ same\ length\ as`,$$

the relation R of equivalence defines a preorder on B, and $B \div R:: \{drawers\ labels\}$ defines a quotient set from B; $x \in B$, $y \in B$ are called equivalents ($x \equiv y$) when they belong to the same class: $(x, y) \in C_m$. If x, y belong to two different equivalence classes: $x \in C_m$ and $y \in C_p$ with $C_m \cap C_p \to \varnothing$ (disjoint classes), two cases can occur depending on (Fig. 6.13):

- $x \in C_m$ and $y \in C_p$ are comparable in terms of length, if we have, for example: $(x' \equiv x) \wedge (y' \equiv y) \Rightarrow x' > y'$, because of the transitivity of the preorder.
- $x \in C_m$ and $y \in C_p$ are not comparable, $(x' \equiv x) \wedge (y' \equiv y) \nRightarrow x' > y'$; it is impossible for x 'to be greater than y'.

By the relation R associated with the preorder on B, we can thus define an order on $B \div R$. A class C_m will be previous to a class C_p if all the elements of C_m are previous to those of C_p. In fact, for a class C_m to be prior to another class C_p, it is necessary and sufficient that an element of C_m is prior to an element of C_p. It is therefore possible to define an order on a quotient set E with the relation R associated with the preorder. By way of illustration, we will take a set E comprising four partially ordered equivalence classes as shown in Fig. 6.14.

Fig. 6.13 Preorder induced on a quotient set

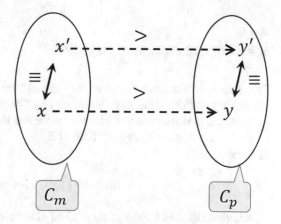

Fig. 6.14 Order induced on a quotient set

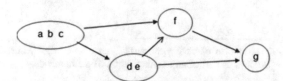

Fig. 6.15 Matrix of an
order induced on a quotient
space

a	b	c	d	e	f	g
1	1	1				
1	1	1	-	-	-	-
1	1	1	-	-	-	-
-	-	-	1	1	-	-
-	-	-	1	1	-	-
-	-	-	-	-	1	-
-	-	-	-	-	-	1

Starting from Fig. 6.14, we can establish its corresponding matrix and highlight
the four equivalence classes (shown in gray in Fig. 6.15). All elements that are
symmetrical with respect to the diagonal belong to the same equivalence class.

6.4   Formal Notion of Partition

The notion of partition refers to frequent attitudes of everyday life that lead to
sharing sets of elements in different classes[4] or cells that are clearly identifiable;
the partition or differentiation operates according to strict rules, so that an element is
in one and only one category at a time. If $E_1, E_2, \ldots E_i, \ldots, E_n$ constitute a partition of
E, then:

$$\exists E_i \neq \varnothing \mid \forall x \in E \Rightarrow x \in E_i$$

Reciprocally, if a family of $E_i \neq \varnothing$ of E is such that:

---

[4]The notion of class generalizes that of the whole, introduced in 1937 by von Neumann and
Bernays as part of an axiomatization of mathematics.

$$\forall x \in E \Rightarrow (x \in E_i) \wedge (x \notin E_i \neq i),$$

then this family E_i is a partition of E. If we take the example of the set:

$$E :: \{E_i \cup E_j\},$$

we call partition $\mathfrak{P}$ of E the set $\mathfrak{P} :: \{E_i, E_j\}$, whose non-empty families are disjoint:

$$E_i \cap E_j = \varnothing$$

and their union equal to $E = E_i \cup E_j$. Thus, in the case of $E :: \{a, b, c\}$, the three parts of E: $\{a\}$, $\{b\}$, $\{c\}$ constitute a partition of E because:

$$\{a\} \cap \{b\} \Leftrightarrow \{b\} \cap \{c\} \Leftrightarrow \{a\} \cap \{c\};$$

all these intersections are empty and $\{a\} \cup \{b\} \cup \{c\} \rightarrow E$.

In the same way, $\{a\}$ and $\{b, c\}$ also constitute parts of E. We call partition $\mathfrak{P}$ of a finite set E, the set of parts: $E_1, E_2, \ldots E_i, \ldots, E_n$, not empty, disjoint two by two and whose union is E:

$$\forall i \in I, \ I = [1, 2, 3, \ldots, n],$$

$$\forall i, \forall j \neq i, E_i \neq \varnothing, E_i \cap E_j = \varnothing,$$

$$\bigcup_{i \in I} E_i = E$$

Any partition $\mathfrak{P}$ is classificatory for the set E on which it applies because of the two preceding properties. This remark will be of great importance later in particular in the notion of refining situations. A partition $\mathfrak{P}$ can be noted by:

$$\mathfrak{P} :: \{E_1, E_2, \ldots E_i, \ldots, E_n\} \Leftrightarrow \{E_i\} \ i = 1, n.$$

A set $A \subset E, A \neq \varnothing$ and its complement A^C also constitute partitions of E. The classes being sets and the partition being a set of classes, the order of the classes as well as that of the elements in a class n is not significant. We can match to any partition of E an equivalence class in a set E, defined according to an equivalence relation R because every element x of a partition of E is related:

- With itself:
  $\forall x \in E, R(x, x) \rightarrow$ property of reflexivity,
- With an element y of a partition of E, then y is in relation with x:
  $\forall x \in E, \forall y \in E, R(x, y) \Rightarrow R(y, x) \rightarrow$ property of symmetry,
- With $y \in E$ which is in relation with $z \in E$ then $z \in E$ is related to $x \in E$:
  $\forall x \in E, \forall y \in E, \forall z \in E, R(x, y) \wedge R(y, x) \Rightarrow R(x, z) \rightarrow$ property of transitivity.

6.4.1 Partitioning of a Situation by an Observer

Consider the case of a United Nations surveillance officer in charge of the surveillance of a disputed area, who to carry out his mission must constantly update a situation in this area. The concept of situation is assimilated here to the whole of permanent knowledge combined with the factual elements collected along the way. The fusion of these elements according to certain rules, notably from AIF processes, must make it possible to apprehend a "certain" reality perceived from the ground ("uncertainty" because rendered by a perceptual system which has its own defects and technological limits). For ease of understanding, this situation is most often associated with ad hoc iconographic representations of situational facts, on representations of digitized or non-digitized terrain, to best allow the interpretation of the reality of the terrain and, thus, facilitate decision-making.

To carry out his mission, this officer will often be required to isolate "observation partitions" within his area of responsibility, if only to arrange and effectively manage the sensor systems. To accomplish his mission, it must also consider the technological qualities of observation devices, their implementation techniques, and even the surveillance requirements imposed by higher authorities. His organization, and especially that of its means, must make it able to react and trigger, if necessary, actions to the extent of events. To meet these expectations, the general approach requires, formally, to consider the following points:

- To divide the universe of observation, according to a coherence dictated by the purposes of the surveillance mission: to bring out a notion of partition
- To carry out groupings by classes necessitated by the nature of the terrain and the setting of technical means: what amounts to establishing equivalence relations

6.4.2 Modalities of Classifying Partitions

Defining a partition $\mathfrak{P}$ on a certain situation S is to give it somehow a beginning of structuring:

- Two elements of S that can then be declared, from a certain point of view (generator of the partition) equivalent or not, but it is especially to be able, as part of a decision-making process, later to define partitions
- To know how to order them according to different criteria. The assessment of a situation S has no other objective than to provide materialized help to the decision-makers.

From a set like $E :: \{a, b, c, d, e, f\}$, we can, for example, make several successive rankings of its possible partitions (Fig. 6.16):

Fig. 6.16 Representation of partitions of a set

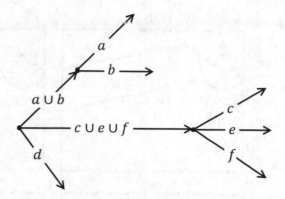

- $\mathfrak{P}_1 :: \{a, b, c, e, f\}$, corresponding to a partition of $a \cup b$, $c \cup e \cup f$, and d.
- $\mathfrak{P}_2 :: \{a, b\}$ corresponding to a partition of $a \cup b$.
- $\mathfrak{P}_3 :: \{c, e, f\}$ corresponding to a partition of $c \cup e \cup f$.

One quickly gets the different parts $\mathfrak{P}_i$ using a classification tree where each arrow of the tree from the root corresponds to a partition of E. It may be interesting to have a way to systematically rank all parts of a family of sets. For instance, if a situation S is considered as a set of classes or enumerable elements (which will often be the case) and if there exists an algorithm knowing that two elements $x \in S$ and $y \in S$ are equivalent, according to some equivalence relation R, then it is possible to find a so-called ranking algorithm which enumerates the classes of the partition: $\{S_i\} \in \mathfrak{P}(S)$ corresponding to the relation R.

6.4.3 Refinement on an Observed Situation

If we take the example of the surveillance officer of United Nations, he may be led according to circumstances, to pay more attention to certain portions of the observed territory. This may lead to a better disposition of the sensors, or even the establishment of new means, specifically adapted to certain conditions of observation. Therefore, it is necessary to be able to identify, compare, and classify the partitions between them because of not having, somehow, the same "degree of refinement" in a specific context of observation (see Fig. 6.17).

Each of the $S_i \in \mathfrak{P}(S_i)$ parts can be considered as a class of a partition. The elements x, y belonging to the same class S_i are said to be equivalent in the situation S relative to an equivalence relation R (which moreover specifies the nature of the partition). Let the partitions $\mathfrak{P} \subset S, \mathfrak{Q} \subset S, R \subset S$ say that the partition $\mathfrak{P}$ is finer than the partition $\mathfrak{Q}$ if:

- Any class of $\mathfrak{P}$ is included in a partition of $\mathfrak{Q}$, inclusion here denoted by "$\ldots \prec \ldots$"
- $\mathfrak{P} \subset \mathfrak{Q} \Leftrightarrow \mathfrak{P} \prec \mathfrak{Q}$ where $\mathfrak{P} \prec \mathfrak{Q}$ is representative of a refinement of $\mathfrak{Q}$.

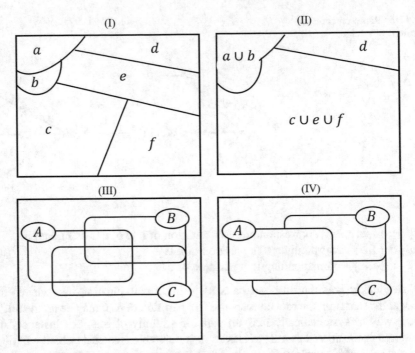

Fig. 6.17 Refinement of a situation: (I) is finer than (II); (IV) is coarser than (III)

Imagine that information about an event (several facts that can be observed at the same time) results from the combination of several sources. It is more difficult to judge the credit of this information as the sources are numerous and noisy and, what is more, can interfere with each other. The difficulty will be less if we succeed in correctly partitioning this set of information sources. Consider the following set of informational facts:

$$E :: \{a, c, d, f, g, h, i, j, k\}$$

and perform, for example, on this set, the partitions (Fig. 6.18):

As the class $\mathfrak{P}$ is finer than $\mathfrak{Q}$, $\mathfrak{P}$ brings more precision to the information collected, so it provides informational details allowing interpretation from another angle of view. Note also that the refinement of partitions of a situation returning in fact, to make inclusions on parts of sets, is like an equivalence relation since it is:

- Reflexive: $\mathfrak{P} \prec \mathfrak{P}$
- Antisymmetric: $(\mathfrak{P} \prec \mathfrak{Q}) \wedge (\mathfrak{Q} \prec \mathfrak{P})(\mathfrak{P}\mathfrak{Q})$
- Transitive: $(\mathfrak{P} \prec \mathfrak{Q}) \wedge (\mathfrak{Q} \prec R) \Rightarrow (\mathfrak{P} \prec R)$

For a set $\mathfrak{Q} :: \{a, b, c\}$ there are two characteristic partitions:

- The finest where all the elements constitute a class apart
- The coarsest where all the elements are in the same class

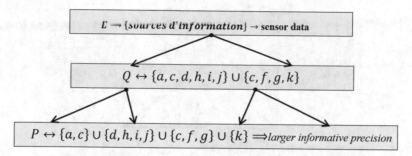

Fig. 6.18 Refinement of partitions

6.4.4   Refining Limits on a Situation

Making several partitions on the same situation can lead to ask, for the later phases of interpretation of the situation, which is the threshold of refinement likely to give the most informational precision. Is it desirable to go to finer scores, by increasing the level of inclusion, or on the contrary, as part of an established partition, group some of them to avoid some unnecessary details that can generate redundancies? It thus appears useful to be able to perform elementary operations on partition classes, in all respects analogous to set operations. Two basic operations are useful to achieve this goal: the *intersection of partitions* or their *merging* or *union*.

Intersection of partitions   We may need to qualify the credibility of subsets of facts from a situation by making a partition according to the following conditions: A corresponding to the statement "*accepted*," H to the statement "*hypothesis*." This amounts to making two 2-class partitions:

$$\{A, \ \overline{A}\} \text{ where } \overline{A} \to no\ A, \ \ \{H, \ \overline{H}\}, \ \overline{H} \to \ no\ H,$$

which implies in S: a partition to making 4 classes:

$$\{A \cap H, A \cap H, A \cap H, A \cap H\}.$$

There is a risk of not retaining good hypotheses and/or accepting as good bad hypotheses: two attitudes that may correspond to errors of appreciation or measure (Fig. 6.19).

In this example, we see that we can somehow compose two partitions of two classes of a situation to obtain a partition of four classes on the same situation. A priori, nothing prevents the generalization of this mode of composition to partitions having any number of classes.

Example 6.5 Situation assessment from a United Nations (UN) surveillance officer   In the example of the UN surveillance officer, it is assumed that the presence of several armed groups on the ground, with their more or less heavy weaponry and

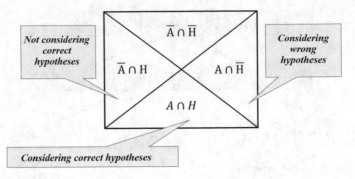

Fig. 6.19 Intersection of two partitions

Table 6.1 Partition-intersection: armed groups versus armament

Partition Intersection		Armament		
		A_1	A_2	A_3
Armed Groups	G_1	$G_1 \cap A_1$	$G_1 \cap A_2$	$G_1 \cap A_3$
	G_2	$G_2 \cap A_1$	$G_2 \cap A_2$	$G_2 \cap A_3$
	G_3	$G_3 \cap A_1$	$G_3 \cap A_2$	$G_3 \cap A_3$
	G_4	$G_4 \cap A_1$	$G_4 \cap A_2$	$G_4 \cap A_3$

their deployment assets, distinguish them from each other, but, above all, demonstrate their intention to undertake aggressive and large-scale actions. To do this, a rough idea, as far as possible, of the intentions of these groups, and especially to see the underlying impact of their supposed behavior, will seek to cross as much as possible the two types of information available to it:

- Type of armament used by an armed group
- Characterization (obtained from analysis of previous information) of a group according to several criteria: numerical importance, mobility, tactics employed, etc.

Knowing that:

- G is the set of known armed groups: $G :: \{G_1, G_2, G_3, G_4\}$.
- To all identified group armaments: $A :: \{A_1, A_2, A_3\}$.

The interesting connections to be made will be in the intersection of all classes G_i with all classes A_j. The intersections obtained, which are the non-empty parts of S, constitute the classes of the partition sought. Given two partitions A and B on the situation S, we will define the intersection-partition of A and B that we will define as $"\wedge_p"$: $A \wedge_p B$ (Table 6.1).

6.4.5   Application to Electronic Warfare

Now let us imagine that armed groups are carrying out, against UN surveillance (observation) officers, sudden interference on their sensor systems, which is basically a go for electronic warfare. It is considered that the observations made for situation S can be summarized as follows:

- At the time starting from t_1: $O(., t_1) :: [\{O_1, O_6, O_8\}; \{O_3, O_5\}; \{O_4, O_7\}; \{O_2\}]$
- At the starting point of t_2: $O(t_1, t_2) :: [\{O_1, O_6\}; \{O_5, O_4\}; \{O_7, O_2\}; \{O_3\}]$, with $t_2 > t_1$

To understand what is the "persistent" information on which we must pay special attention, to adapt the level of vigilance of surveillance officers, at the appropriate level (avoid generating irritant alarms), it becomes interesting to realize the intersection-partition:

$$O(., t_1) \wedge_p O(t_1, t_2) \Leftrightarrow [\{O_1, O_6\}; \{O_2\}; \{O_3\}; \{O_4\}]$$

Suppose now that this state of electronic warfare continues with increased aggression by more targeted attacks from t_3 and $t_4 > t_3$, $(t_3 > t_2)$ on some surveillance optronic systems. Surveillance systems identify the presence of several signals, represented here by iconic symbols:

$$E :: \{*, \Delta, O, T, \nabla, \perp, \circ\}.$$

For technical reasons (collection and ease of analysis), we have made several different classes for the periods $O(t_2, t_3)$, $O(t_3, t_4)$ which themselves constitute two partitions of S:

$$O(t_2, t_3) :: \{T, \Delta, O\} \cup \{\perp, \nabla\} \cup \{\circ, *\},$$

$$O(t_3, t_4) :: \{O, \perp\} \cup \{\circ\} \cup \{\nabla\} \cup \{T, *, \Delta\}.$$

If we want to understand now what this electronic warfare phase can mean, we must look at whether there are persistent signals, or formally to examine whether the classes $O(t_3)$ and $O(t_4)$ have elements in common. Making an intersection-partition of classes $O(t_3)$ and $O(t_4)$ can help achieve this.

By direct reading of Table 6.2 of the intersection-partition $O(t_2, t_3) \wedge_p O(t_3, t_4)$, the signals whose persistence can support the presumption of a particular threat are easily identified. The intersection-partition operation can be useful when it comes to knowing what level of detail one is with respect to the situation. For example, if we go back to previous observation periods, we notice that (Fig. 6.20):

Table 6.2 Intersection-partition: 7 classes

Partition Intersection		Observation period, $O(t_3)$		
$O(t_2,t_3) \wedge_p O(t_3,t_4)$		$T,\Delta,0$	$\perp,\nabla$	$\Diamond, *$
Observation period, $O(t_4)$	$0,\perp$	0	$\perp$	
	$\Diamond$			$\Diamond$
	∇		∇	
	$T,*, \Delta$	T,Δ		$*$

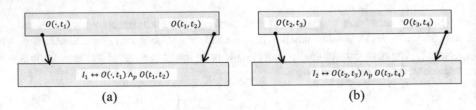

Fig. 6.20 Interest of an intersection-partition

In Fig. 6.20a, one observes that I_1 is finer than $O(.,t_1)$ and $O(t_1,t_2)$:

$$I_1 \prec O(., \ t_1) \ \&| \prec O(t_1, \ t_2).$$
$$\text{If } O(.,t_1) \prec O(t_1,t_2) \text{ then}$$
$$I_1 \leftrightarrow O(., \ t_1) \wedge_p O(t_1, \ t_2) \equiv O(., \ t_1).$$

In Fig. 6.20b, one observes that I_2 is finer than (t_2,t_3) and $O(t_3,t_4)$:

$$I_2 \prec O(t_2,t_3) \ \&| \prec O(t_3,t_4).$$

If it becomes possible to order I_1 and I_2, we stand in a comfortable situation where the meaning to be conferred to the observations is reinforced by the durability of certain distinctive signals.

Union of two partitions If now we are confronted with an abundance of details that no longer allows us to manage properly, then try to gather more information to obtain a better overall signified on an observed situation. Performing a grouping characterizes the so-called union-partition operation that we designate as "$\vee_\mu$." Thus for

$$A \subset E, B \subset E, \quad (A\vee_p B) \in \mathfrak{P}(E),$$

the partition $U \Leftrightarrow A \vee_p B$ is at the same time less fine than the partitions A and B of E:

$$U \prec A \text{ and } U \prec B.$$

We see that U is induced in E by an equivalence relation called a relation of *relatedness* defined with the conventions following two elements $a \in E$, $b \in E$ that are called:

- *Neighbors* if they are elements of the same class either $(a, b) \in A$, or $(a, b) \in B$, designated by $(a_{n=}b)$
- *Parents* if they constitute ancestors common to a third $c \in E$ designated by $(a_{p=}b)$

Let relations R:: '*be neighbor of*', R' :: '*be parent of*'; we notice that R is a binary relation that is reflexive, symmetrical, and non-transitive, while R' is reflexive, symmetric, and transitive. Consider the case where $A \subset E, B \subset E, \; (A\vee_p B) \in \mathfrak{P}(E)$, if A, B correspond to the two following partitions:

$$A :: \{(a,f,h); (c,e); (d,g); (b)\},$$
$$B :: \{(a,f); (e,d); (g,b); (c); (h)\}.$$

In terms of neighborhood relations, we have:

$$(a_{n=} f); (a_{n=} h); (f_{n=} h); (d_{n=} g); (c_{n=} e); (e_{n=} d); (g_{n=} b),$$

since $(a_{=n} f)$ belongs to the 1st class of A and the 1st class of B, likewise, $(c_{=n}e)$,

$$(c, e) \subset A, (d_{n=} g), (d, g) \subset A, (e_{n=} d), (e, d) \subset B, (g_{n=} b), (g, b) \subset B.$$

In terms of *parents* relations, we have:

$$R' :: \{(c_{p=} e); (d_{p=} g); (e_{p=} d); (g_{p=} b); (c_{p=} d)(*)\},$$

(*) $(c_{p=} d) \in R'$ because of the transitivity: $(c, e) \wedge (e, d) \rightarrow (c, d)$.

To perform the union-partition $A\vee_p B$ (Fig. 6.21) which is both thinner than the partitions A and B of E: $(A\vee_p B) \prec A$ and $(A\vee_p B) \prec B$, we will connect all the classes of A and B which are not disjoint to each other. In practice, this amounts to drawing a path between all the classes of A and B which, being not disjoint, have common elements and then constitute resultant classes in $(A\vee_p B)$.

Fig. 6.21 Union of partitions

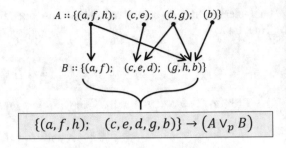

$$\{(a,f,h); \quad (c,e,d,g,b)\} \rightarrow \left(A \vee_p B\right)$$

6.5   Composition of Relations for the Fusion/Merging AIF Process

One is often led in everyday life to "infer" additional knowledge, from knowledge previously acquired. The search for filiations or foundation of ancestries in genealogy illustrates this need. When one begins to have well-characterized information elements or knowledge items, it becomes interesting to know if by associating them we can gain additional information, in other words, whether their combination leads to an increase in the signified of the situation. On the relational side, when there are neighborhood links between domains expressed by:

- A relation R between domains A and B
- A relation S between domains B and C

it is often interesting to build "filiations" by explicitly defining a third relation T, formed from R and S, called a relation composed of R and S:

$$T \subset A \subseteq C \mid [\exists(x,y) \in R] \wedge [\exists(y,z) \in S] \Rightarrow (x,z) \in T.$$

This amounts to writing[5]: $T :: S \circ R$ and not $R \circ S$, since S applies first to the elements of B, themselves obtained after application of R to the elements of set A. Example of composition from the following relations:

$$R :: \quad \{(a_1, \ b_2)\}, \ (a_2, \ b_1), \ (a_2, \ b_3), \ (a_3, \ b_2)\};$$

$$S :: \quad \{(b_1, \ c_1)\}, \ (b_1, \ c_2), \ (b_2, \ c_2), \ (b_3, \ c_1), \ (b_3, \ c_2)\},$$

From the previous representation and by successive applications, returning, in fact, to make cuts, it results in:

$$R(a_1) \rightarrow \{b_2\}, \ R(a_2) \rightarrow \{b_1, b_3\}, \ R(a_3) \rightarrow \{b_2\},$$

---

[5]The composition operation is indicated by the symbol "$\circ$".

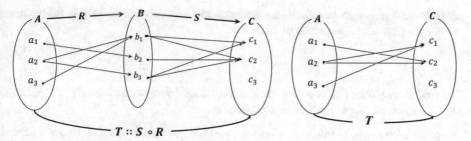

Fig. 6.22 Composition of relations

$$S(b_1) \rightarrow \{c_1, c_2\}, \quad S(b_2) \rightarrow \{c_2\}, \quad S(b_3) \rightarrow \{c_1, c_2\},$$
$$S(b_2) \rightarrow S(R(a_1) \rightarrow \{c_2\},$$

With $\{b_1, b_3\} \rightarrow R(a_2)$, we have:

$$S[R(a_2)] \rightarrow S[\{b_1, b_2\}] \Rightarrow [S(b_1) \cup S(b_2)] \Rightarrow [\{c_1, c_2\} \cup \{c_2\}], \quad S[R(a_2)] \\ \rightarrow \{c_1, c_2\}.$$

Using the composed relation T,

$$S{\circ}R(a_1) \rightarrow T(a_1) \rightarrow \{c_2\},$$
$$S{\circ}R(a_2) \rightarrow T(a_2) \rightarrow \{c_1, c_2\},$$
$$S{\circ}R(a_3) \rightarrow T(a_3) \rightarrow \{c_2\}.$$

It is possible to compose the inverse relations (Fig. 6.22).
If starting from $T^{-1} \subset C \times A$, one makes, for example:

$$T^{-1}(c_2) \rightarrow \{a_1, a_2, a_3\},$$

then,

$$R^{-1}[S^{-1}(c2)] \rightarrow [R^{-1}(\{b_2, b_3\})] \rightarrow [R^{-1}(b_2) \cup R^{-1}(b_3)] \rightarrow \{a_1, a_3\} \cup \{a_2\} \\ \rightarrow \{a_1, a_2, a_3\},$$

therefore:

$$T^{-1} \Leftrightarrow R^{-1} {\circ} S^{-1} \Leftrightarrow (R{\circ}S)^{-1}$$

Knowing that a relation can be represented by a Boolean matrix, the composition of two relations can be obtained directly by the multiplication of two matrices.

Matrix multiplication can facilitate understanding as illustrated in the following example. Consider a sequence of elements:

$$E :: \{a, b, c, d, e\}$$

And if one wants to organize their successions by two relations $R_1 \subset E$ and $R_2 \subset E$, according to the following modality:

- to an element x_i correspond elements x_j which are such that (x_i, x_j) are in the relation R_i:

$$\forall x_i \exists x_j \mid (x_i, x_j) \in R_i,$$

R_1 and R_2 are specified by the following tables (Tables 6.3, 6.4, and 6.5).

Table 6.3 Table of the successors

	R_1		R_2
a	c,e	a	$\emptyset$
b	$\emptyset$	b	b,c
c	a,d	c	b,d
d	e	d	a,e
e	c	e	c

Table 6.4 Composition of the successors

Elements	$[R_1 \circ R_2]$
a	b, c, e
b	$\emptyset$
c	a, e
d	c
e	b, d

Table 6.5 Fusion of successor lists

Elements	$[R_1 + R_2]$
a	c, e
b	Ø
c	a, d
d	e
e	c

To have an idea, via these relations, about the successors of the successors, it is enough either to compose the relations R_1 and R_2 or to carry out the product of their respective matrix:

$$
{}^1[R_1] = \begin{bmatrix} 0 & 0 & 1 & 0 & 1 \\ 0 & 0 & 0 & 0 & 0 \\ 1 & 0 & 0 & 1 & 0 \\ 0 & 0 & 0 & 0 & 1 \\ 0 & 0 & 1 & 0 & 0 \end{bmatrix} \quad [R_2] = \begin{bmatrix} 0 & 0 & 0 & 0 & 0 \\ 0 & 1 & 1 & 0 & 0 \\ 0 & 1 & 0 & 1 & 0 \\ 1 & 0 & 0 & 0 & 1 \\ 0 & 0 & 1 & 0 & 0 \end{bmatrix} \quad [R_1] \times [R_2]
$$

$$
= \begin{bmatrix} 0 & 1 & 1 & 1 & 0 \\ 0 & 0 & 0 & 0 & 0 \\ 1 & 0 & 0 & 0 & 1 \\ 0 & 0 & 1 & 0 & 0 \\ 0 & 1 & 0 & 1 & 0 \end{bmatrix}
$$

$[R_1] \times [R_2]$ directly gives the desired result.

We can also be led to look for a resulting table of successors that is in fact a fusion, which is then easily obtained by adding $M[R_1] + M[R_2]$.

$$
\begin{bmatrix} 0 & 0 & 1 & 0 & 1 \\ 0 & 0 & 0 & 0 & 0 \\ 1 & 0 & 0 & 1 & 0 \\ 0 & 0 & 0 & 0 & 1 \\ 0 & 0 & 1 & 0 & 0 \end{bmatrix} + \begin{bmatrix} 0 & 0 & 0 & 0 & 0 \\ 0 & 0 & 1 & 0 & 0 \\ 0 & 1 & 0 & 1 & 0 \\ 1 & 0 & 0 & 0 & 1 \\ 0 & 0 & 1 & 0 & 0 \end{bmatrix} = \begin{bmatrix} 0 & 0 & 1 & 0 & 1 \\ 0 & 0 & 1 & 0 & 0 \\ 1 & 1 & 0 & 1 & 0 \\ 1 & 0 & 0 & 0 & 1 \\ 0 & 0 & 1 & 0 & 0 \end{bmatrix}
$$

corresponding to the resulting list:

6.5.1   *Properties of a Composition of Relations*

Since the compositions of relations themselves are sets, the properties of the sets are applicable to them.

Commutative and associative It is easy to verify that the composition is not commutative:

$$R_1 \circ R_2 \neq R_2 \circ R_1.$$

To show the associativity between three relations R_1, R_2, R_3, it is enough to verify the equivalence:

$$R_3 \circ (R_2 \circ R_1) \Longleftrightarrow (R_3 \circ R_2) \circ R_1$$

Consider Fig. 6.23 with $R_3 \circ (R_2 \circ R_1)$,
$R_2 \circ R_1(\{a\}) \rightarrow \{c\}$ and $R_3(\{c\}) \rightarrow \{d\}$, where

$$R_3[R_2 \circ R_1(\{a\})] \leftrightarrow R_3 \circ [R_2 \circ R_1(\{a\})] \rightarrow \{d\},$$

Considering now $(R_3 \circ R_2) \circ R_1$:

$$R_3 \circ R_2 (\{b\}) \rightarrow \{d\} \text{ and } \{b\} \rightarrow R1(\{a\}),$$

where

$$R_3 \circ R_2(R_1(\{a\})) \rightarrow (R_3 \circ R_2) \circ R_1(\{a\}) \rightarrow \{d\},$$

therefore

$$R_3 \circ (R_2 \circ R_1) \leftrightarrow (R_3 \circ R_2) \circ R_1.$$

Fig. 6.23 Associativity of composition of relations

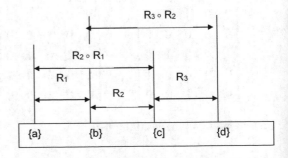

Distributivity of composition with respect to union   The composition is distributive with respect to the union:

$$R_1 \circ (R_2 \cup R_3) \leftrightarrow (R_1 \circ R_2) \cup (R_1 \circ R_3).$$

If we consider that (a, b) satisfies $R_1 \circ (R_2 \cup R_3)$, this implies that:

$$(a, b) \in R_1 \circ (R_2 \cup R_3)$$

if and only if:

$$\exists c \mid [(a, c) \in R_1] \wedge [(c, b) \in R_2 \cup R_3], (c, b) \in R_2 \cup R_3$$

which amounts to saying (logically) that either (c, b) satisfies R_2, or (c, b) verifies R_3, then it means that (a, b) satisfies:
either $R_1 \circ R_2$, or $R_2 \circ R_3$, $(a, b) \in (R_1 \circ R_2) \cup (R_1 \circ R_3)$;therefore,

$$R_1 \circ (R_2 \cup R_3) \leftrightarrow (R_1 \circ R_2) \cup (R_1 \circ R_3).$$

Absorbing element for relations composition   Let $A \subset E \times E$ be a relation never verified,

$$\forall (x_i, y_j) \in A \mid (x_i, y_j) = 0.$$

All the components of the matrix A are therefore equal to 0. A is an absorbing element for the composition, while A is a neutral element for the union. We can actually check using matrix multiplication:

$$\forall R \subset E \times E, (R \circ A) = (A \circ A) = 0.$$

Neutral element for relations composition   Let $N \subset E \times E$ be a verified relation if:
$$\forall (x_i, y_j) \in N \mid x_i \leftrightarrow y_j$$

the components of the main diagonal of A are equal to 1, the other components being equal to 0. We verify by matrix multiplication that N is a neutral element for the composition:

$$\forall R \subset E \times E, R \circ N = N \circ R = R.$$

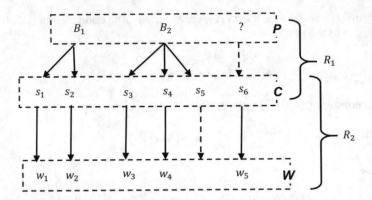

Fig. 6.24 Application of composed relations

6.5.2 Application to the Formalization of Parentage

The use of composition of relations can be a convenient means for modeling the relations between different levels of relatedness in a manner analogous to what is done in genealogical research. It can be illustrated from a simple example, that of a family of two brothers B_1 and B_2 that have the following relationships:

- Brother B_1 has two sons: s_1, s_2.
- Brother B_2 has three sons s_3, s_4, s_5 and an adopted son, s_6.

The children got married under the following conditions ($w_i \rightarrow$ wife): (see Fig. 6.24)

- Son s_1 with w_1.
- Son s_2 with w_2.
- Son s_3 with w_3.
- Son s_4 with w_4.
- Son s_5 stays single.
- Son s_6 with w_5.

From the situation described above, it becomes interesting to represent certain relations and to observe the consequences they induce. We can consider, for example:

$$R_1 :: \text{''to be father of''} \text{ with } R_1 \subset P \times C.$$

$$R_2 :: \text{''to be a husband of''} \text{ with } R_2 \subset C \times W.$$

Starting from R_1, the application of cuts gives:

$$R_1(B_1) \rightarrow \{s_1, s_2\}$$

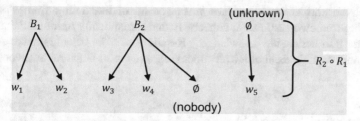

Fig. 6.25 Resultant relation from composition

$$R_1(B_2) \rightarrow \{s_3, s_4, s_5\}$$

$R_1(\varnothing) \rightarrow \{s_6\}$ (unknown father)

$$R_1(B_1) \cup R_1(B_2) \cup R_1(\varnothing) \rightarrow C; R_1(P) \rightarrow C.$$

With R_2, we get the same:

$$R_2(s_1) \rightarrow \{w_1\}, R_2(s_2) \rightarrow \{w_2\}, R_2(s_3) \rightarrow \{w_3\}, R_2(s_4) \rightarrow \{w_4\}, R_2(s_5)$$
$$\rightarrow \{\varnothing\}, R_2(s_6) \rightarrow \{w_5\}; R_2(C) \rightarrow W.$$

If one wants to know now who is "the father of the husband of," in other words, to find out who is "the father-in-law of," that amounts in fact to composing the relations R_1 and R_2, that is to say:

$$R_2 \circ R_1$$

as:

$$R_2(C) \rightarrow W \text{ and } R_2(R_1(P)) \rightarrow W,$$

we have:

$$R_2 \circ R_1(P) \rightarrow W$$

that is immediately verified on the sagittal representation of Fig. 6.25.

6.6   Conclusion

Analytics and information fusion (AIF) processes support the generation of action-able knowledge. This chapter proposes a discussion on the usage of relational calculus when applied to the AIF core processes. AIF processes are being implemented via an assemblage of techniques and methods that should ideally be guided by a sort of transcendence principle or an integrating framework. This integrating framework allows to coherently and semantically frame AIF that "transform data into actionable knowledge." Relations and its calculus make AIF processes more capable technologically speaking to support situation awareness.

References

1. M. Barès, *Maîtrise du savoir et efficience de l'action*: Editions L'Harmattan, 2007.
2. M. Barès, *Pratique du calcul relationnel*. Paris: Edilivre, 2016.
3. É. Bossé, J. Roy, and S. Wark, *Concepts, models, and tools for information fusion*: Artech House, Inc., 2007.
4. E. Blasch, E. Bosse, and D. A. Lambert, *High-level Information Fusion Management and Systems Design*. Boston & London: Artech House, 2012.
5. É. Bossé and B. Solaiman, *Information fusion and analytics for big data and IoT*: Artech House, 2016.
6. B. Solaiman, É. Bossé, L. Pigeon, D. Gueriot, and M. C. Florea, "A conceptual definition of a holonic processing framework to support the design of information fusion systems," *Information Fusion,* vol. 21, pp. 85-99, 2015.
7. E. Bossé and B. Solaiman, "Fusion of information and analytics: a discussion on potential methods to cope with uncertainty in complex environments (big data and IoT)," *International Journal of Digital Signals and Smart Systems,* vol. 2, pp. 279-316, 2018.
8. B. Solaiman and É. Bossé, *Possibility Theory for the Design of Information Fusion Systems*: Springer, 2019.
9. P. Zikopoulos and C. Eaton, *Understanding big data: Analytics for enterprise class hadoop and streaming data*: McGraw-Hill Osborne Media, 2011.
10. W. Sulis, "Archetypal dynamics, emergent situations, and the reality game," *Nonlinear dynamics, psychology, and life sciences,* vol. 14, pp. 209-238, 2010.
11. C. R. Mesle, *Process-relational philosophy: an introduction to Alfred North Whitehead*: Templeton Foundation Press, 2008.
12. W. Sulis, "A process model of non-relativistic quantum mechanics," Ph.D. Ph.D., Physics, Waterloo, Waterloo, Ontario, Canada, 2014.
13. J. Le Moigne, N. S. Netanyahu, and R. D. Eastman, *Image registration for remote sensing*: Cambridge University Press, 2011.
14. J. Modersitzki, *Numerical methods for image registration*: Oxford University Press on Demand, 2004.
15. A. A. Goshtasby, *Image registration: Principles, tools and methods*: Springer Science & Business Media, 2012.
16. M. Liggins, D. Hall, and J. Llinas, *Handbook of Multisensor Data Fusion: Theory and Practice, Second Edition*: Taylor & Francis, 2008.
17. A. D. Baxevanis, G. D. Bader, and D. S. Wishart, *Bioinformatics*: John Wiley & Sons, 2020.
18. P. K. Varshney, *Distributed detection and data fusion*: Springer Science & Business Media, 2012.
19. G. L. Foresti, C. S. Regazzoni, and P. K. Varshney, *Multisensor surveillance systems: the fusion perspective*: Springer Science & Business Media, 2012.
20. O. Maimon and L. Rokach, "Data mining and knowledge discovery handbook," 2005.
21. A. Azevedo, "Data mining and knowledge discovery in databases," in *Advanced Methodologies and Technologies in Network Architecture, Mobile Computing, and Data Analytics*, ed: IGI Global, 2019, pp. 502-514.
22. J. T. Wang, M. J. Zaki, H. T. Toivonen, and D. Shasha, "Introduction to data mining in bioinformatics," in *Data Mining in Bioinformatics*, ed: Springer, 2005, pp. 3-8.

23. K. G. Mehrotra, C. K. Mohan, and H. Huang, *Anomaly detection principles and algorithms*: Springer, 2017.
24. A. El-Baz, X. Jiang, and J. S. Suri, *Biomedical image segmentation: advances and trends*: CRC Press, 2016.
25. D. Dubois and H. Prade, "Possibility theory in information fusion," in *Proceedings of the third international conference on information fusion*, 2000, pp. PS6-P19 vol. 1.
26. D. Dubois, W. Liu, J. Ma, and H. Prade, "The basic principles of uncertain information fusion. An organised review of merging rules in different representation frameworks," *Information Fusion*, vol. 32, pp. 12-39, 2016.
27. D. Dubois and H. Prade, "On the use of aggregation operations in information fusion processes," *Fuzzy sets and systems*, vol. 142, pp. 143-161, 2004.
28. E. P. Blasch, D. A. Lambert, P. Valin, M. M. Kokar, J. Llinas, S. Das, *et al.*, "High Level Information Fusion (HLIF): Survey of models, issues, and grand challenges," *Aerospace and Electronic Systems Magazine, IEEE*, vol. 27, pp. 4-20, 2012.
29. É. Bossé and G. L. Rogova, *Information Quality in Information Fusion and Decision Making*: Springer, 2019.

Chapter 7
Conclusion

Information overload and complexity are core problems to most organizations of today. One of the major challenges of a newly created scientific domain called "data science" is to turn data into actionable knowledge to exploit the increasing data volumes and deal with their inherent complexity (Big data and IoT). The advances in networking capabilities have created the conditions of complexity by enabling richer, real-time interactions between and among individuals, objects, systems, and organizations. Networking involves relations of all kinds and presents challenges of complexity specially when the objective is to provide technological supports to human decision-making.

Two main poles have been treated in this book: computations with relations (relational calculus) and actionable knowledge. Actionable knowledge has been qualitatively and intensively studied in management, business, and social sciences [1], but in this book, the focus has been on computer sciences and engineering, i.e., technological and technical (data science). We have explored basic properties of knowledge, knowledge representations, and knowledge processes from scientific and practical perspectives emphasizing existing directions and areas in knowledge studies. We also have examined the fundamental role of information. We endorsed the views expressed in conclusion of Burgin's book on general theory of information: "*Thus, we have seen that information is not merely an indispensable adjunct to personal, social and organizational functioning, a body of facts, data and knowledge applied to solutions of problems or to support actions. Rather it is a central and defining characteristic of all life forms, manifested in genetic transfer, in stimulus-response mechanisms, in the communication of signals and messages and, in the case of humans, in the intelligent acquisition of knowledge, understanding and achieving wisdom.*"

In Chap. 1, we have set up the scene with a discussion on actionable knowledge and its related concepts often differently labeled, such as situation awareness, analytics, and information fusion. In addition, we have positioned "*relational calculus*" with respect to these related notions of actionable knowledge. What is knowledge and what is information are the two main questions introduced in this

© The Author(s), under exclusive license to Springer Nature Switzerland AG 2022
M. Barès, É. Bossé, *Relational Calculus for Actionable Knowledge*, Information
Fusion and Data Science, https://doi.org/10.1007/978-3-030-92430-0_7

chapter. Actionable knowledge is not a new term. It has been qualitatively and intensively studied in management and social sciences. It illustrates the relationship between theory and practice. Actionable knowledge is linked with its user: the practitioner. It has been positioned as a response to the relevance of management research to management practice. Is the generated knowledge actionable by the users whom it is intended to engage (business practitioners, policymakers, researchers)? Actionable knowledge should advance our understanding of the nature of action as a phenomenon and the relationship between action and knowledge (modes of knowing) in organizations. But what if the user is a machine? Or a system?

Actionable knowledge is explicit symbolic knowledge that allows the decision-maker to perform an action, for instance, select customers for a direct marketing campaign or select individuals for population screening concerning high disease risk. Its main connection is with data mining, an emerging discipline that has been booming for the last two decades. Data mining seeks to extract interesting patterns from data. However, it is a reality that the so-called interesting patterns discovered from data have not always supported meaningful decision-making actions. This has motivated the evolution of data mining next-generation research and development from data mining to actionable knowledge discovery and delivery [2–4].

Knowledge remains the domain of philosophers since the nature of the issues and questions challenge so deeply the human mind. The issues treated in Chap. 2 are apprehension and knowledge representation prior to defining a suitable formalism that remains, in effect, the required essential for any subsequent artificial reasoning. Any automatic process geared to support human decision-making must be indeed endowed with reasoning ability or interpretation depending on the circumstances and the context of its employment. Most often, that capacity will be achieved through an inference mechanism. Inference cannot operate without a priori properly formalized knowledge. For data science and engineering, an important direction is the measurement theory of knowledge and information. It is necessary to build efficient measures of knowledge providing effective tools for evaluation of knowledge and information quality. It is also necessary to have more measures for evaluation of different knowledge and information properties based on sound theoretical foundations. All these subsequent actions are in part conditioned by two aspects:

- A sort of permanent knowledge, originally initialized and maintained over time
- Timely knowledge to be acquired continuously, depending on needs

Humans must not only learn but also understand to act. It requires not only to possess a thinking capacity but also to have the following skills:

- To attach a meaning to shapes and relations connecting them
- To make the necessary bridges between data, information, and knowledge
- To take advantage of doing in the process of knowledge enrichment

Chapter 2 studied the properties and multiple dimensions of knowledge, initially introduced in Barès [5]. We treat knowledge in the context of epistemic structures and from a semiotic basis. Symbols and signs and their interpretations are necessary to acquire meaning. Knowledge items are epistemic structures. Beliefs are epistemic structures as well but associated with descriptive knowledge. Knowledge representation and acquisition are key issues. To understand and process knowledge, it is important to know that there are various types, sorts, dimensions, and kinds of knowledge. This chapter emphasizes the multidimensionality of knowledge and discusses dimensions such as ontological, semantic, temporal, and reference. Formalization and quantification of knowledge are discussed as well.

Mainly, the technological processes that support the generation of knowledge for action can be described by the following two categories: analysis (analytics) and synthesis (fusion) of information. Chapter 3 brought a discussion on a knowledge processing chain that under numerous sub-processes transforms data into actionable knowledge. Analytics and fusion of information (AIF) processes are the technological key enablers to the knowledge processing chain. The concepts of symbolic fusion and semantic growth have been treated with representative examples. AIF offers through its processes the capability of numerous transformations of data into actionable knowledge. Relations and their calculations are essential to these AIF processes. Transformations are made possible using information. This is the reason of the label "*infocentric knowledge chain.*" Let us repeat one of the citations of Chap. 3: "*Data, under the influence (action) of additional information, become Knowledge. That is, information is the active essence that transforms data into knowledge. It is similar to the situation in the physical world, where energy is used to perform work, which changes material things, their positions and dynamics.*"

Chapter 3 has addressed the information-processing aspect of a knowledge chain. We have discussed the relationship between data, information, and knowledge with their associated imperfections (i.e., imperfect world). Distinctions in terms of semantic status, role, and processing are important since they have a significant impact on their conceptual and implementation approaches. The notion of "quality" is of prime importance to any information processing, and particularly for a knowledge processing chain, since the objective is to improve, through processing, data, information, and knowledge quality. Various aspects of quality of information (QoI) have been discussed with emphasis on uncertainty-based information.

The contributions of the first three chapters can be summarized as follows:

1. What is actionable knowledge from a technological perspective?
2. What are data, information, and knowledge? What are the interactions between these three entities?
3. A discussion on formalization of multiple dimensions of knowledge
4. Context representations and reasoning
5. A semiotic basis to define data, information, and knowledge
6. What is an infocentric knowledge processing chain?
7. How to get sense out of data along a knowledge processing chain?
8. How to cope with information or knowledge imperfections?

Chapter 4 presented preliminaries of crisp and fuzzy relational calculus to support the discussion in the subsequent Chaps. 5 and 6. The basic elements and notions to perform different operations on the calculation of the mathematical relations and to examine the interest of their properties according to their respective contexts of use have been presented. The material in Chap. 4 is essentially a short synthesis of the "*existing*" coming from the following sources [6–11]. Chapter 4 is necessary to really appreciate the next two chapters (5 & 6). The relationship between the elements is a crucial thing to address. Relations may exist between objects of the same set or between objects of two or more sets. Often, we do not have a complete knowledge about these relations, so comes the notion of "*fuzziness.*" Fuzzy relations appear as a generalization of crisp relations. While a crisp relation determines the presence or absence of interconnectedness between the elements of two or more sets, fuzzy relations supply additional information for degrees of membership between elements. Zadeh [12] was the first to look at relations as fuzzy sets [13] on the universe $X \times X$. Zadeh [13] introduced the concept of fuzzy relation, defined the notion of equivalence, and gave the concept of fuzzy ordering. Compared with crisp relations, fuzzy relations have greater expressive power and broader utility. They are considered as softer models for expressing the strength of links between elements. They also permit to manipulate values that can be specified in linguistic terms.

The importance of the theory of fuzzy relational equations is best described by Zadeh in the preface of the monograph by Di Nola et al. [14]:

> *Human knowledge may be viewed as a collection of facts and rules, each of which may be represented as the assignment of a fuzzy relation to the unconditional or conditional possibility distribution of a variable. What this implies is that the knowledge may be viewed as a system of fuzzy relational equations. In this perspective, then, inference from a body of knowledge reduces to the solution of a system of fuzzy relational equations.*

Since the 1960s, fuzzy relations have been defined, investigated, and applied in many ways, e.g., in fuzzy modeling, fuzzy diagnosis, and fuzzy control. In systems, the relationship between input and output parameters can be modeled by fuzzy relation between input and output spaces [15]. Fuzzy relational calculus is a powerful tool to study the behavior of such systems. Fuzzy relations and fuzzy relational calculus have many applications in pure and applied mathematics, artificial intelligence, psychology, medicine, economics, and sociology. They are implemented in all inference forward or backward chain reasoning schemes. The enumeration of applications listed in Peeva and Kyosev [6] advocates strongly on the importance considering fuzzy system science in systems design:

> *The most valuable implementations are in expert systems and in artificial intelligence areas— approximate reasoning, inference systems, knowledge representation, knowledge acquisition and validation, learning, in information processing, in pattern analysis and classification, in fuzzy system science for fuzzy control and modelling, in decision making, in engineering for fault detection and diagnosis, in management, etc.*

Knowledge is the culmination of a chain of different types of entities whose essence is determined by the perception of signs and signals in the perceived world participating in action control according to the level of signified conveyed (i.e., semantic status). Chapter 5 has examined the couple (knowledge, action). Knowledge is a prerequisite to taking any reasoned action or course of action according to rational rules. It can be described as a superset of domains of knowledge aggregated according to a specific application. Its complex structure is formed through several levels of aggregation:

- Aggregation of signs from the perceived world that are considered relevant for future decision-making contexts
- Aggregation of data serving as a basis for enactment
- Aggregation of knowledge expanded and structured according to the cultural level and expertise of the acting entity

Chapter 5 has examined what facilitates the relevant decision-making and the modalities that can make the action (effect) more efficient. There is a strong dependency between the notion of knowing about a given world and the decisions that can be made and consecutively the potential actions that can be undertaken [5]. For the sake of efficiency, one cannot act on the constitutive objects of the world without first gathering useful knowledge related to this world, in other words, to acquire knowledge on it. If the disposition of a piece of knowledge favors the conduct of an action, it does not constitute a sufficient condition. A book or a knowledge base, although it contains knowledge, is not able to act accordingly to the information content it conveys. This knowledge will only be used appropriately to have both a good interpretation of the situations encountered in real world and to consciously take the attitudes driving to the good conduct (effecting) of an action. In the absence of such dispositions, inherent to an actor endowed with intelligence, knowledge cannot be "mastered."

In Chap. 5, we brought the notion of "*mastering knowledge*" for efficient actions [5]. Mastering knowledge amounts to having a coherent set of means to represent the most useful knowledge in the context of the action and to know how to resort as necessary to the appropriate formalizations to model the situations. Chapter 5 brought contributions for a better understanding of the couple (knowledge, action). By questioning the deeper meaning of the notion of efficiency attached to an action, we proposed an approach organized around the following points:

- To partition the universe in line with the aims of the projected actions, which leads to the proper discrimination of exogenous and endogenous data by proposing a bipartition of the universe of action
- To establish a typology of knowledge useful for action and their concomitant attributes
- To examine the impact that the "quality" of information can have on the conduct of an action, and seek to improve it, particularly through the so-called concept of enrichment of semantics
- To master imperfect knowledge, by looking for the models most adapted to the situations in which their control can have a positive impact on the effective actions
- To situate the notion of relevant information in relation to the efficiency of an action: the search for efficiency will in fact, in many circumstances, lead to

partitioning the exogenous universe or refining its partitioning to consider, for example, certain criteria or categorization of elements

- To examine what facilitates the relevant decision-making and the modalities that can make the action more efficient
- To examine the notions of cooperability and interoperability in a complex world of actions

Analytics and information fusion (AIF) processes are what we call the technological dimension to support the generation of actionable knowledge. Chapter 5 discussed the complexity of the universe decision-action and the efficiency of actions. Chapter 6 addressed the usage of relational calculus [16] when applied to the AIF core processes. The following eight AIF core processes have been identified and listed:

1. *Alignment* (spatial and temporal)

    Description: World is observed in time and space. To make sense out of observations, one needs to understand the context and common referencing under time and space.

2. *Detection/Mining*

    Description: Multi-sensor detection and data processing are well known in engineering and physics. Applying the term "*detection*" to all levels of the data-information-knowledge processing chain implies to also use the term "*knowledge discovery.*" Knowledge discovery is the process of automatically searching large volumes of data to search for complex patterns.

3. *Partition/Classification*

    Description: Facing the complexity of the world, partitioning eases its apprehension and understanding. Partitioning is the act of dividing a unit into its components. Clustering and classification are examples of partitioning.

4. *Fusion/Merging*

    Description: Aggregation, integration, combination, merging, and fusion of information elements have the objective to gain better awareness. It is generally known that merge is joining together of two flows while fusion is the merging of similar or different elements into a union.

5. *Truthfulness/Veracity* (true, untrue, degree of truth) Elements, objects, relations

    Description: The action of putting forward some statement or proposition as true. The statement, or proposition, can also be put forward not as assertion, but as a supposition or hypothesis, as possibly true, and so on. Assertion using logical approaches is an example of that process. Reliability and relevance are also notions that can be assessed under that process.

6. *Inference/Reasoning*

    Description: Inference is using observation and background to reach a logical conclusion. An inference is the process of drawing a conclusion from supporting evidence. This is the way to understand the world. Inferences are steps in reasoning. Inference is theoretically divided into three main categories: deduction, induction, and abduction.

7. *Prediction/Forecasting*

Description: Prediction is concerned with estimating the outcomes for unseen data. Forecasting is a sub-discipline of prediction in which we are making predictions about the future, based on time-series data. Thus, the only difference between prediction and forecasting is that we consider the temporal dimension. Simple object tracking as well as more complex objects like event tracking or group tracking or situation tracking can be seen as examples of this AIF process.

8. *Association/Linking*

Description: This important AIF process is about the identification and characterization of any link or relation between objects, knowledge items, and information elements considering all dimensions. Correlation between information elements, association between detections and object tracks, and application such as link analysis or relations in social networks are all examples of implementation of that process. The links can be of three types—one-to-one, one-to-many, many-to-many.

For each of the above eight generic AIF processes, an analysis has been conducted of how to apply relational calculus. Some of the AIF core processes are currently implemented in the literature through a plethora of methods and techniques. However, there is still no intelligent AIF-based system designed according to some coherent integration rules or principles which can frame the transformation of data to actionable knowledge. The notion of relation and its calculus, treated in this book, is fundamental for that framework. This book was a starting point, but an enormous effort is still required.

References

1. P. Meusburger, B. Werlen, and L. Suarsana, *Knowledge and action*: Springer Nature, 2017.
2. L. Cao, "Actionable knowledge discovery and delivery," *Wiley Interdisciplinary Reviews: Data Mining and Knowledge Discovery,* vol. 2, pp. 149-163, 2012.
3. R. Batra and M. A. Rehman, "Actionable Knowledge Discovery for Increasing Enterprise Profit, Using Domain Driven-Data Mining," *IEEE Access,* vol. 7, pp. 182924-182936, 2019.
4. K. De Smedt, D. Koureas, and P. Wittenburg, "FAIR digital objects for science: from data pieces to actionable knowledge units," *Publications,* vol. 8, p. 21, 2020.
5. M. Barès, *Maîtrise du savoir et efficience de l'action*: Editions L'Harmattan, 2007.
6. K. Peeva and Y. Kyosev, *Fuzzy relational calculus: theory, applications and software (with CD-ROM)* vol. 22 World Scientific, 2004.
7. I. Beg and S. Ashraf, "Fuzzy relational calculus," *Bulletin of the Malaysian Mathematical Sciences Society (2),* vol. 37, pp. 203-237, 2014.
8. G. J. Klir and B. Yuan, "Fuzzy sets and fuzzy logic: theory and applications," *Upper Saddle River,* p. 563, 1995.
9. G. J. Klir and R. V. Demicco, *Fuzzy logic in Geology*: Elsevier academic press, 2004.
10. L. Seymour and L. Marc, "Schaum's Outline of Discrete Mathematics, Revised," ed: McGraw-Hill, 2009.
11. B. Bede, *Mathematics of Fuzzy sets and Fuzzy logic*. New York: Springer, 2013.
12. L. A. Zadeh, "Fuzzy sets," *Information and control,* vol. 8, pp. 338-353, 1965.

13. L. A. Zadeh, "Similarity relations and fuzzy orderings," *Information sciences,* vol. 3, pp. 177-200, 1971.
14. A. Di Nola, S. Sessa, W. Pedrycz, and E. Sanchez, *Fuzzy relation equations and their applications to knowledge engineering*: Springer Science & Business Media, 1989.
15. L. Zadeh and C. Desoer, *Linear system theory: the state space approach*: Courier Dover Publications, 2008.
16. M. Barès, *Pratique du calcul relationnel*. Paris: Edilivre, 2016.

Index

Printed in the United States
by Baker & Taylor Publisher Services